INVERTEBRATE TISSUE CULTURE

Volume II

ADVISORY BOARD

INVERTEBRATE TISSUE CULTURE

Edited by C. Vago

STATION DE RECHERCHES CYTOPATHOLOGIQUES
INRA-CNRS, SAINT-CHRISTOL
UNIVERSITÉ DES SCIENCES
MONTPELLIER, FRANCE

VOLUME II

ACADEMIC PRESS   New York and London   1972

ACADEMIC PRESS, INC.
111 Fifth Avenue, New York, New York 10003

United Kingdom Edition published by
ACADEMIC PRESS, INC. (LONDON) LTD.
24/28 Oval Road, London NW1

LIBRARY OF CONGRESS CATALOG CARD NUMBER: 75-154371

PRINTED IN THE UNITED STATES OF AMERICA

CONTENTS

I. ORGAN CULTURES

1 Organ Culture of Insects

J. Demal and A. M. Leloup

10   Use of Invertebrate Cell Culture for the Study of Plant Viruses

Jun Mitsuhashi

III. CELL LINES

11   A Catalog of Invertebrate Cell Lines

W. Fred Hink

LIST OF CONTRIBUTORS

Numbers in parentheses indicate the pages on which the authors' contributions begin.

GORDON H. BALL, Department of Zoology, University of California, Los Angeles, California (321)

CLAUDIO BARIGOZZI, Istituto di Genetica, Università di Milano, Milan, Italy (163)

JOSETTE BERREUR-BONNENFANT, Laboratoire de Génétique Evolutive, CNRS Gif-sur-Yvette, France (181)

J. DAVID, Service d'Entomologie Expérimentale et de Génétique, Université Lyon, Lyon, France (211)

J. DEMAL, Laboratoire de Morphologie Animale, Université Catholique de Louvain, Louvain, Belgium (3)

L. GOMOT, Laboratoire de Zoologie, Université Besançon, Besançon, France (41)

W. FRED HINK, Department of Entomology, The Ohio State University, Columbus, Ohio (363)

A. M. LELOUP, Laboratoire de Morphologie Animale, Université Catholique de Louvain, Louvain, Belgium (3)

JUN MITSUHASHI, Division of Entomology, National Institute of Agricultural Sciences, Nishigahara, Kita-ku Tokyo, Japan (343)

JOSEF ŘEHÁČEK, Institute of Virology, Czechoslovak Academy of Sciences, Bratislava, Czechoslovakia (279)

M. ROUGIER, Service d'Entomologie Expérimentale et de Génétique, Université Lyon, Lyon, France (211)

TAKEO TAKAMI, Sericultural Experiment Station, Tokyo, Japan (137)

C. VAGO, Station de Recherches Cytopathologiques, INRA-CNRS, Saint-Christol, Université des Sciences, Montpellier, France (245)

PREFACE

The contribution of cell and tissue culture to the development of medical and biological sciences is universally known. Most of the recent progress in cellular biology and pathology has been achieved by *in vitro* investigations. Fifteen years ago this was true for biological experimentation dealing with man and other vertebrates. The use of tissue culture in invertebrate research in physiology and pathology was being considered, but only a few attempts were made to achieve such cultures, and then in only a small number of insect and mollusk species.

The First International Colloquium on Invertebrate Tissue Culture held in 1962 in Montpellier (France) enabled us to evaluate results of previous research and opened new vistas for future research. Since then rapid advances have occurred in cell and organ culture of invertebrates— Arthropoda, Mollusca, Echinoderma, Nematoda, and Coelenterata. Such cultures, often barely developed, were also used in studies involving genetics, physiology, and pathology.

At present, investigations and results are so numerous and diversified that the publication of a treatise to collate this material seemed desirable. With the aid of the foremost specialists in the field of invertebrate cell and organ culture, this two-volume treatise was made possible.

This work was organized so that the techniques utilized and their applications to the various biological disciplines are developed, accompanied by the results and characteristics of the resulting cultures. Some overlap was necessary to ensure linkage of information from chapter to chapter.

Volume I includes general methodology concerning both cell and organ cultures and their preparation from aseptic conditions. Methods for the examination of cultures are also developed, particularly those concerning ultrastructure studies by electron microscopy. Cell cultures obtained from different groups of invertebrates are then discussed with emphasis on peculiarities specific to each group.

Organ cultures of different invertebrates are the subject of several chapters in Volume II, including an important section on the use of cell and organ cultures in physiology, genetics, and pathology. The study of the effect of pathogens is distinguished from that of microorganisms transmitted by vectors.

I wish to thank the members of the advisory board—C. Barigozzi, P. Buchner, J. de Wilde, M. Florkin, E. Hadorn, P. Lepine, K. Maramorosch, N. Oker-Blom, R. C. Parker, G. Ubrizsy, and E. Wolff—for their aid in the preparation of this treatise. The continuous assistance of M. Laporte is gratefully acknowledged.

It is hoped that this treatise will not only evaluate the present state of research and the problems in invertebrate tissue culture but will also serve as a guide to those working in this field and as a technical and scientific introduction to those intending to culture invertebrate tissues or to use cultures in pathology, physiology, or other biological disciplines.

C. Vago

CONTENTS OF VOLUME I

I

Organ Cultures

I

ORGAN CULTURE OF INSECTS

J. Demal and A. M. Leloup

I. INTRODUCTION

The *in vitro* technique of explantation, whether applied to vertebrates or to invertebrates, is used in two fairly distinct areas of research. The first, described in the preceding chapters, aims at obtaining from explants of varying size and nature both survival and cell proliferation. More often than not, the cultured cells lose the specific characteristics which they had at the time of explantation, but this is without importance for the aim of the investigation. The second area seeks to maintain in culture the structural and functional integrity of the explant; this calls for survival and growth, as well as for the maintenance, and eventually, an evolution of the object's differentiation. Since Maximov's time (1925), the technique of explantation used in the first area of research has usually been called tissue or histiotypic culture, and that in the second, organ or organotypic culture. Truth to say, this terminology is inadequate. It would seem preferable to speak, in the first case, of cell culture since it is the cell which is the biological unit cultured *in vitro;* this technique could also be termed unorganized culture (Thomson, 1914). In the second instance, that of the method which aims at maintaining the entire explant unimpaired *in vitro*, it would be better to speak of organized culture. This broader term would cover experiments carried out on anatomical bodies lower or higher than the organ, i.e., tissues or fragments of organs, embryos, or whole organisms. Thus, when the term organ culture is used, we consider that it must be given this broader meaning.

II. HISTORICAL SURVEY

A. Preliminary Works

Since the *in vitro* technique was first used in the study of vertebrates, it is not surprising to find that the first attempts at organotypic culture of invertebrates should have been made in laboratories generally studying vertebrates. Goldschmidt (1915, 1916, 1917), working in the Harrison Laboratory at Yale University, was the first to attempt to set up an *in vitro* culture of testicular cysts of *Samia cecropia* pupae. The explants, set in hanging drops of the blood of the insect, showed all the phases of spermatogenesis during a period of three weeks. The death of the sex cells in no way hindered survival and a certain cellular growth of the follicle membranes and of blood cells during a period lasting more than six months. Goldschmidt immediately grasped the value of the method he had just experimented with, and he was not satisfied with a mere passive observation of the evolution of these cultures. Organ culture presents the advantage of allowing the explant to be observed independently of the organism from which it has been taken and freed from internal correlation and the many influences which it receives from the rest of the animal. The investigator is therefore free to vary systematically the experimental conditions of the culture in order to determine exactly which factors favor or hinder survival, growth, and differentiation of the explant. Once Goldschmidt had become convinced that the evolution of his cultures *in vitro* was analogous to evolution *in vivo*, he began experimental and causal study of spermatogenesis. He varied the chemical composition of the culture medium, and studied the effects of heat, cold, and osmotic conditions on his explants. He was able, in particular, to point out the importance of the action of the follicle membrane in maintaining the physical conditions necessary for the normal evolution of spermatogenesis. Shortly after Goldschmidt's first experiments, other authors, such as Lewis (1916) and Lewis and Robertson (1916), working on *Chorthippus curtipennis*, and Chambers (1925), working on *Dissosteira carolina*, applied this technique to the cytological study of spermatogenesis.

However, the technique, in spite of promising beginnings, showed itself to be of delicate and often fruitless application in the use of the study of insects. This explains why no further publication was devoted to the culture of insect organs until 1928, when Frew (1928) set out to study by *in vitro* culture the formation of the mesoderm in the leg imaginal discs of *Calliphora erythrocephala*. Since the explants showed but very limited differentiation *in vitro*, he realized his aim only partially. But his systematic research for improving culture conditions allowed him to point out a number of important problems besetting the would-be research worker in *in vitro* organ culture.

In the first place, he drew attention to the necessity of severe aseptic conditions in obtaining a good *in vitro* culture; he showed how difficult it is to maintain such conditions in the course of explantation of insect organs and in blood collecting. The majority of types of explants, with the exception of those from the digestive tract and its annexes, may be sterilized by repeated washings. There are various methods for filtration of collected hemolymph, but it always undergoes some modification while passing through the filter. It might be possible to avoid these difficulties by aseptic rearing of experiment animals, wherever this can be done. A second problem confronting the investigator concerns the knowledge of the composition of the insect's body fluid, and since this composition varies in the course of metamorphosis, it is necessary to have a minimum of knowledge of the physiology of metamorphosis in order to be able to estimate the variations of the hemolymph composition during the development.

Efforts were made in many laboratories to realize all the conditions favorable to a good *in vitro* culture of insect explants. However, it was a good ten years after the work carried out by Frew before any satisfactory results were obtained, but from that time onward, the aim of the investigations carried out by means of this technique, as well as the forms which it took, varied. We shall therefore now attempt to describe briefly the principal types of work done. The results will be commented upon at greater length in later paragraphs or in other chapters. Nevertheless, it must be remembered that any such systematization is relative, because on the one hand, certain works correspond to several fields of research, and on the other hand, results obtained in one field may prove to be very precious for the progress of other studies.

B. Cellular Substrata by Organ Culture

A certain number of animal and plant parasites live part, or even the whole, of their life cycle within vector insects. The study of these parasites is of great importance, and the method of *in vitro* culture of insect organs, tissues, or cells has raised the hope of being able to analyze better the factors conditioning the development and the transmission of certain viruses or pathogenic protozoa. It was with this end in view that Trager (1937, 1938, 1959a,b) undertook the culture of the walls of the ovaric tubules of *Bombyx mori* and of different organs of *Aedes aegypti* and *Glossina palpalis*. Gavrilov and Cowez (1941) assured *in vitro* survival of organs of *Stegomyia fasciata* and *Anopheles maculipennis*, and Nauck and Weyer (1941) obtained the same result with different organs of *Cimex lectularius*, *Ctenopsyllus segnis*, and *Pediculus humanus*. Ball (1947, 1948, 1954) and Ball and Chao (1960) maintained an *in vitro* culture of the digestive organs of *Culiseta incidens*, *Culex tarsalis*, and *Culex stigmatosoma*. Pursuing a similar aim, Ragab (1949) cultivated in hanging drops or by perfusion the digestive tract of *Anopheles maculipennis* and *Aedes aegypti*. Vago and Chastang (1960) maintained in *in vitro* culture for several weeks ovaries and fragments of ovaries of *Antheraea pernyi*, *Bombyx mori*, *Galleria mellonella*, and *Lymantria dispar*. Similar works were carried out by Hirumi and Maramorosch (1963) on different organs of *Agallia constricta* and *Macrosteles fasci-*

frons. Lastly, Peleg and Trager (1963) obtained cell survival from larval imaginal discs of *Aedes aegypti* and *Culex molesus.*

The different authors we have just mentioned have, for the most part obtained *in vitro* cultures of insect organs. However, their aim was not to obtain an organized culture but rather a cellular substratum which would favor the development of protozoa or viruses.

C. Morphological Studies

Experimenting with the *in vitro* technique, many investigators have tried to obtain a real organized culture for studies in histogenesis or morphogenesis. They met with the greatest of difficulties for they expected the explant to show not only *in vitro* survival but also growth and differentiation. Fischer and Gottschewski (1939) and Gottschewski and Fischer (1939) explanted wing, leg, and eye imaginal discs of *Drosophila melanogaster,* and they were able to observe a certain *in vitro* differentiation of these organs without, however, proceeding with histological control. Similarly, Stern (1940), limited himself to the external morphological aspect, reproduced *in vitro* a stage of the differentiation of pupal testes of *D. melanogaster.* Goodchild (1954) had very little success with *in vitro* culture of the epiderm, heart, and testis of *Rhodnius prolixus* and of *Blatta orientalis.* Demal (1955, 1956), using first Wolff and Haffen's culture medium (1952), then a liquid medium, obtained *in vitro* a restricted differentiation, histologically controlled, of imaginal buds of *Calliphora erythrocephala* legs and of *D. melanogaster* eyes. Adult ovaries of *Aedes hexodontus, A. aegypti,* and *A. communis* were kept in survival by Beckel (1956), but the eggs did not mature and no mitosis was observed. Pursuing his research on the development of the *D. melanogaster* eye, Gottschewski (1960) was able to observe *in vitro,* in different strains, distinct bud growth and differentiation. In the years 1956–1960 the Japanese school made several important contributions in favor of organized *in vitro* culture of insect organs. Their investigations were chiefly of the genetical order and will therefore be mentioned under that heading. One important study of morphogenesis must, however, be mentioned here, i.e., that of Horikawa and Sugahara (1960) on the effects of *in vivo* irradiation on the behavior of larval organs of *Drosophila melanogaster,* Oregon strain, cultured *in vitro.* Hanly (1961) signaled a good *in vitro* survival of third stage larval brains of *D. melanogaster.* Demal (1961) obtained *in vitro* survival and continued contraction of the aorta of *C. erythrocephala* as well as limited differentiation of the pronymphal ovaries and testes of the same insect

and testes of *D. melanogaster*. Gottschewski and Querner (1961) noticed
that the second larval stage eye bud of *D. melanogaster* does not develop
in vitro. They were able, however, to show, by means of fluorochromes,
the possibility of restricted migration of substances within a brain *in
vitro*. Lender and Duveau-Hagège (1962, 1963a,b, 1965) obtained an
in vitro differentiation of the gonads of full-grown larva and nymphs of
Galleria mellonella. Duveau-Hagège (1963, 1964) extended to *Periplaneta
americana* her investigations on *in vitro* ovogenesis. Finally, Lender and
Laverdure (1967, 1968) and Laverdure (1967, 1969) were able to reproduce *in vitro* a more advanced differentiation of the ovaries of *Tenebrio
molitor*, including vitellogenesis. Similar results were obtained by
Ittycheriah and Stephanos (1969) for ovaries of *Iphita limbata*. Very
long *in vitro* survivals were observed by Larsen (1963) for different
organs of *Blaberus craniifer;* even whole embryos were seen to develop.
Cannon (1964) showed an *in vitro* DNA synthesis in larval salivary
glands of *Sciara coprophila*. Schneider (1964, 1965, 1966) obtained *in
vitro* cultures of the eyes, antennae, brains, and ovaries of *D. melanogaster* with varied results concerning differentiation; the best results,
histologically controlled, were obtained for the eyes and, especially,
antennae of explants of well-defined ages. Other workers (Hanly *et al.,*
1967; Burdette *et al.,* 1968; Sengel and Mandaron, 1969; Mandaron,
1970) obtained similar results for organs of the same insect. Demal and
Leloup (1963), Leloup (1964, 1969, 1970), and Leloup and Demal (1968)
set up cultures of the gonads of *C. erythrocephala* and carried out histological verification of *in vitro* survival, growth, and differentiation of explants of well-defined stages. Martin and Nishiitsutsuji-Uwo (1967)
describe cellular proliferation and *in vitro* tracheolar formation from
fragments of the trachea of *Locusta migratoria*. Similarly, Pihan (1969)
set up cultures of trachea from *C. erythrocephala;* Courgeon (1969),
cephalic discs and brains of the same insect; and Demaure (1968),
cardias, brains, and salivary glands of *Lucilia sericata*. With the aim
of investigating wound healing processes, nerve regeneration, cuticular
formation, and endocrine interaction in *Leucophaea maderae*, Marks
(1968, 1969, 1970) and Marks *et al.* (1968; Marks and Leopold, 1970;
Marks and Reinecke, 1964, 1965) cultivated leg regenerates and the
different constituents of the cephalic complex.

D. Endocrinology

   Since *in vitro* culture requires that the explant be withdrawn from
the influence of the many factors which interact *in vivo* among the

different organs of the same animal, this method has rendered great services to those investigators who wished to analyze endocrinological factors. As these play an active part in insects in the phenomena of growth and differentiation, it is evident that the results obtained by the *in vitro* experiments in the field of endocrinology are of the greatest interest for those who work on histogenesis and morphogenesis. The works of Schmidt and Williams (1949, 1953) on *Platysamia cecropia* and *Samia walkeri* established, by the *in vitro* culture of spermatic cysts of pupae, that at certain stages of development the blood contains a factor favorable to differentiation. This factor would appear to be the growth and differentiation hormone. Schneiderman *et al.* (1953) made *in vitro* analysis, on the same material, of the action of certain gases, metabolic inhibitors, temperature, and light on spermatogenesis. The *in vitro* culture of insect organs carried out by different Japanese authors also brings out the endocrinological factors. We can mention here the researches of Kuroda and Yamaguchi (1956) on *Drosophila melanogaster*, Bar and Oregon strains, which show the *in vitro* action of the cephalic complex on the differentiation of larval buds of eyes and antennae. Horikawa (1960) working on similar material, obtained comparable results but observed a variation of the endocrine activity of cephalic complexes in the different strains. Larval organs cultured *in vitro* may show, by the extent of the differentiation produced *in vitro* in relation to the moment of explantation, the hormonal variations of the insect's body fluid. Such are the results established by the researches of Miciarelli *et al.* (1967) on *Schistocerca gregaria*. Studies on the continuance of neurosecretion among brains isolated in *in vitro* culture, have been made by Leloup and Gianfelici (1966) and by Gianfelici (1968a,b) working on *Calliphora erythrocephala* and by Schaller and Meunier (1967) on *Aeschna cyanea*. More recently, the *in vitro* effects of synthetic hormone analogues on organs from different insects were reported (Agui *et al.*, 1969a,b; Burdette *et al.*, 1968; Judy, 1969; Laverdure, 1969; Mandaron, 1970; Marks and Leopold, 1970; Oberlander and Fulco, 1967; Oberlander, 1969a,b; Pihan, 1969; Sengel and Mandaron, 1969; Williams and Kambysellis, 1969; Yagi *et al.*, 1969).

E. Cell Proliferation from Organs in Culture

The finest result recently obtained by the technique of *in vitro* insect culture is certainly the achievement of long survival of stabilized cell strains. This, of course, does not concern organized *in vitro* culture, but a certain number of investigators, in their attempts to obtain cell proliferation, have, in fact, sometimes realized cultures which reply to certain characteristics of organized cultures. Thus, Grace (1954) obtained an *in vitro* survival of the gonads, imaginal discs, and intestines of *Bombyx mori, Periplaneta americana,* and *Drosophila melanogaster* of varying ages. Loeb and Schneiderman (1956) maintained in prolonged survival, but without true proliferation, different organs or fragments of organs of *Samia cynthia* and *Antheraea polyphemus.* Vago and Chastang (1958) obtained a good survival and *in vitro* cell migration from fragments of ovaries of *B. mori.* Chabaud *et al.* (1960) were able to observe *in vitro* the contractions of thoracic imaginal discs of *Aedes aegypti* as well as cell migration. Castiglioni and Rezzonico Raimondi (1961, 1963) set up *in vitro* cultures of the larval brains and lymph glands of different strains of *D. melanogaster* with the view to obtaining the liberation of cells suitable for cytological and genetical study. Stanley and Vaughn (1968), using histological procedures to determine the origin of migrating cells from cultured ovaries of various insects, also observed that the primary explants had survived *in vitro.* Greenberg and Archetti (1969) reported that intestinal segments from *Musca* larva which did not give rise to cell migration showed the persistence of contractions for seven weeks. They also pointed out that brains plus ring glands, releasing lymphocyte-like cells during the first week, underwent no differentiation.

F. Genetic Experiments

The chain reactions which lead to the phenotypic expressions of genes are particularly suitable for analysis by the method of *in vitro* culture which allows of an experimental breaking up of these chains and, eventually, an exchange of links between chains of neighboring strains. These researches are carried out on *Drosophila melanogaster,* which offers a great variety of mutant strains allowing comparison and *in vitro* interaction between organs of different stocks. Gottschewski (1958) compared *in vitro* the formation of the ommatidian field in different mutants distinguishable by the eye form. Horikawa (1958) attributes the differences observed *in vitro* in the deposit of eye pigment of different strains to either varying interaction between metabolism of tryptophan and of the pteridines or qualitative and quantitative differences in the hormonal activity of cephalic complexes. Kuroda (1959) explanted wing buds of the third larval stage of certain strains. The conditions of *in vitro* culture show that the influence of temperature on wing development varies from strain to strain and that determination of the type of development is made before the third larval stage. Fujio (1962), using the experimental *in vitro* techniques, poses the hypothesis of the secretion of B substances by the cephalic complexes of *Drosophila* of Bar strains; these substances would seem to induce the characteristic eye development of these strains.

G. Physiology

Many aspects of insect physiology can be more rigorously analyzed when the functioning of a single organ is examined independently of the remainder of the organism. It is normal, then, that numerous physiologists should have explanted organs and kept them in *in vitro* survival for the time necessary for measurements or for the action of substances modifying function to take place. Here again, it is not, strictly speaking a matter of organized organ cultures including survival, growth, and differentiation, but it is sometimes difficult to draw the boundary between these two types of research. By way of examples of such *in vitro* physiological investigations, we can mention the work carried out by Ramsay (1954) on the functioning of the Malpighian tubules of *Dixippus morosus*, as well as the studies of heart functioning of various insects carried out by Barsa (1955), Butz (1957), Ludwig *et al.* (1957a,b), and David and Rougier (1965); the works of Ozbas and Hodgson (1958) on the influence of neurosecretion on the nervous system; the researches of Wayne Wiens and Gilbert (1965) on the influence of the corpus cardiacum on the metabolism of the fat body of *Leucophaea madorae;* and the study made by Daillie and Prudhomme (1966) on respiratory intensity and thymidine incorporation in the DNA of the sericigenous gland of the silkworm.

III. RESULTS AND DISCUSSION

In the rather long historical notice which we have just given we have tried to sketch a synthesis of all research work carried out on insects by means of organ culture in its broadest sense. The results of the researches carried out in the fields of genetics, endocrinology, pathology, and physiology will be commented upon in other chapters. Here, we shall limit ourselves to analyzing those studies which, in general, are connected with morphogenesis. We shall consider successively the cultures of different types of organs. (A synopsis of this section is given in Table I.)

TABLE I
A GUIDE TO SPECIES AND REFERENCES

Origin of explants	Location in Section III	References
Coleoptera		
Cybister lateralimarginalis	D	David and Rougier (1965)
Tenebrio molitor	A,3; E,2	Lender and Laverdure (1967, 1968)
	A,3; E,2	Laverdure (1967, 1969)
Diptera		
Aedes aegypti	B; D; E,2	Trager (1938)
	D	Gavrilov and Cowez (1941)
	D	Ragab (1949)
	E,2	Beckel (1956)
	B	Chabaud et al. (1960)
Aedes communis	E,2	Beckel (1956)
Aedes hexodontus	E,2	Beckel (1956)
Anopheles maculipennis	D	Gavrilov and Cowez (1941)
	D	Ragab (1949)
Calliphora erythrocephala	B	Frew (1928)
	A,1; B	Demal (1961)
	B; C; E,1; E,2	Demal and Leloup (1963)
	A,3; E,1; E,2	Leloup (1964, 1969, 1970)
	A,1	Leloup and Gianfelici (1966)
	A,1	Gianfelici (1968a,b)
	A,3; C; E,2	Leloup and Demal (1968)
	A,2	Courgeon (1969)
Culex pipiens	D	Ball (1947, 1948, 1954)
Culex tarsalis	D	Ball and Chao (1960)
Culiseta incidens	D	Ball and Chao (1960)
Drosophila sp.	B	Fischer and Gottschewski (1939)
	A,2	Gottschewski and Fischer (1939)
Drosophila melanogaster	E,1	Stern (1940)
	A,2; B; D	Demal (1955, 1956)
	A,2; A,3	Kuroda and Yamaguchi (1956)
	A,2; A,3	Gottschewski (1958, 1960)
	A,2; A,3	Horikawa (1958, 1960)
	A,3; B	Kuroda (1959)
	A,2; A,3; B; D; E,1	Horikawa and Sugahara (1960)
	A,2; A,3	Fujio (1960, 1962)
	C; E,1; E,2	Demal (1961)
	A,3	Gottschewski and Querner (1961)
	A,1; A,3	Hanly (1961)
	A,2	Hanly et al. (1967)

TABLE I
(Continued)

Origin of explants	Location in Section III	References
Diptera (con't.)	D	Demal and Leloup (1963)
	A,1; A,2; C; D	Schneider (1963, 1964, 1966)
	E,2	Schneider (1965)
	C	Castiglioni and Rezzonico Raimondi (1961, 1963)
	C	Rezzonico Raimondi and Ghini (1963)
	A,2	Burdette et al. (1968)
	A,1; A,2; B	Sengel and Mandaron (1969); Mandaron (1970)
Drosophila virilis	E,1	Stern (1940)
	A,2; C; D	Schneider (1963, 1964)
Glossina palpalis	A,1; A,2; B; D	Trager (1959b)
Lucilia sericata	A,1; D	Demaure (1968)
Musca domestica	A,1	Greenberg and Archetti (1969)
Musca sorbens	A,1	Greenberg and Archetti (1969)
Sciara coprophila	D	Cannon (1964)
Stegomyia fasciata	C; D	Gavrilov and Cowez (1941)
Hemiptera		
Iphita limbata	E,2	Ittycheriah and Stephanos (1969)
Homoptera		
Agallia constricta	A,1; D; E,1; E,2	Hirumi and Maramorosch (1963)
Macrosteles fascifrons	A,1; D; E,1; E,2	Hirumi and Maramorosch (1963)
Lepidoptera		
Cecropia silkworm	E,1	Bowers (1961)
	E,1	Schneiderman et al. (1953)
	E,1	Ketchel and Williams (1955)
	E,1	Williams and Kambysellis (1969)
Chilo suppressalis	B	Agui et al. (1969a)
	E	Yagi et al. (1969)
Galleria mellonella	E,1; E,2	Lender and Duveau-Hugège (1962, 1963a,b, 1965)
	E,1; E,2	Duveau-Hagège (1964)
	D	Plantevin et al. (1968)
	B	Oberlander (1969a,b)
Philosamia cynthia	E,1	Laufer and Berman (1961)
Platysamia cecropia	E,1	Schmidt and Williams (1949, 1953)
Platysamia walkeri	E,1	Laufer and Berman (1961)

(continued)

TABLE I (Continued)

Origin of explants	Location in Section III	References
Lepidoptera (con't.)		
Samia cecropia	E,1	Goldschmidt (1915, 1916, 1917)
Samia cynthia	E,1	Laufer and Berman (1961)
Samia walkeri	E,1	Schmidt and Williams (1949, 1953)
	E,1	Laufer and Berman (1961)
Odonata		
Aeschna cyanea	A,1	Schaller and Meunier (1967)
Orthoptera		
Blaberus craniifer	C; D	Larsen (1963)
Blaberus discoidalis	C	Larsen (1967)
Chorthippus curtipennis	E,1	Lewis (1916)
	E,1	Lewis and Robertson (1916)
Periplaneta americana	E,2	Duveau-Hagège (1963)
	E,1; E,2	Duveau-Hagège (1964)

A. Brain and Cephalic Buds

A preliminary distinction must be made: certain explanted organs are the direct object of a study; others are used for the purpose of enriching or modifying the culture medium, especially from a hormonal point of view. In this latter case we are dealing with the cephalic complex, which includes the brain and adjacent endocrine structures. We shall deal first with brain and cephalic buds and then with the cephalic complexes.

1. The Brain

In vitro brain survivals have been signaled by different authors, carrying out various researches. Demal (1955, 1956) explanted pupal brains of *Calliphora erythrocephala;* Trager (1959b), brain fragments of *Glossina palpalis; Hanly* (1961), larval brains of *Drosophila melanogaster;* Hirumi and Maramorosch (1963), nymphal and adult brains of *Agalia constricta* and *Macrosteles fascifrons.*

Schneider (1966) carried out histological and comparative examination of third stage larval brains of *Drosophila melanogaster* cultured *in vitro.* First, the cerebral lobes spread laterally and the two lobes fused medio-dorsally. The optic glomeruli differentiated; the external one protruded outwards. Between the fifth and the eighth days of culture the external glomerulus assumed a fan-shaped appearance and was in close apposition to the ommatidia of the unevaginated eye. All the constituents of the adult brain were soon to be recognized, but they differed from those of the controls by their topographical localization; the general shape of the brain was also more flattened than *in vivo*. Finally, although the organ as a whole showed good survival, some centers of vacuolization

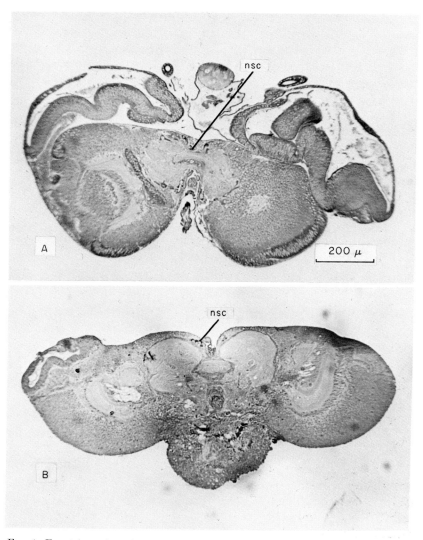

FIG. 1. Frontal section through the pronymphal cephalic complex of *Calliphora erythrocephala* showing the neurosecretory cells (nsc) of the protocerebrum (stained with paraldehyde fuchsin). (A) Control; (B) explant after 11 days in medium M_1 supplemented with chick embryo extract. (From E. Gianfelici, unpublished.)

were observed peripherically from the third or fourth day of culture. The ventral ganglions, more susceptible to injury at the moment of explantation, did not survive well.

Leloup and Gianfelici (1966) and Gianfelici (1968a,b) explanted pronymphal brains of *Calliphora erythrocephala.* The organs gave evidence of survival for more than three weeks and grew by mitosis even in the protocerebrum (Fig. 1). Differentiation continued to take place in culture but with a time-lag, in comparison with the controls, of about eight days. The brain spread laterally and increased in volume; histological control revealed medullary growth and cortical thinning; morphogenetic movements were observed within the organ. Neurosecretory phenomena continued *in vitro* but with a time-lag of about five days (in comparison with the controls); the neurosecretion diffused toward the corpora cardiaca, where it accumulated.

Larval brains of *Lucilia sericata* (Demaure, 1968), *Musca domestica, M. sorbens* (Greenberg and Archetti, 1969), *Drosophila melanogaster* (Sengel and Mandaron, 1969; Mandaron, 1970), and *Calliphora erythrocephala* (Courgeon, 1969) were also reported to be successfully maintained in culture in different types of media for several days.

Schaller and Meunier (1967) kept in *in vitro* culture for a period of two to three weeks the brain and subesophagal ganglion of different aged larvae of *Aeschna cyanea.* The optic ganglions gave evidence of numerous mitoses. The neurosecretory cells showed an increase of the amount of neurosecretory product; the hormonal antagonism between the brain and the subesophagal ganglion also showed itself *in vitro.*

2. Eyes and Antennae

Dealing with the pigmentation of the eye, Gottschewski and Fischer (1939) explanted prepupal eye discs of *Drosophila;* they obtained *in vitro* pigmentation of these buds. They also showed that this differentiation depends upon the genetic constitution of the explant and that substances extracted from certain mutants can favor *in vitro* pigmentation.

Demal (1955, 1956), using histological techniques, examined the different aspects of limited differentiation shown *in vitro* by prepupal explants of about 100-hour-old *Drosophila melanogaster.* The explants included the cerebral lobes bearing eye discs and the subesophagal ganglion. At the time of putting into culture, the ommatidial zone was represented by a thin band in which the separation of cells into ommatidial groups was scarcely perceptible. After two days of culture, the ommatidial cap was closely apposed to the optic lobe, the cells of each ommatidium were clearly grouped in clusters, the eye hairs were present, the ommatidia had doubled in thickness, the cornea and lens were to be clearly seen, and finally, the characteristic pigmentation of the wild strain had appeared.

For the purpose of verifying if the form of the eye is controlled by genic action working in the cephalic complex, Kuroda and Yamaguchi (1956) undertook cultures of eye–antennal discs of 95-hour-old larvae of *Drosophila melanogaster*, Bar and Oregon strains. The cultures were maintained for 17 hours, and the explants were judged on their external appearance. The *in vitro* eye–antennal differentiation of the Oregon strain was increased by the presence of one, and even two, homologous cephalic complexes. The association of a cephalic complex of the Bar strain with Oregon explants in no way modified *in vitro* differentiation of the antenna, but the eye differentiated according to the Bar type. Inversely, Bar explants associated with an Oregon cephalic complex differentiated according to the Oregon type. The genic action seems to show itself, then, first in the brain and the ring gland which appear to influence the differentiation of the eye disc.

Gottschewski (1958) examined the *in vitro* differentiation of the corneal facets of different mutants of *Drosophila*. The larval explants, of unspecified age, included the brain, the eye discs, and the ring gland. A corneal bud first developed on the whole of the ommatidial field in the explants of all strains used. This phenomena seems, then, to be independent of the genes. It was only afterwards that that portion of the corneal bud which, in certain mutants, does not change into facets, differentiated in a chitinous envelope with distinctive characteristics in comparison with the surrounding chitin. The genes which influence the shape of the facets seemed to act very early. Gottschewski (1960) continued his *in vitro* experiments of third larval stage eye discs of *D. melanogaster* of different strains. Fifteen percent of the explants survived for 170 hours. When the eye discs are cultured without the brain, the presence of the ring gland is necessary for a certain atypic development *in vitro* to take place, but the inductive influence of the brain seems to be indispensable for more pronounced normal development. In this case, the retinal fibers multiply, the optic ganglions differentiate, the external chiasma form, and the eye disc shows mitosis and takes on a hemispherical form in adhering to the optic lobe. A membrane develops on the eye disc and the adjacent antennal disc; it appears to be at the origin of the cornea, which, then, does not seem to originate from a secretion of the primary pigment cells. The facet buds form on the whole ommatidial field for all mutants. Preommatidia then appear, and pigmentation begins 72–80 hours after the start of the culture. In comparison with the controls, a lesser development of the retinal fibers and a reduced number of facets is observed and differentiation shows a time-lag of 6–12 hours. From the onset of pigmentation the ommatidial field modifies in all mutants (in particular, by the degeneration of the facets in the case of eyeless mutants) and takes on the form corresponding to the genetic type; this is, then, independent of the primitive ommatidial field.

Horikawa (1958, 1960) analyzed the factors influencing *in vitro* development of eye discs of approximately 96-hour-old larva of different strains of *Drosophila melanogaster*. The first of his studies dealt mainly with certain chemical and enzymatic factors. The eye–antennal discs cultured together with the cephalic complexes in various media containing tryptophan, kynurenine, or 3-hydroxykynurenine showed more distinct pigmentation than the eye–antennal discs alone. These facts indicate that the tryptophan enzyme activities in the eye discs were either directly or indirectly increased by the metamorphic hormone, in the latter case, through the growth and differentiation of the eye discs. The amounts of pigment deposited in the eye discs seem to be controlled mainly by enzyme activities in the chemical process converting kynurenine to 3-hydroxykynurenine, even though the substratum of this enzyme is in abundance. The smaller amount of pigment in the eye discs of double recessive mutants, v bw and bw, seems to indicate that there may be either some interaction between tryptophan and pteridine metabolism or some quantitative or qualitative differences among the metamorphic hormones secreted from the brains of several strains. The second work of the same author deals with the hormonal factor. The explants of 92-hour-old larva were kept in culture for 72 hours, and the differentiation was estimated by direct microscopic examination of the explants. Ten cephalic complexes of Oregon-R larvae provide the optimum concentration of the metamorphic hormone in a hanging drop culture for obtaining growth and differentiation of the explants. This indicates that the metamorphic hormone had already been secreted into the body fluid from the cephalic complexes of the mature third instar larvae, and its secretion at this stage was decreased. The cephalic complex of various eye-color mutant larvae has a similar effect upon growth and differentiation of eye–antennal discs. The metamorphic hormone secreted from cephalic complex of other mutants [B, bar-3, or Dp/In(3L)p,In(3R)c,sbel(3)e] seems to be qualitatively different. These facts suggest that the genic action appears simultaneously in both cephalic complexes and eye–antennal discs.

In another type of research, Trager (1959b) set up a culture of the eyes, optic lobes, and associated head structures of 6-day-old pupae of *Glossina palpalis*. He observed an *in vitro* differentiation of the omma-

tidia and pigmentation; the explants were kept alive for about 35 days.
The consequences of the irradiation of 80- to 85-hour-old larvae of *Drosophila melanogaster* on the *in vitro* evolution of their eye–antennal discs were examined by Horikawa and Sugahara (1960). The cultures were suspended after 72 hours. The experiments showed that when eye discs irradiated with a dose of less than 20 krad were cultured together with unirradiated cephalic complexes, the eye discs showed normal growth and differentiation. When eye discs were irradiated with a dose of 25 krad, degeneration appeared. On the other hand, when the cephalic complexes were irradiated with 0.5 or 3 krad, the associated eye discs showed normal growth and differentiation; but when cephalic complexes were irradiated at a dose level of 5–15 krad, no differentiation appeared in the eye discs although its growth remained normal. Thus, in eye discs the effect of radiation on either growth or differentiation appeared to be separable by the dose–effect relations. In conclusion, it seems that the primary effect of irradiation of the larva in a disturbance of the hormonal function of the cephalic complex would be to provoke an inhibition of the differentiation of the discs.

Fujio (1960, 1962) used *in vitro* culture for analyzing, on the one hand, the effects of the substances which increase the number of eye facets, and on the other, the effects of the cephalic complexes on the development of eye–antennal discs of different strains of *Drosophila melanogaster*. The explants were dissected from approximately 95-hour-old larvae and cultured for 48 hours. The quantity of metamorphic hormone secreted by five cephalic complexes in hanging drops seems to be sufficient to assure the growth and the differentiation of eye–antennal discs, this differs from the quantities given by Horikawa (1960). But on the other hand, and this is in keeping with preceding authors, the cephalic complexes of different eye-color mutants seem to liberate substances orienting the eye differentiation according to the characteristic type of each strain. Nevertheless, facet-increasing substances (ammonium lactate, acetamide, urea, and polypeptone) would only seem to have some effect by the intermediary action of the cephalic complexes; they have no effect upon isolated discs. The comparison of different associations of eye discs and heterologous cephalic complexes in media containing facet-increasing substances indicates that the genetic action controlling the growth and differentiation of eye–antennal discs appears first on the cephalic complexes in eye–mutant strains. The eye–antennal discs of the Dp strains seems to be different from other eye-mutant strains in the responsibility to the cephalic complexes of the wild type. In another set of experiments the same author cultured eye–antennal discs in a medium containing some substances secreted from

the cephalic complexes which were previously cultured in the rotating
tubes. In order to do this, 10 to 60 cephalic complexes were cultured
during 24 hours in 1 ml of synthetic medium and then withdrawn. Eye–
antennal discs were then cultured for 48 hours in this medium. The substance secreted, per milliliter of medium, by at least 20 cephalic complexes of Oregon strain is sufficient to bring on differentiation of Oregon
eye discs. Sixty cephalic complexes of Oregon strain are enough for
starting differentiation of eye discs of other strains except the Dp strain.
Moreover, Bar strain cephalic complexes seem to liberate B substances,
other than the metamorphic hormone, which appear to bring on characteristic eye growth and differentiation of the Bar strains. Urea, a facet-
increasing substance, seems to have specific effects in inhibiting the secretion of B substances by the cephalic complexes of the mutants. We
might conclude, then, that the genic action operates in the secretion of
B substances which promote the characteristic growth and differentiation
for eye-mutant strains through the cephalic complexes of the eye-mutant
larvae.

In vitro pigmentation of larval and pupal eye discs of Drosophila
melanogaster obtained under different experimental conditions should
also be mentioned (Hanly, 1961; Hanly et al., 1967; Burdette et al.,
1968; Sengel and Mandaron, 1969; Mandaron, 1970).

Schneider (1963, 1964) carried out further studies, by in vitro culture,
of the differentiation of larval eye–antennal buds of Drosophila melanogaster and D. virilis. The explants came from 95-hour-old larvae reared
at 25°; they included brain and ventral ganglion, eye–antennal discs,
ring gland, aorta, lymph gland, esophagus, and proventriculus. The cultures were examined each day, and the ommochrome pigments were
analyzed by the redox indicator method of Ephrussi and Herold, the
pteridins by chromatography. Immediately after being set in culture,
segmentation of the antennal disc occurred. After 4 days the eye discs
underwent morphogenetic movements, i.e., they shifted upwards and
outwards and 10% of the eye discs took on a normal hemispheric configuration. After 7–12 days, the cornea, lens, and ommatidia were noticed.
Pigmentation began after 5–8 days, and it passed from yellow to red
(drosopterins) in 5–12 days. The phenomena began around the margin
of the eye. Chemical analysis revealed the normal succession of all the
pigments. Four to 6 days after explantation, the three parts of the
antenna were seen; the hairs formed, and the aristae were often present.
This differentiation occurred for the explants of both D. melanogaster
and D. virilis, but with a time-lag of 1–3 days for the latter. It should
be noticed that replacing the medium is favorable to survival but arrests
differentiation. If the explants come from 72- to 85-hour-old larvae,

there is no development. Schneider repeated the experiments of Fujio (1962) but obtained only very limited differentiation in these experimental conditions; similarly, the repetition of the cultures made by Horikawa (1958) was equally disappointing. The author attributes the time-lag of *in vitro* development which she observed in comparison with *in vivo* development, to the imperfect composition of the culture medium. Schneider's regret of the lack of precision of many earlier authors in the estimation of larval age at the time of explantation and the lack of objective criteria with many modern authors for judging *in vitro* growth and differentiation is to be shared, for it makes comparison of results difficult, if not impossible. It was in order to define these criteria that Schneider's second study (1966) included histological analysis of organs cultivated *in vitro*. This analysis revealed good survival up to 14 days for the antennal discs, an average survival for the brain and eye discs, and a short survival, 24–48 hours, for the ventral ganglion. The eye-antennal discs did not evaginate but shifted only to a position normally attained in the late prepupal stage. According to Schneider, reports of evagination of larval discs explanted *in vitro* must be regarded with scepticism. The histological examination of eye discs cultured *in vitro* for one day showed a differentiation of the upper layer but little or no organization of the lower layer. After 5–6 days the cornea formed, the facets were defined, and the hairs appeared but in more than normal numbers. If the upper part of the ommatidia showed more or less normal differentiation, the retinulae showed reduced and atypic development. After 10–15 days, the explants degenerated. The antennal disc underwent normal histological differentiation and reached the adult stage about seven days after explantation. It can be concluded, with Schneider, that in general, if a developmental sequence takes place within 24 hours after puparium formation, this same sequence will occur *in vitro* without a serious time-lag; if such a sequence normally takes place after this period, the time-lag not only increases considerably but there is also much greater diversity among the explants with respect to degree of development. The fact that some organs in culture differentiate more fully than others can probably be attributed to three main factors: (a) the relative extent of injury during dissection, (b) vulnerability to adverse conditions in an artificial environment, and (c) inherent capacity for autonomous development.

Courgeon (1969) reported some observations concerning the survival of *Calliphora erythrocephala* larval eye discs maintained in culture alone or in association with the brain and the ring gland for a time up to three weeks.

On the whole, the works which we have just recalled have made important contributions in pointing out different factors favoring differentiation of eye–antennal discs of insects. There are some divergences in the results, but these are of little importance. It is to be hoped, however, that more precise criteria will be given in order to allow a better estimation of the progress made and so that still further progress may be made.

3. Cephalic Complexes

In most of the investigations of *in vitro* culture of eye–antennal discs which we have just mentioned, the explants included the cerebral lobes, the ventral ganglion, and the ring gland. Certain authors have emphasized the necessity for the whole of the cephalic complex to be present in order to assure growth and differentiation of cephalic or other buds (Kuroda and Yamaguchi, 1956; Horikawa, 1958; Kuroda, 1959; Gottschewski, 1960). In 1960 Horikawa estimated at 10 the number of cephalic complexes necessary in the culture medium for assuring differentiation of eye–antennal discs; Horikawa and Sugahara (1960) confirmed this inductive action of the cephalic complex by experiments in irradiation. It seems that the genic action proper to certain mutants is carried out on the differentiation of the buds by the intermediary action of the cephalic complex; likewise, the facet-increasing substances seem to act first on the cephalic complex (Fujio, 1960, 1962). Gottschewski and Querner (1961), remarking the absence of *in vitro* differentiation for eye buds not yet attached to the brain at time of explantation, made experimental study on the possibility of the diffusion of substances within the brain and from the brain towards eye–antennal buds. They proceeded by *in vitro* injection of fluorochromes in larval brains of *Drosophila melanogaster*. This showed that substances can diffuse towards the eye–antennal discs from the medulla externa, but it seems to be separated from the remainder of the brain by some sort of barrier. The brain may well serve, then, as an inductor for the differentiation of cephalic buds.

The presence of the cephalic complex in the culture medium in association with organs other than the cephalic buds also seems to be favorable. This has been shown by the experiments on *in vitro* parabiosis between the ovaries and cephalic complexes of *Tenebrio molitor* made by Lender and Laverdure (1967). Unfortunately, the poor *in vitro* survival of the cephalic complex obliged them to change this organ daily. Leloup and Demal (1968) have shown the favorable action of prolonged *in vitro* parabiosis with the cephalic complex on the differentiation of the prepupal ovary of *Calliphora erythrocephala*. But if in certain cases (Leloup and Demal, 1968; Leloup, 1970) its action seemed to be of hormonal nature, in other cases (Leloup, 1969) it was shown to be acting by enriching the medium in subtances which may be considered as essential nutrients for certain organs *in vitro*.

B. Thoracic Imaginal Discs

The works carried out by Frew (1928), the histological importance of which has been mentioned above, concern the explantation of leg imaginal discs of *Calliphora erythrocephala* in the larval hemolymph and pupal extract filtered on collodion. The author observed the evagination of discs arising from old larva when cultivated in pupal extract. This phenomena does not occur for younger explants nor for explants cultivated in larval hemolymph. The fact that evagination occurs *in vitro* proves that it is not due to an increase of internal pressure or to body turgescence. The use of salt solutions of differing concentrations shows that the evagination does not result from a mere variation of osmotic condition but rather from changes in the composition of the hemolymph.

Following these works, other investigators report the *in vitro* evagination of wing or leg imaginal discs of different insects, i.e., the last larval stage of *Aedes aegypti* (Trager, 1938; Chabaud *et al.*, 1960), third larval stage of *Drosophila* (Fischer and Gottschewski, 1939; Demal, 1955, 1956; Kuroda, 1959; Horikawa and Sugahara, 1960; Sengel and Mandaron, 1969; Mandaron, 1970), young prepupae of *Calliphora erythrocephala* (Demal, 1955, 1956; Demal and Leloup, 1963), pupae of *Glossina palpalis* (Trager, 1959b), and last larval stage of *Galleria mellonella* (Oberlander and Fulco, 1967; Oberlander, 1969a,b). Injuring the explants brings on the formation of cell migrations (Fischer and Gottschewski, 1939; Chabaud *et al.*, 1960).

Certain authors have carried their investigations further. Demal (1955, 1956) and Demal and Leloup (1963) verified survival and differentiation of leg buds of *Calliphora erythrocephala* by means of histological techniques. Demal described the bud evolution as thinning of the epithelium, disappearance of reserve cells in the light of the bud, formation of a mesenchymatic membrane and tarsal tendon, and development of tracheolar structures. Finally, he signaled that the imaginal buds of *Drosophila melanogaster* behaved in a comparable way *in vitro*. Similar observations were also reported by Sengel and Mandaron (1969) and Mandaron (1970) for the same insect.

Kuroda (1959) and Horikawa and Sugahara (1960) used Kuroda and Tamura's technique for cultivating imaginal discs of *D. melanogaster*. Kuroda explanted wing buds of wild and vg strains. He observed different behavior of buds of vg strains from the point of view of growth according to the temperature at which they were kept. On the other hand, from the moment of parabiosis with the cerebral complex of either, homologous or heterologous strain, the wing buds develop in a similar way, which suggests that the realization of the wild or vestigial phenotype of the wing is determined at this stage.

Horikawa and Sugahara (1960) study the action of *in vitro* behavior of their imaginal discs in parabiosis with the cephalic complex. The radiosensitivity of wing buds is greater than that of leg buds. It has been noticed that the growth of wing buds is inhibited when they have undergone an irradiation of 10 krad, that of leg buds for doses above 20 krad. Furthermore, parabiosis with cephalic complexes which have received 20 krad, inhibits growth of wing buds but not of leg buds, which, however, cease to differentiate. The differentiation of wing buds is not considered.

Oberlander and Fulco (1967) and Oberlander (1969a,b) successfully studied *in vitro* the action of ecdyson and other hormones on wing discs of the last instar larva of *Galleria mellonella*. They succeeded in initiating the metamorphosis of the organs with ecdyson even when they were explanted at a very early stage. Similarly, Agui *et al.* (1969a) cultivating wing discs of the last instar larva of *Chilo suppressalis* observed their moulting. This result was obtained only when a certain stage of differentiation was already reached at the time of explantation, and ecdyson was stated to be no longer required for the moulting of older discs. This suggested the existence of a critical stage which required the intervention of the hormone to promote further differentiation.

C. Circulatory Organs

Since the majority of explantations of the dorsal vessel were made for physiological studies, there are but few works for us to consider, and more often than not, they concern the dorsal vessel only incidentally. Thus, in the publication of Gavrilov and Cowez (1941), the key to a figure shows that these authors were able to maintain contractions for 10 days in the heart of *Stegomyia fasciata*. The culture medium consisted of chick plasma to which had been added "embryonic juice of mosquito larvae."

Demal (1961) and Demal and Leloup (1963) explanted prepupal hearts of *Calliphora erythrocephala* in different media and observed, for one of them, the continuance of contractions for more than 10 days. The lymph elements and the nephrocytes which accompanied the aorta were shown in perfect survival. The prolongation of the experience without renewal of the medium allowed the observation of continued contractions for two months but not of the survival of adjacent formations (unpublished results). Other media used by these same authors yielded similar results (Leloup and Demal, 1968).

David and Rougier (1965) explanted the dorsal vessel of *Cybister lateralimarginalis* in a medium of varying pH for the purpose of testing this factor. This will be dealt with at greater length in another chapter.

In a medium made up of TC 199 and for shorter periods in other media, Larsen kept in contraction fragments of the embryonic heart of *Blaberus craniifer* for 260 days (1963) and of *Blaberus discoidalis* (1967) for over 2½ years.

Other authors explanted *Drosophila melanogaster* larval heart with lymph glands (Castiglioni and Rezzonico Raimondi, 1961, 1963; Rezzonico Raimondi and Ghini, 1963; Schneider, 1964). But the results obtained, i.e., liberation of lymph cells, concern histiotypic culture and do not, therefore, come within the scope of this chapter.

D. Digestive Organs

The digestive tube and the salivary glands of vector insects have been the object of numerous *in vitro* explantations in order to serve as substrata for parasites. The insects used were almost all Diptera. The explanted organs were salivary glands (Gavrilov and Cowez, 1941) or fragments of the digestive tube of different mosquitoes (Trager, 1938; Ball, 1947, 1948, 1954; Ball and Chao, 1960, 1963; Ragab, 1948, 1949) or of *Glossina palpalis* (Trager, 1959a,b). Often, the explants gave rise to cell migrations (Trager, 1959b), and this was sometimes the authors' main objective (Hirumi and Maramorosch, 1963). The survival of the explants was generally estimated from the maintenance of contraction.

Apart from these works, which will be considered in another chapter, we have very little information concerning the culture of digestive organs. Demal (1955, 1956) signaled the maintenance for four days of proventricular contractions of larval *Drosophila melanogaster*. Horikawa and Sugahara (1960) observed a distinct increase in size of the salivary glands of the same insect in Kuroda and Tamura's medium (1956). An irradiation of the larvae with a dose less than 15 krad does not affect the behavior of the glands. Demal and Leloup (1963) obtained survival of these glands in Vago and Chastang's medium (1958) and in Trager's A₂ medium (1959b). Larsen (1963) observed continued contractions of fragments of the digestive tract of *Blaberus craniifer*

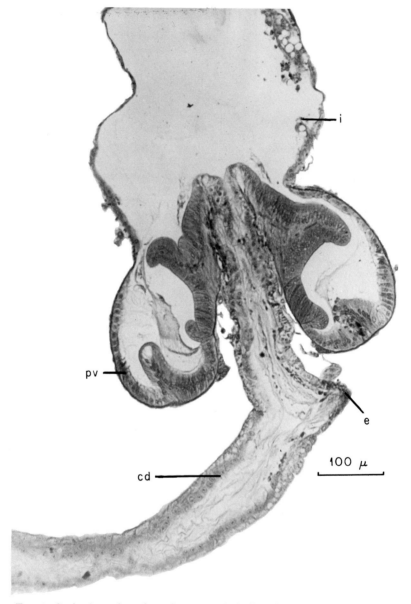

Fig. 2. Sagittal section through a nymphal digestive tube of *Calliphora ery-throcephala* maintained *in vitro* for seven days in medium M_1. The explant consisted of the esophagus (e), the crop and its duct (cd), the proventriculus (pv), and the beginning of the intestine (i). It was still contracting at the time of fixation, and histological control showed perfect survival (stained with hematoxylin-eosin).

during 67 days in TC 199 medium. Demaure (1968) observed a good survival of larval salivary glands and cardias of *Lucilia sericata* for two weeks and six days, respectively. During that time, the salivary glands were secreting and the cardias contracting. Nymphal digestive tracts of *Calliphora erythrocephala* were occasionally successfully maintained *in vitro* in our laboratory for seven days (Leloup, unpublished) (Fig. 2).

Cannon (1964) explanted larval salivary glands of *Sciara coprophila* for the purpose of studying the modification of chromosome development under the influence of the cerebral complex in parabiosis. The cells of the front area were not affected by the presence of the complex, whereas those behind only developed "puffs" in its presence. The use of tritiated thymidine shows that the synthesis of DNA continues in culture.

Plantevin *et al.* (1968) cultured the mesenteron of fifth larval stage *Galleria mellonella*. They used four different culture media, the precise composition of which was not stated. The length of the experiment varied from one case to another. Certain explants were kept *in vitro* for 28 days. By means of histological techniques these authors were able to observe, for one of these media, the onset of metamorphosis in culture. The larval epithelium became detached from the basal membrane and began to degenerate, and there was a beginning of proliferation of regenerative cells. The histological observations were compared with enzymatic analyses of the explants. Putting intestines into the culture prolonged the preservation of a certain number of enzymes, which disappeared very quickly during *in vivo* metamorphosis. This time-lag is accompanied with the degeneration of the larval epithelium. Measurements of the pH of the medium during the experiments also showed a close connection between the decrease, or the disappearance, of certain enzymes and observed pH variations. One of the chief merits of this work was to have brought out the possibilities of enzymatic control of *in vitro* organ cultures.

E. Gonads

The culture of genital organs is generally limited to the explantation of gonads severed from the genital ducts and annexes. In certain cases, it is not the whole organ which is cultured, but fragments making up well-defined morphological units, i.e., ovarioles, follicles, or even cysts.

They are used mainly for studies in morphogenesis and physiology. This implies a precise knowledge of both the structure of the gonads at the time of explantation, and their normal *in vivo* development, as well as experiments controlled by cytological or histological techniques. The observation of continued contractions, or the maintenance of tissue transparence, gives only the preliminary indications concerning the survival and functioning of the explants, and it seems to be happily completed by thorough microscopic examination. This would have the added advantage of giving more valid support to the interpretation of the growth and development of the explants.

1. Testes

We shall not refer again here to the works which Goldschmidt (1915, 1916, 1917), Lewis (1916), and Lewis and Robertson (1916) carried out on spermatogenesis. They have been dealt with at the beginning of this chapter.

In 1940, Stern published preliminary results on the cultures of whole testes from *Drosophila melanogaster* and *D. virilis* in diluted seawater. Withdrawing these organs from the action of the internal body medium, the author attempted to show the influence of genital ducts on their growth and spiralization. Testes taken from young nymphae in which the testes–genital ducts junction had not yet taken place showed no change in shape, and far from showing any sort of growth, gave evidence of shrinking. Older explants grew longer and thinner as they curved. Spiralization was more pronounced in *D. melanogaster* than in *D. virilis*. This development, which is similar to that observed *in vivo*, occurs equally in the presence or in the absence of the genital ducts. The survival of the explants was estimated according to their transparence. Unfortunately, this work was without future.

Within the framework of research on the physiology of diapause, Williams and his collaborators used the technique which Goldschmidt had drawn up some 30 years earlier. Schmidt and Williams (1949, 1953) explanted, in hemolymph, cysts from diapausal pupae of *Platysamia cecropia* and *Samia walkeri*, and they observed that under certain conditions spermatogenesis started again. According to these authors, the power which hemolymph has for helping the development of germ cells would result from its hormonal fraction, and the rapidity with which spermatogenesis takes place would be a test for detecting the presence, in the hemolymph, of the hormone influencing growth and differentiation during development.

Schneiderman *et al.* (1953), using this same technique, studied the action of different factors on spermatogenesis. In this way they were able to determine that it is an aerobic process in which the Krebs cycle and oxidative phosphorylations play some part, that the optimal tem-

perature lies between 25 and 30°, and that the pressure of oxygen satura-
tion of the terminal oxidase mediating the respiration is lower than
8 mm Hg. Among the metabolic inhibitors which were experimented with,
it was remarked that the phenylthiourea used for inhibiting blood
tyrosinase is not able to block spermatogenesis.

The importance of the standardization of technique and a strict control of experimental conditions is emphasized by the accident which
befell Ketchell and Williams in 1955. They had noticed that results
were influenced by the number of cultures per chamber, and this led
them to suppose that there existed some volatile influence, the properties
of which they were able to study without, however, being able to identify
it. Bowers (1961) and Laufer and Berman (1961) determined simultaneously that this factor is the water vapor formed from the culture
media. The same vapor pressure is set up in each case, from a greater
or lesser volume of medium, and this brings about an inversely propor-
tional modification of the concentration of the dissolved constituents.

Horikawa and Sugahara (1960), using Kuroda and Tamura's medium
(1956), carried out cultures of the testes of *Drosophila melanogaster*
in parabiosis with the cephalic complex. The organs were taken from
larvae, some of which had been given varying doses of irradiation and
some, none at all. When there had been no irradiation, the testis in
culture showed growth and differentiation, manifested by a distinct
increase in volume and a slight change of shape. When the testis had
been given a dose of less than 10 krad, or the cephalic complex a dose
of more than 5 krad, there was inhibition of both growth and differentiation of the testis. This showed that these organs are highly radiosensitive
and indicated a direct action of the cephalic complex on the testis in
culture.

From 1961 onwards, Demal and Leloup published results concerning
cultures of the testis of *Drosophila melanogaster* (Demal, 1961) and of
Calliphora erythrocephala (Demal, 1961; Demal and Leloup, 1963;
Leloup, 1964, 1969). Almost simultaneously, Lender and Duveau-
Hagège (1962, 1963a,b, 1965) and Duveau-Hagège (1964) published
results of work on other insects, *Galleria mellonella* and *Periplaneta
americana*. These experiments were undertaken with the aim of studying certain problems of morphogenesis. The different authors report
that they were able to observe good survival of explants during varying
periods of culture. They obtained a continuation of spermatogenesis
(Fig. 3) but not its onset. These results, histologically controlled, are
similar if we take into consideration, not the stage of the insects, but
the state of the differentiation of the gonads at the time of explantation,
i.e., with *Drosophila* spermatogenesis starts at the time of pupation;

30   J. DEMAL AND A. M. LELOUP

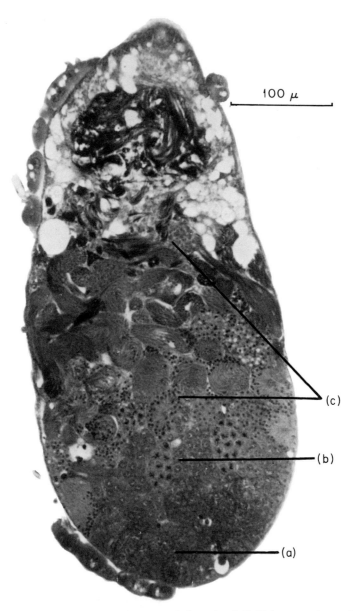

Fig. 3. Axial section through a nymphal testis of *Calliphora erythrocephala* after 72 hours culture in medium M₁H, showing (a) the zone of primary spermatogonia, (b) the zone of germ cell growth and multiplication, and (c) the zone of spermiogenesis (stained with hematoxylin-eosin).

with *Calliphora*, at the phanerocephalic nymphal stage, and with *Galleria*, at the end of the larval period. Except for *Calliphora*, in certain conditions, testes explanted before this moment usually show poorer *in vitro* survival. The same may be said of the primary spermatogonia zone even at more advanced stages. But recently, we succeeded in obtaining a better survival of prepupal testes of *Calliphora* (Leloup, 1969), and further, we were able to determine that, from this stage, for this insect the cerebro–endocrine complex played no, or no further, direct role in the starting and continuation of spermatogenesis. On the contrary, Williams and Kambysellis (1969) reported that ecdyson was necessary for obtaining spermiogenesis *in vitro* in intact testes of diapausing pupae of *Samia cynthia*, but not in free spermatocytes. The authors suggested that the hormone was acting by altering the permeability of the testicular walls to allow a macromolecular blood factor to enter the testes. Similarly, Yagi *et al.* (1969), cultivating testes of diapausing larvae of *Chilo suppressalis*, reported that ecdysterone promoted spermiogenesis. In our opinion, the apparent contradiction between these last two studies and the previous one may arise from the fact that, in the last two cases, the insects were in a diapausing stage at the time of explantation.

Before ending this section, we must point out that Hirumi and Maramorosch (1963) explanted in different media nymphal and adult testes of Cicadellidae. They signaled survival for more than three months as well as mitosis and cell migrations. These cultures were to serve as substrata for viral studies.

2. Ovaries

Trager (1938) carried out the explantation of ovaries from young females of *Aedes aegypti* in a physiological solution richened in organic matter by the addition of vertebrate plasma. The somatic part of the explant remained in good condition for several weeks, but the germ cells degenerated within a few days. This is of secondary importance for the author since his cultures were to be used as the cellular foundation necessary for virus cultures.

Beckel (1956) had quite another aim in view. He set out to discover the nature of the development of the egg in the adult female mosquito. The explants were taken from autogenous and anautogenous species, *Aedes aegypti, A. hexodontus,* and *A. communis.* They gave evidence of contraction and good survival in Morgan's synthetic medium for periods varying between 7 and 60 days but did not show the slightest sign of differentiation. The author signaled some cell migrations.

In order to study the phenomena of histogenesis occurring during metamorphosis of Diptera, Demal and Leloup cultured ovaries of *Calliphora erythrocephala* (Demal, 1961; Demal and Leloup, 1963; Leloup, 1964; Leloup and Demal, 1968). The works undertaken by Lender and his collaborators have to do with the differentiation of the ovaries and with vitellogenesis in *Galleria mellonella* and *Periplaneta americana* (Lender and Duveau-Hagège, 1962, 1963a,b, 1965; Duveau-Hagège, 1963, 1964) and, later, in *Tenebrio molitor* (Laverdure, 1967, 1969; Lender and Laverdure, 1967, 1968). Ovaries of *Iphita limbata* were cultivated by Ittycheriah and Stephanos (1969).

The explants used for these works were taken from insects at different stages of development. Here again, as for the testes, the different stages showing comparable development of the ovaries must be determined.

The stages used were as follows: for *Calliphora,* the pronymph and the pigmented phanerocephalic nymph; for *Galleria,* the end of the last larval stage, the nymph, and the adult; for *Periplaneta,* a preimaginal stage and the adult; for *Tenebrio,* different nymphal stages and the young unfed adult; for *Iphita,* late fifth instar.

The pronymphal ovary of *Calliphora* shows but very little differentiation. There is no sign of the ovarioles, and the germen is composed of primary gonia. The division of the ovary into ovarioles starts within the first hours following the formation of the puparium. The larval ovaries of the other species considered show greater differentiation, and their degree of development is that observed in the phanerocephalic nymph of *Calliphora.* Vitellogenesis occurs during nymphosis for *Galleria* (Lender and Duveau-Hagège, 1963a,b), before the imaginal change for *Periplaneta* (Duveau-Hagège, 1963), and *Iphita* (Ittycheriah and Stephanos, 1969), and 24–48 hours after this change for *Tenebrio* (Lender and Laverdure, 1967). For *Calliphora,* it occurs later on, at the adult stage (Thomsen, 1942). Having thus determined the points of comparison, we shall now consider successively the works dealing with the pronymphal ovaries of *Calliphora,* then those concerning the stages preceding vitellogenesis, and finally, those studying the stages of vitellogenesis.

Demal (1961) obtained pronymphal ovary survival of *Calliphora* in a medium containing extract of chick embryo. But the differentiation was insignificant and was not improved by parabiosis with the brain or change of medium. These results were confirmed in 1963 (Demal and Leloup, 1963). The use of new culture media based on blood composition led to prolonged survival accompanied with more or less pronounced growth during the first few days and varied development according to the case (Leloup, 1964; Leloup and Demal, 1968). These are parallel experiments, in which use is made of a basic medium supplemented, in certain cases with hemolymph, and in which the ovary was sometimes put in parabiosis with the cephalic complex or the imaginal discs of thoracic appendages. The results may be summarized as follows: When the ovary alone was cultured the basic medium allowed survival, but differentiation and growth were insignificant. The addition of hemolymph favored survival and growth but frequently brought about abnormal development. Parabiosis with the brain was favorable to growth and differentiation in the basic medium (Fig. 4) but delayed growth and stimulated differentiation in the medium to which blood had been added. Parabiosis with imaginal buds in the basic medium favored slight growth; in the presence of hemolymph it delayed growth but hindered abnormal development. In these experiments, it is difficult to distinguish the factors resulting from nutritional influence and those which are caused by hormonal action.

Concerning the ovaries, which show more advanced development, it seems that a fairly good survival was obtained for all the insects. The germarium seems to be the ovariole zone giving the least good results.

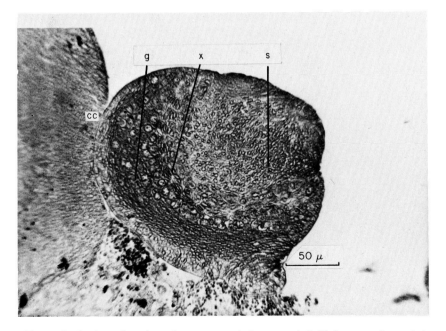

Fig. 4. Sagittal section through a pronymphal ovary of *Calliphora erythrocephala* after culture for three days in the medium M₁, in parabiosis with the cephalic complex (cc), showing the first signs of ovarioles formation (x) at the limit of the somatic (s) and germinal (g) areas of the explant.

In 1961 Demal obtained the survival of the nymphal ovary of *Calliphora* for a period of at least 12 days in a medium containing chick embryo extract. Following this it was shown that it is possible to prolong these cultures for a period as long as 15 days either in the liquid medium or in this same medium gelosed (Leloup, unpublished). But no trace of differentiation was observed. The use of new media (Leloup, 1964) gave slightly less good results for survival since there occurred some necrosis of the youngest germ cells, but a certain differentiation was obtained and was shown in some cases by the separation of the last egg chamber, or by a change in colorability of the oldest germ cells.

The larval ovaries of *Galleria* showed a high survival percentage on gelosed chick embryo medium to which had been added mammal serum (Lender and Duveau-Hagège, 1962, 1963a,b; Duveau-Hagège, 1964). When, after seven days in culture, these were grafted in the abdomen of last larval stage caterpillars, they are able to continue normal development (Lender and Duveau-Hagège, 1965). The basic medium also allowed for an increase of oocytes, differentiation of the nurse cells (some of which showed division), organization of the follicles, and previtellogenesis. Different modifications of the medium, as well as parabiosis with the cephalic complex, were tried out in view of bettering these results. The addition of crushed head brought on greater growth of the oocytes.

Lastly, the results concerning ovaries during the period of vitellogenesis may be summarized as follows. Lender and Duveau-Hagège (1963) obtained no survival of the adult gonad containing eggs ready for laying, nor of isolated nymphal ovarioles of *Galleria*. But Duveau-Hagège (1963, 1964) managed to keep alive ovarioles of *Periplaneta* for seven days, except for the adult vitellarium. From the point of view of differentiation, there was no notable growth of oocytes, but the follicular cells actively multiplied and produced a substance having the same characteristics as the vitellus.

Ovarian cultures of *Tenebrio* (Lender and Laverdure, 1967, 1968; Laverdure, 1967) brought about, for the first time, simultaneous survival of different ovarian structures and a certain differentiation. The basic medium, containing trehalose, allowed the survival of the nurse cells and medium-sized oocytes; when glucose replaced trehalose, deposit of vitelline droplets occurred (Fig. 5). In the adult ovaries, the addition of the ether-insoluble fraction of the fat body of the adult females assured the survival of young oocytes. The other fraction brought on growth of the ovocytes. Finally, the addition of both fractions together not only verified the results obtained separately but also assured simultaneous survival of different parts of the ovariole, growth of oocytes, and vitellogenesis. These results were obtained for cultures lasting from 4 to 6½

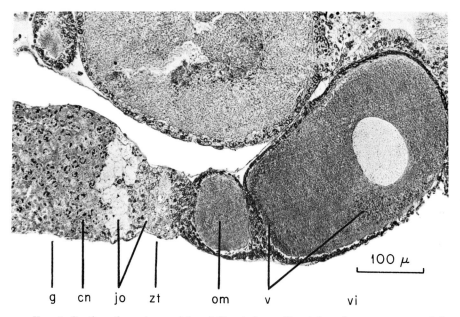

g   cn   jo   zt   om   v   vi

Fig. 5. Section through ovariole of *Tenebrio molitor* taken from a young unfed adult female, at most 48-hours old, after four or five days culture in standard medium, with glucose in presence of cephalic complex: cn, nurse cells; g, germarium; jo, young oocytes; om, medium-sized oocytes; v, vitellus; vi, vitellarium; zt, zone of transition. (From Lender and Laverdure, 1968.)

days. In nymphal ovaries, ecdysone was shown to influence the growth of the oocytes and vitellogenesis (Laverdure, 1970).

Similar results were obtained by Ittycheriah and Stephanos (1969) for *Iphita*. The basic medium did not allow any differentiation to be seen. The brain and corpus allatum from a mating female influenced the growth of the germarium. An extract of fat body from mating females permitted the initiation of oogenesis and deposition of small yolk droplets in the oocytes. In the presence of both the fat body extract and the brain corpus allatum, oogenesis and vitellogenesis were obtained.

Hirumi and Maramorosch (1963) mention ovaries among the organs of Cicadellidae explanted for cellular substrata to virus.

IV. CONCLUSIONS

There is no need to supply further proof of the interest which *in vitro* organ culture holds for the study of insects. The various results which wc have just summarized and the long list of authors quoted are witness enough. The obstacles which confronted the workers in *in vitro* insect organ culture were many and redoubtable; i.e., the small size of the material, difficulties in obtaining aseptic organs and hemolymph, the scanty knowledge of the internal body medium, the important variations which it undergoes in the course of metamorphosis, the very relative stability of the concentration of its components, and finally, discontinuity in development involving alternation of very rapid phases with slower ones.

Most of these difficulties have now been overcome, at least in part, thanks to the recent progress made in physiology, biochemistry, and the knowledge of insect development, thus, allowing the preparation of more adequate culture media. Nevertheless, still further research is necessary before the very precious tool which is *in vitro* culture can be used with maximum efficacy. Many investigations would profit by being concentrated on one stage or one well-defined problem. It is also greatly important that rigorous experimental conditions should be observed in the use of this technique: the setting up of a parallel between organ behavior *in vitro* and *in vivo;* mention of the exact stage at which explantation was made; and exact control of explants by histological, enzymatic, or biochemical techniques. The fulfillment of these conditions will have the double advantage of allowing investigators to communicate their experimental results with greater precision and of creating the possibility of greater coordination and efficacy in the drawing up of research programs.

REFERENCES

Agui, N., Yagi, S., and Fukaya, M. (1969a). *Appl. Entomol. Zool.* 4, 158–159.
Agui, N., Yasi, S., and Fukaya, M. (1969b). *Appl. Entomol. Zool.* 4, 156–157.
Ball, G. H. (1947). *Amer. J Trop. Med.* 27, 301–307.
Ball, G. H. (1948). *Amer. J. Trop. Med.* 28, 533–536.
Ball, G. H. (1954). *Exp. Parasitol.* 3, 358–467.
Ball, G. H., and Chao, J. (1960). *Exp. Parasitol.* 9, 47–55.
Ball, G. H., and Chao, J. (1963). *Ann. Epiphyt.* 14, 205–210.
Barsa, M. C. (1955). *J. Gen. Physiol.* 38, 79–92.
Beckel, W. E. (1956). *Nature (London)* 177, 534–535.
Bowers, B. (1961). *Science* 133, 42–43.
Burdette, W. J., Hanley, E. W., and Grosch, H. (1968). *Tex. Rep. Biol. Med.* 26, 173–180.
Butz, A. (1957). *J. N. Y. Entomol. Soc.* 65, 22–31.
Cannon, G. (1964). *Science* 146, 1063.
Castiglioni, M. C., and Rezzonico Raimondi, G. (1961). *Experientia* 17, 88–90.
Castiglioni, M. C., and Rezzonico Raimondi, G. (1963). *Experientia* 19, 527–529.

Chabaud, M. A., Chelelovitch, S., and de Lalun, E. (1960). *Bull. Soc. Pathol. Exot.* **53**, 170–172.

Chambers, R. (1925). *Cellule* **35**, 105–124.

Courgeon, A. M. (1969). *C. R. Acad. Sci.* **268**, 950–952.

Daillie, J., and Prudhomme, J. C. (1966). *Ann. Nutr. Aliment.* **20**, 353–360.

David, J., and Rougier, M. (1965). *C. R. Acad. Sci.* **261**, 1394–1396.

Demal, J. (1955). *Bull. Acad. Roy. Belg. Cl. Sci.* [5] **41**, 1061–1071.

Demal, J. (1956). *Ann. Sci. Nat. Zool.* **18**, 155–161.

Demal, J. (1961). *Bull. Soc. Zool. Fr.* **86**, 522–533.

Demal, J., and Leloup, A. M. (1963). *Ann. Epiphyt.* **14**, 91–93.

Demaure, J. C. (1968). *C. R. Soc. Biol.* **162**, 224–227.

Duveau-Hagège, J. (1963). *C. R. Acad. Sci.* **256**, 5429–5430.

Duveau-Hagège, J. (1964). *Bull. Soc. Zool. Fr.* **89**, 66–69.

Fischer, I., and Gottschewski, G. (1939). *Naturwissenschaften* **27**, 391–392.

Frew, J. C. H. (1928). *J. Exp. Biol.* **6**, 1–11.

Fujio, Y. (1960). *Jap. J. Genet.* **35**, 361–370.

Fujio, Y. (1962). *Jap. J. Genet.* **37**, 110–117.

Gavrilov, W., and Cowez, S. (1941). *Ann. Parasitol.* **18**, 180–186.

Gianfelici, E. (1968a). *Ann. Endocrinol.* **29**, 495–500.

Gianfelici, E. (1968b). *Proc. Colloq. Int. Invertebr. Tissue Cult., 2nd, 1967*, pp. 155–167.

Goldschmidt, R. (1915). *Proc. Nat. Acad. Sci. U. S.* **1**, 220–222.

Goldschmidt, R. (1916). *Biol. Zentralbl.* **36**, 160–167.

Goldschmidt, R. (1917). *Arch. Zellforsch.* **14**, 421–450.

Goodchild, A. J. P. (1954). *Nature (London)* **173**, 504–505.

Gottschewski, G. (1958). *Naturwissenschaften* **45**, 400.

Gottschewski, G. (1960). *Wilhelm Roux' Arch. Entwicklungsmech. Organismen* **152**, 204–229.

Gottschewski, G., and Fischer, I. (1939). *Naturwissenschaften* **27**, 584.

Gottschewski, G., and Querner, W. (1961). *Wilhelm Roux' Arch. Entwicklungsmech. Organismen* **153**, 168–175.

Grace, T. D. C. (1954). *Nature (London)* **174**, 187.

Greenberg, G., and Archetti, I. (1969). *Exp. Cell. Res.* **54**, 284–287.

Hanly, E. W. (1961). *Diss. Abstr.* **22**, 980.

Hanly, E. W., and Hemmert, W. H. (1967). *J. Embryol. Exp. Morphol.* **17**, 501–511.

Hanly, E. W., Fuller, C. W., and Stanley, M. S. M. (1967). *J. Embryol. Exp. Morphol.* **17**, 491–499.

Hirumi, H., and Maramorosch, K. (1963). *Ann. Epiphyt.* **14**, 77–79.

Horikawa, M. (1958). *Cytologia* **23**, 468–477.

Horikawa, M. (1960). *Jap. J. Genet.* **35**, 76–83.

Horikawa, M., and Sugahara, T. (1960). *Radiat. Res.* **12**, 266–275.

Ittycheriah, P. I., and Stephanos, S. (1969). *Indian J. Exp. Biol.* **7**, 17–19.

Judy, K. L. (1969). *Science* **165**, 1374–1375.

Ketchel, M. M., and Williams, C. M. (1955). *Biol. Bull.* **109**, 64–74.

Kuroda, Y. (1959). *Med. J. Osaka Univ.* **10**, 1–16.

Kuroda, Y., and Tamura, S. (1956). *Med. J. Osaka Univ.* **7**, 137–144.

Kuroda, Y., and Yamaguchi, K. (1956). *Jap. J. Genet.* **31**, 98–103.

Larsen, W. (1963). *Life Sci.* **8**, 606–610.

Larsen, W. (1967). *J. Insect Physiol.* **13**, 611–619.

Laufer, H., and Berman, R. H. (1961). *Science* **133**, 34–36.

Laverdure, A. M. (1967). *C. R. Acad. Sci.* **265**, 505–507.

Laverdure, A. M. (1969). *C. R. Acad. Sci.* **269**, 82–85.
Laverdure, A. M. (1970). *Année Biol.* **9**, 455–463.
Leloup, A. M. (1964). *Bull. Soc. Zool. Fr.* **89**, 70–77.
Leloup, A. M. (1969). *Ann. Endocrinol.* **30**, 852–856.
Leloup, A. M. (1970). *Année Biol.* **9**, 447–453.
Leloup, A. M., and Demal, J. (1968). *Proc. Int. Colloq. Invertebr. Tissue Cult., 2nd, 1967,* pp. 126–137.
Leloup, A. M., and Demal, J. (1970). Unpublished.
Leloup, A. M., and Gianfelici, E. (1966). *Ann. Endocrinol.* **27**, 506–508.
Lender, T., and Duveau-Hagège, J. (1962). *C. R. Acad. Sci.* **254**, 2825–2827.
Lender, T., and Duveau-Hagège, J. (1963a). *Ann. Epiphyt.* **14**, 81–89.
Lender, T., and Duveau-Hagège, J. (1963b). *Develop. Biol.* **6**, 1–22.
Lender, T., and Duveau-Hagège, J. (1965). *C. R. Soc. Biol.* **159**, 104–106.
Lender, T., and Laverdure, A. M. (1967). *C. R. Acad. Sci.* **265**, 451–454.
Lender, T., and Laverdure, A. M. (1968). *Proc. Int. Colloq. Invertebr. Tissue Cult., 2nd, 1967,* pp. 138–146.
Lewis, M. R. (1916). *Anat. Rec.* **10**, 287–299.
Lewis, M. R., and Robertson, W. R. (1916). *Biol. Bull.* **30**, 99–114.
Loeb, M. J., and Schneiderman, H. A. (1956). *Ann. Entomol. Soc. Amer.* **49**, 493–494.
Ludwig, D., Tefft, E. R., and Suchyta, M. D. (1957a). *J. Cell. Comp. Physiol.* **49**, 503–508.
Ludwig, D., Tracey, K. M., and Burns, M. L. (1957b). *Ann. Entomol. Soc. Amer.* **50**, 244–246.
Mandaron, P. (1970). *Develop. Biol.* **22**, 298–320.
Marks, E. P. (1968). *Gen. Comp. Endocrinol.* **11**, 31–42.
Marks, E. P. (1969). *Biol. Bull.* **137**, 181–188.
Marks, E. P. (1970). *Symp. Arthropod Cell Cult. Appl. Study Viruses, 1970* Abstract No. 15.
Marks, E. P., and Leopold, R. A. (1970). *Science* **167**, 61–62.
Marks, E. P., and Reinecke, J. P. (1964). *Science* **143**, 961–963.
Marks, E. P., and Reinecke, J. P. (1965). *Gen. Comp. Endocrinol.* **5**, 241–247.
Marks, E. P., Reinecke, J. P., and Leopold, R. A. (1968). *Biol. Bull.* **135**, 520–529.
Martin, B., and Nishiitsutsuji-Uwo, J. (1967). *Biochem. Z.* **346**, 491–495.
Maximov, A. (1925). *Contrib. Embryol. Carnegie Inst. Wash.* **16**, 49–113.
Miciarelli, A., Sbrenna, G., and Colombo, G. (1967). *Experientia* **23**, 64–66.
Nauck, E. G., and Weyer, F. (1941). *Zentralbl. Bakteriol., Parasitenk. Infektionskr., Abt. 1* **147**, 365–367.
Oberlander, H. (1969a). *J. Insect Physiol.* **15**, 297–304.
Oberlander, H. (1969b). *J. Insect Physiol.* **15**, 1803–1806.
Oberlander, H., and Fulco, L. (1967). *Nature (London)* **216**, 1140–1141.
Ozbaz, S., and Hodgson, E. S. (1958). *Proc. Nat. Acad. Sci. U. S.* **44**, 825–830.
Peleg, J., and Trager, W. (1963). *Ann. Epiphyt.* **14**, 211–212.
Pihan, J. C. (1969). *C. R. Acad. Sci.* **268**, 2074–2077.
Plantevin, G., Nardon, P., and Laviolette, P. (1968). *Proc. Int. Colloq. Invertebr. Tissue Cult., 2nd, 1967.*
Ragab, H. A. (1948). *Trans. Roy. Soc. Trop. Med. Hyg.* **41**, 434–441.
Ragab, H. A. (1949). *Trans. Roy. Soc. Trop. Med. Hyg.* **43**, 225–230.
Ramsay, J. A. (1954). *J. Exp. Biol.* **31**, 104–113.
Rezzonico Raimondi, G., and Ghini, C. (1963). *Ann. Epiphyti.* **14**, 153–159.
Schaller, F., and Meunier, J. (1967). *C. R. Acad. Sci.* **264**, 1441–1444.

Schmidt, E. L., and Williams, C. M. (1949). *Anat. Rec.* **105**, 487.
Schmidt, E. L., and Williams, C. M. (1953). *Biol. Bull.* **105**, 174–187.
Schneider, I. (1963). *Genetics* **48**, 908.
Schneider, I. (1964). *J. Exp. Zool.* **156**, 91–104.
Schneider, I. (1965). *Genetics* **50**, 284.
Schneider, I. (1966). *J. Embryol. Exp. Morphol.* **15**, 271–279.
Schneiderman, H. A., Ketchel, M., and Williams, C. M. (1953). *Biol. Bull.* **105**, 188–199.
Sengel, P., and Mandaron, P. (1969). *C. R. Acad. Sci.* **268**, 405–407.
Stanley, S. M., and Vaughn, J. L. (1968). *Ann. Entomol. Soc. Amer.* **61**, 1064–1067.
Stern, C. (1940). *Growth* **4**, 377–382.
Thomsen, E. (1942). *Vidensk. Medd. Dansk Naturh. Foren. Kjobenhavn* **106**, 320–405.
Thomson, D. T. (1914). *Proc. Roy. Soc. Med.* **7**, 71–75.
Trager, W. (1937). *J. Parasitol.* **23**, 226–227.
Trager, W. (1938). *Amer. J. Trop. Med.* **18**, 387–393.
Trager, W. (1959a). *Nature (London)* **184**, 30–31.
Trager, W. (1959b). *Ann. Trop. Med. Parasitol.* **53**, 473–491.
Vago, C., and Chastang, S. (1958). *Experientia* **14**, 426–427.
Vago, C., and Chastang, S. (1960). *C. R. Acad. Sci.* **251**, 903–905.
Wayne Wiens, A., and Gilbert, L. I. (1965). *Science* **150**, 614–616.
Williams, C. M., and Kambysellis, M. P. (1969). *Proc. Nat. Acad. Sci. U. S.* **63**, 231.
Wolff, E., and Haffen, K. (1952). *Texas Rep. Biol. Med.* **10**, 463–472.
Yagi, S., Kondo, E., and Fukaya, M. (1969). *Appl. Entomol. Zool.* **4**, 70–78.

2

THE ORGANOTYPIC CULTURE OF
INVERTEBRATES OTHER
THAN INSECTS

L. Gomot

I. INTRODUCTION

The contributions of the culture technique for cells, tissues, or organs are considerable in the field of biological research.

Experiments in the culture of animal or vegetable tissues were attempted as early as the end of the 19th century: Weltner (1893) tried to obtain cultures of the internal part of the larva of the sponge *Spongilla;* Loeb (1897) cultured fragments of adults rabbit organs for three days in a test tube containing a little serum or a clot of plasma; Haberlandt (1902) tried to obtain cultures of isolated cells of the palisade tissue of leaves, of medullary parenchyma, and of the epidermis of various plants. Haberlandt did not attain his object but his working hypotheses were prophetic. He attempted to culture isolated cells of higher plants in order to see if they could function as elementary organisms and to determine the reciprocal influences that the individual cells of a multicellular plant may undergo. Haberlandt foresaw that if isolated plant cells were successfully cultured it would be possible to prove experimentally the supposed *totipotence* of all the living cells of higher plants and that one would then have the means of experimentally directing the course of cell differentiation. Verification of his hypothesis was only proved within the past few years.

In animals, on the other hand, differentiation of frog embryo neuroblasts (Harrison, 1907) and the indefinite multiplication of chicken organ cells (Carrel and Burrows, 1910) were the starting point of experiments on *in vitro* explantation. In amniote vertebrates the problems treated covered a wide field, as already reviewed by Carleton in 1923. These methods gave rise to two main tendencies, depending on the conditions of culture. One of them, *tissue culture,* aims at obtaining abundant and indefinite multiplication of cells. The other maintains the organization of the explants, and at any moment of the culture, the structure and the function of the explanted tissues or organs are recognizable. After the first differentiations with early organ primordia of chick embryos were obtained, this type of culture, called "organotypic" by Maximow (1925), was widely utilized in the vertebrates. This technique made it possible to obtain the differentiation of organ rudiments removed at various stages of embryonic development. Explantation is achieved in different ways, depending on the authors, the media, and the vessels (see Volume 1, Chapter 2).

Cultures of adult vertebrate organs have also been achieved [Trowell (1952, 1961), various organs of adult mice; Prop (1959), mammary glands of adult mice; and Petrovic (1960), guinea pig hypophyses].

The problems solved by organ culture have been analyzed in several publications. In the book "Les cultures organotypiques" edited by Thomas (1965), Et. Wolff and his collaborators show the importance of the results achieved in vertebrates and announce the possibilities offered by this technique in invertebrates. Information on techniques and applications is to be found in the work "Cells and Tissues in Culture" by Wilmer (1965, 1966).

In vitro culture has made it possible to deal with a large number of problems of animal and plant physiology. However, up to now, most of the experiments in the culture of animal organs have been performed with vertebrate material. The vertebrates are, in fact, the most independent of animals (with regard to their environment) as a result of the acquisition of a stable internal medium. The presence of numerous organs implies correlations which are precise and diversified and which produce modifications that cause the cells to be different from the cells from which they originate.

In order to establish this fact it is necessary to do as diversified as possible research on animals other than vertebrates. Hence, the scope for study represented by the invertebrate phyla should be an interesting source for the understanding of progressive specialization which has gained the interest of so many contemporary biologists.

The results obtained by the application of *in vitro* invertebrate organ culture techniques have become more and more numerous over the past few years. The general themes developed in vertebrate research work are again encountered. The culture may be simply a means of "keeping alive," thus allowing physiological investigations to be made. It is also a process of analysis of the metabolism of the organs isolated on synthetic media. Above all, it represents a good means for the study of the capacity of autodifferentiation of isolated embryo or larvae organ primordia and for the analysis of differentiation and organogenesis. Lastly, since this technique has been extended to explantations of very young amniote vertebrate embryos made up of a few cells, it is very useful for studying the embryonic development of invertebrates which are dependent for their development on particular conditions of the environment that are generally provided either by their parents' organs or by the tissues of intermediary hosts.

Two general lines of research seem to be developing for invertebrates, due to the fact that cultures are as feasible with embryo organs as with adult organs. These are, on the one hand, the search for correlations between organs and, on the other hand, the dissociation of simple organs which should enable the study of dedifferentiation or of the irreversibility of cell differentiation in animals.

The different phyla of invertebrates represent types of organization that are so different from one another that we shall first of all review the main research work performed in each one of them and, subsequently, give details of the essential general results.

In invertebrates it is sometimes difficult to distinguish tissue culture from organ culture since certain invertebrates such as sponges and coelenterates do not have clearly differentiated organs. Hence, we shall report the results of the explantations with respect to the structural units and their functions since, in some types of invertebrate organisms, certain cells are the equivalent of vertebrate organs.

II. REVIEW OF ORGAN CULTURE IN THE VARIOUS PHYLA

A. Protochordates

In the course of experimental segmentation of the ascidian *Clavelina lepadiformis* Müller into four parts, Driesch (1902) and, later, Schultz (1907) observed regeneration by epimorphosis, that is, by reparation of the missing parts from the visceral sac or from the branchial corbeilla maintained in seawater.

When they placed fragments of the branchial corbeilla in small vessels containing seawater, they observed both the regeneration and the transformation of the cells, which finally formed a cellular mass in which morphogenetic processes, comparable with those of embryonic development, took place. The detailed study of the *in vitro* organogenesis of these "reduction bodies" of Driesch has never been resumed experimentally. However, Brien (1930, 1932, 1969), taking inspiration from Driesch's experiments, showed the part played by polarity in the course of regeneration. When he isolated the esophagus portion of a *Clavelina*, he discovered that there is heteromorphosis, and the segment in culture became bithoracic. But if the sectioning is done successively, the first preceding the second by 72 hours, then regeneration becomes monothoracic.

More recently, new ascidian organs have been cultured *in vitro*.

P. Sengel (1961) adapted the embryo organ culture technique of Et. Wolff and Haffen (1952) to the culture of marine invertebrate organs by replacing the Tyrode's solution in the medium with seawater which had been previously filtered and sterilized. He explanted the ovary and the pericardial organ of young *Ciona intestinalis* Fleming that had not yet reached sexual maturity. These organs showed little modification in structure during seven days in culture.

P. Sengel and Kieny (1962, 1963) utilize this technique for studying the maturation of the genital products of the solitary ascidian *Molgula manhattensis* De Kay. Among the culture media (see Volume 1, Chapter 2) found to be the most suitable for the ovary and the testis that have been removed at various stages of their development are the mediums CP and HS, containing respectively chicken plasma and horse serum. The P medium with Morgan 199 solution and EJ medium with chicken embryo extract have a beneficial action on the ovocytes, while the LS medium, consisting basically of Limule serum, suits the spermatocytes; but neither liquid gives really satisfactory results.

On the other hand, juxtaposition in the same medium of a gonad and a GNV complex (neural gland–nerve ganglion–vibratile organ) improves the viability of the germinal elements of the gonads and provokes the differentiation of certain ovocytes. The determination of the maturation of the gametes does not seem to depend directly on this complex (see this volume, Chapter 5).

The Et. Wolff and Haffen (1952) medium has also been adapted to the study of morphogenesis and secretory activity of the endostyle of *Ciona intestinalis* Fleming by Ghiani *et al.* (1964). These authors explanted the endostyle of immature and mature animals to gel media (see Volume 1, Chapter 2).

An SM medium, consisting basically of seawater and agar gel, is suitable for the culture of explants of mature animals while the organs of immature animals grow better on an SE medium containing chick embryo extract diluted with Hanks' solution. For immature explants the authors observe *in vitro* differentiation by way of a thickening of the lateral cells of space 7 defined and studied by Barrington and Franchi (1956a,b) and Barrington (1957) which appears to be the result of an inhibitory action being suppressed. Lastly, the structure of the endostyle is better conserved and its viability made to be of longer duration. The authors generally used the neutral SN medium for studying *in vitro* the action of thyrotropic substances on the histological structure of lateral zone 7, which was capable of fixing iodine.

On a TSHM medium with thyrotropic hormone the endostyle of immature specimens undergoes cytological modifications in the region of segment 7, which increases in length. On the PTM medium containing propylthiouracil the cells of segment 7 of the endostyle of mature specimens are taller and more globular than those of the controls.

The action of these substances is difficult to interpret at the cellular level, but the modification of the secretions as well as their change in rate of production by substances having a specific action on the thyroid cell seem to plead in favor of the hypothesis of an endocrine function.

B. Echinoderms

1. Experiments in Survival and Study of Structural Dedifferentiation

As early as 1911 Wilson (1911a) dissociated immature gonads of *Asterias arenicola* by pressing them through gauze so as to determine whether cells which have not yet achieved specialization remain capable of dedifferentiating. The dissociated cells group together and form cellular masses which survive for two days in seawater but do not undergo any other changes.

Thomas (1941a) observed that the ovary of the sea urchin *Paracentrotus lividus* Lamark survives for a few days in seawater after spawning. The ovary fragment heals and swims by means of the coordinated movement of its ciliature. The author proposed a technique for experimental survival in a nutrient medium that is renewed every day. This survival technique that Thomas (1963) qualified as an intermediary between culture and breeding is applicable to the small external organs of aquatic invertebrates and lower vertebrates.

In echinoderms, as in other marine invertebrates, certain organs remain in their state of differentiation, whereas others undergo structural dedifferentiation.

The experiments carried out by Thomas (1941b,c) involved the external organs (pedicels, ambulacral tubes, tentacles, and pinnules) of various classes of echinoderms.

In echinoids, holothuroids, crinoids, and ophiuroids, most of the organs kept alive undergo extensive structural dedifferentiation. This varies with the different types of pedicels of echinoids and is more pronounced in the ambulacral tubes of echinoids and ophiuroids, in fragments of holothuroid tentacles, and in the crinoid pinnules. This regression in shape is particularly striking in the ambulacral tubes of *Sphaerechinus granularis* Lamark and in the fragments of tentacles of *Cucumaria planci* Brandt. The histological study of these tentacles, on the 15th and 23rd days of survival reveals considerable growth with dedifferentiation of the epidermis and of the canals while the fibrillary structures of the central zone are disordered and undergo slow autolysis. In general, dedifferentiation is more rapid and pronounced in small fragments. This led Thomas (1941c) to put forward the hypothesis that the main cause of structural dedifferentiation, during survival, is probably due to the gradual depletion of a supply of substances, which are perhaps specific and indispensable for maintaining differentiation.

On the other hand, the ambulacral tubes and, particularly, the pedicels of asteroids survive for a long time without any important modification in structure. For instance, the small right dorsal and lateral pedicels of the *Marthasterias glacialis* L., isolated in filtered seawater, maintain their appearance and functions for over three months. After rapid healing of the sectioned surface of the pedicle they continue to contract, to twist, and to open and close their valves, either spontaneously or after stimulation by contact. Histological study shows that on the 60th day of survival, the structure of these pedicels remains practically normal. The musculature, connective tissue, and calcareous skeleton of the valves retain their normal appearance. In the epidermis, the mucous glands are empty, but in the poison glands the secretion accumulates and the remains of various victims that are supposed to have been phagocytosed may be observed. These organs are also capable of regenerating parts of their clarareous framework, which seems to indicate that calcium metabolism has taken place in the surviving organ. Moreover, as these organs frequently expend muscular energy, Thomas (1941b) presumed that these organs are not reduced to total fasting leading to autophagia but that they feed by absorbing dissolved substances through their teguments. Incubation experiments of these organs in seawater containing either ferrous citrate or trypan blue carried out by this author in 1941 prove that these substances pass into the epidermal cells and into the free cells of the subjacent connective tissue. Attempts to incorporate nutrient protein substances were made, but they considerably increase microbial pollution and only provided preliminary indications in the invertebrate organs studied by Thomas, who improved his technique by devising an antiseptic synthetic nutrient medium which enables organs of Urodela to survive (Thomas and Borderioux, 1948).

2. Survival and Study of Metabolic Processes

So as to ascertain the part played by the coelomic fluid of the starfish *Asterias forbesi* Desor in the process of transferring nutrient substances from the supply area of the digestive glands to other organs, Ferguson (1964) explanted the main organs of the body (digestive glands, stomach, gonads, and rectal coeca) in coelomic fluid or seawater. The long survival of these organs under these conditions reveals the relatively autonomous existence that they are able to lead. These organs are capable of absorbing amino acids and glucose labeled with ^{14}C added to the seawater or the coelomic fluid. Analyses were made over only 12 hours, during which time the explants remained in good health and showed spontaneous movements. The analysis of the rate of absorption of the labeled substances by the isolated digestive glands enables the author to conclude that the labeled substances are not only taken from the liquid medium but that there is a turn over of these substances which are replaced by other unlabeled molecules.

Measurements of the passive diffusion of nitrogenous substances from the isolated digestive glands revealed that a rapid flow of organic food material took place between the internal liquid and the tissues of the starfish. Recently, Ferguson (1968) kept the starfish (*Echinaster spinulosus*) digestive glands alive in seawater and found that 15 amino acids may cross the peritoneal surface.

3. Organ Culture of Gonads

a. Sexual Differentiation in *Asterina gibbosa* Penn. Experiments by Delavault and Bruslé (1965) have shown that gonads of the hermaphrodite starfish *Asterina gibbosa* Penn. survive for about 40 days in *in vitro* culture on a nutrient medium derived from that of Et. Wolff and Haffen (1952). Organ culture of these gonads enabled these authors to pursue research on the determination of the differentiation of the two sexual types in this hermaphrodite species.

The experiments on explantation in an agar medium, the composition of which is given in Volume 1, Chapter 2, are carried out at 14°, 16°, or 18°C on animals coming from Banyuls (Mediterranean) or from Roscoff (Channel). They remained alive for 30, 40, 50, or 60 days.

The presence in the *A. gibbosa* of five pairs of gonads in a similar genital state enabled one or two gonads to be removed from the same animal for *in vitro* culture, one or several autografts to be performed, and control gonads to be kept by rearing the operated animal at the same temperature as the culture.

The differentiations observed by Bruslé (1966, 1967, 1968) depended on the genital state of the donor protandric hermaphrodite which had spermatogenic activity in autumn and in winter.

In gonads explanted while in dominant ovogenetic activity and maintained in culture for 40 days, the young ovocytes remained intact, the more mature ovocytes degenerated, but the "vesiculous tissue" characteristic of the gonads of *A. gibbosa* in functional ovogenesis developed considerably. Bruslé (1967, 1968) considered this increase of the vesiculate tissue to be a form of recuperation of the ovogenetic material capable of ensuring the survival and growth of a new female generation which appeared at the periphery of the gonad in the form of numerous young ovocytes.

When the gonads are explanted while in dominant spermatogenic activity, if spermatogenesis is well established, it continues normally and without the occurrence of vesiculous tissues. If spermatogenesis is at its beginning, it is reduced and ovogenetic development appears at the periphery of the gonad at the base of the spermatic colonnettes.

According to Bruslé (1968) "these facts obviously reflect that in *A. gibbosa* the tendency towards ovogenetic evolution is always preponderant." Also, the fact that the gonads behave in the same way in both *in vitro* culture and in autografts seems to indicate that the differentiation factors of the gamete types are not dispersed in the internal medium of the animal. The gonad of these animals has a constant tendency to evolve in the female direction, but the determination of the induction of spermatogenic activity has yet to be explained.

The fact that the evolution of autografted gonads differs from that of the control gonads would seem to warrant the attention of research workers interested in this question. Indeed, is this difference due to a surgical stress or does the operation separate the gonad from the localized influence of a sexual determining factor?

b. Meiotic Maturation and the Spawning Mechanism in the Starfish. In order to study the maturation and spawning mechanisms of the starfish, Kanatani (1964) and Kanatani and Ohguri (1966) isolated gonads from *Asterias amurensis* and *Asterina pectinifera* in artificial seawater (Van't Hoff's) to which they had added an active substance extracted from the nerve ring and the radial nerves. The active substance seemed to be identical in the two sexes, and this gamete shedding substance was found in the radial nerve at the same concentration in both the male and female starfish. This radial nerve factor is a polypeptide (Kanatani, 1967). Kanatani and Shirai (1967) have established that this polypeptide (probably a neurosecretory substance) acts on the ovary and induces the production of a second active substance which is responsible for ovocyte maturation and shedding of the gametes. The second substance has recently been isolated and identified as 1-methyladenine in *Asterias amurensis* by Kanatani *et al.* (1969). Synthetic 1-methyladenine was also found to be very effective in inducing ovocyte maturation and spawning *in vitro* in *Asterina pectinifera, Asterias amurensis, Marthasterias glacialis, Astropecten aurantiacus,* and *Ceramaster placenta.* The effects of various adenine derivatives were also studied *in vitro* with *Asterias forbesi.* The experiments of Kanatani and Shirai (1969) on *A. amurensis* indicate that the ovarian factor can act through the wall of a ligated ovarian fragment to produce ovocyte germinal vesicle breakdown and dispersion of the follicular cells without shedding. The effect on spawning of deficiency of bivalent cations in seawater was also examined. Ovarian fragments immersed in magnesium-free water released their eggs after a certain interval while those treated with calcium-free seawater for an appropriate period spawn after subsequent addition of calcium. At present the exact effect of various ions on starfish spawning is far from clear. The maturation of oocytes and spawning control may be regulated by several other mechanisms. Chaet (1964) observed that high concentration of the radial nerve factor (RNF) of the starfish *Patiria miniata* added to the ovaries *in vitro* inhibited shedding, thus suggesting that a second neurosecretory substance was present in the radial nerve which counteracted the RNF. Ikegami *et al.* (1967) isolated a gonadal substance that inhibits the action of the spawning factor. This material is present in both the ovary and testis of *Asterias amurensis*. This spawning inhibitor was purified by using the testis of *A. pectinifera* and chemically identified as L-glutamic acid.

C. Arthropods

1. Crustaceans

In crustaceans attempts at organ culture were made by Lewis (1916) in the hermit crab by explantation of fragments of the first chela undergoing regeneration in a liquid medium consisting of crab broth, and this showed only the emigration of small connective or epithelial cells and of a few muscle fibers.

Later, Fischer-Piette (1929, 1933), by explanting fragments of adult lobster lymph gland in a drop of coagulated lobster plasma, noted that the architecture of the lymphocytogenic follicles was preserved and that they continue functioning in a normal manner for over four days, producing lymphocytes by mitosis, which are contained in the medium. Fischer-Piette interpreted this result by the fact that the lobster's lymph gland does not possess blood vessels and that already in the organism the nutrition of the cells and the elimination of waste material takes place through the parenchyma cells.

In the light of knowledge acquired since then the difference in behavior between the fragments of lymphatic gland and vertebrate tissue cultures to which the author compares them is also due to different properties of the media. In fact, the media used in vertebrate tissue cultures are often made up of plasma and embryo fluid which causes rapid cell proliferation. Pantin (1934), by perfusing the walking legs of the crab *Carcinus moenas* Leach, removed by autotomy, with just a physiological saline solution (see Volume 1, Chapter 2), kept the muscles alive for eight hours at temperatures varying from 12° to 17°C and studied the response of crustacean muscles to stimulations by alternating electric current.

The inconsistent results obtained in invertebrate organ culture in several experiments were doubtless the cause of it being little used. In crustaceans it was only in 1962 that new attempts at culture were made by Berreur-Bonnenfant on the gonads of an amphipod crustacean, *Orchestia gammarella* Pallas, using Et. Wolff and Haffen's medium (1952) which had been adapted to crustaceans by replacing the Gey solution by seawater so as to reproduce the internal medium of the crustaceans studied (see Volume 1, Chapter 2).

Testes of adult *O. gammarella* survive 14–18 days on the "Standard C" medium containing chick embryo juice, and the activity of the gonad is seen by the gonial mitoses. However, if the gametogenesis in progress is completed, the germinative zone of the isolated testes degenerates and no further gametogenesis takes place. Gametogenesis is not renewed. Later Berreur-Bonnenfant (1963a,b, 1964) undertook experiments for the purpose of determining precisely the effect on the germinal cells of the hormone elaborated by the androgenic gland discovered by Charniaux-Cotton (1954) (this volume, Chapter 5). On the basis of endocrinological data obtained from ablation or organ graft experiments (Charniaux-Cotton, 1954, 1955; Arvy *et al.*, 1954), Berreur-Bonnenfant (1963a) explanted androgenic glands of *O. gammarella*. The survival of this gland is less than that of the testis. By association of the testis and androgenic gland in the same culture medium gametogenesis, once it has begun, continues just as in the isolated gonads and a new generation of spermatogonia are derived from the germinal area, which afterwards degenerates.

The influence of the androgenic gland on the evolution of the germinal area of the gonad was also shown in other species. In particular, in the *Talitrus saltator* Montagu, in the testis cultured singly, the appearance of ovocytes as early as the 14th day of culture is noted, while in the presence of the androgenic gland (Berreur-Bonnenfant, 1963b, 1964) no onset of ovogenesis was observed. The same observation was made in the protandrous hermaphrodite isopod *Anilocra physodes* L. The hormone released by the androgenic gland prevents the onset of ovogenesis, and though it seems necessary for the maintenance of the germinal area, it does not suffice to prevent its degeneration.

The same author (1963b) then studied the explantation of various organs and their associations with the testis of *O. gammarella*. The moult gland of *Carcinus moenas* survived for eight days in culture but became vacuolated, and this hastened the degeneration of the testicular germinal area of *Orchestia;* the viability of the androgenic gland was not improved.

The brains of animals undergoing male sexual activity cultured together with the testis and renewed every other day maintain the integrity of the germinal area; those of the female do not have an effect on this area.

The association of testis, androgenic gland, and male brain enable the germinal area to maintain its activity, and often two successive waves of spermatogenesis were released. The female brain has no action on the maintenance of the germinal area of the testis in culture (Berreur-Bonnenfant, 1966, 1967). The analysis of the interactions of associated organs has revealed up to now that the androgenic gland acts on the brain of *Orchestia gammarella* and provokes the secretion of the factor which ensures the maintenance of the integrity of the germinal area of the testis. (This factor is present only in the brains of males in which the androgenic gland has not been ablated and in females where the androgenic gland has been implanted.)

Hence, the technique of *in vitro* cultures coupled with ablation and graft techniques shows that male sexual differentiation of higher crustaceans depends on the action of the androgenic gland on the testis and on the brain, which in turn also acts on the testis. The problems of invertebrate endocrinology therefore show many similarities with those of vertebrates, and in the particular case of male crustaceans, where the gonad is distinct from the secretory organ of the androgenic hormone, it is possible to imagine new culture experiments combined with castration, for example, which will perhaps make it possible to demonstrate the feed-back phenomena so important in vertebrate endocrinology.

Oyama and Kamemoto (1970) maintained ovarian explants of the portunid crab *Thalamita crenata* H. Milne Edwards in organ culture in defined media. The organ culture method employed was the floating lens paper technique introduced by Chen (1954) for liquid media. The use of either artificial seawater or crab salines in conjunction with Medium 199 without serum gave similar and favorable results. Cultures were incubated for seven days in a constant temperature chamber set at 25°.

The culture media promoted the proliferation of interstitial cells and some growth of young ovocytes. The germinative zone undergoes cell division and is encompassed by many adherent ovogonia. Maturing ovocytes in later stages of development, however, were not maintained.

2. Myriapods

The technique of *in vitro* culture has recently been attempted in the chilopod myriapod *Lithobius forficatus* L. by Zerbib (1966b). Zerbib observed a phase of sexual undifferentiation covering the first three intermoultings of the larval stage. He showed (Zerbib, 1966a) that all the undifferentiated gonads, grafted into specimens, whether sexually differentiated or not, develop into ovaries. This seems to indicate that the realization of the ♀ sex is governed by autodifferentiation. Up to now this author has explanted sections of young larvae of *L. forticatus* containing the undifferentiated gonad. The culture process is performed according to the Wolff and Haffen technique, but the physiological solutions were replaced by diluted seawater (see Volume 1, Chapter 2). In this medium, consisting basically of seven-day-old chick embryo extract, the explants survived for approximately 12 days without cell proliferation and no sexual differentiation took place.

3. Merostomes (*Limulus*)

a. Cultures in Liquid Medium. The first explantations of fragments of the hypodermis, the heart, and muscle of *Limulus* performed by Lewis (1916) in the medium used to culture hermit crab tissues simply produced the emigration of cells of undetermined nature.

A little later on, Loeb (1922, 1927) obtained *in vitro* the agglutination of the amoebocytes of *Limulus* blood which formed a tissue structure without proliferation. Loeb undertook an interesting study of the influence of the environmental factors on cell agglutination and cohesion. Even today this is still a captivating point of interest with the experiments in organ dissociation and reorganization.

Limulus serum was used as a basic medium in a certain number of Loeb's experiments. In particular, he showed that the presence of proteins in the medium is necessary for the preservation and maintenance of the amoeboid movements of the amoebocytes. Loeb extensively studied the influence of the osmotic pressure and the pH of various saline solutions, and he noted that these factors have the same effect on the agglutination of the amoebocytes of the king crab and on the archaeocytes derived from the dissociation of sponges.

b. Cultures on Agar Gel Medium. Em. Wolff (1962, 1963) explanted fragments of the dorsal heart, liver, nervous system, intestine, and gonads of the king crab *Xiphosura polyphemus*. These were removed from a young king crab 3- to 6-cm long, carefully washed in sterile seawater with the addition of penicillin to obtain aseptic conditions. The behavior of the various organs was observed in media consisting basically of agar and filtered sterile seawater to which various natural nutrient substances had been added (e.g., king crab lymph, king crab serum, chick embryo extract, horse serum, chicken plasma) or synthetic substances (e.g., solution 199 by Morgan *et al.*, 1950) (see Volume 1, Chapter 2).

All the media used enabled the ventral nerve cord to survive for seven days, during which time the neurons retained a perfect structure. It was the same for the liver, where the tubules retained their structure and functions for the same length of time and even longer on the HS medium containing horse serum.

The segments of intestine were kept alive for five days on the HS medium containing horse serum, LS medium containing king crab serum, and EE medium containing chick embryo fluid.

The dorsal heart explanted on the HS medium containing horse serum or on medium P containing Morgan's synthetic solution survived on an average for 10 days, and most of the explants continued to contract spontaneously. The other media were less suitable.

The media used by Wolff hence enabled the differentiated explanted organs to preserve their structural and functional integrity outside the organism when no infection was present.

4. Arachnids

To our knowledge, few experiments have been carried out on arachnid organ culture. Up to now only acarid and opilionid organs have been explanted. These experiments were undertaken for two different purposes: either to study the behavior of viruses or protozoans in the tissues of the vector animal or for a morphogenetic and endocrinological study. Although the purposes of these two research studies were different, it is to be noted that the observations made in each case yielded data that is useful in both fields of research.

a. **Explantations Performed for the Purpose of Studying the Behavior of Protozoans or Viruses.** Here the explantations are used in two ways. Whole organs of adults or of larvae at various stages of their development are placed in culture, and depending on their nature or on the composition of the culture medium, they tend either to retain their organization or else their cells proliferate and emigrate from the explant.

A detailed analysis of the utilization of organ culture in pathology is dealt with in a separate chapter (see this volume, Chapter 7).

To precisely determine the developmental cycle of the protozoan *Theileria parva* Theiler in the vector, Martin and Vidler (1962) cultured all the organs of the cavity of nymph and adult ticks (*Rhipicephalus appendiculatus* Neumann): the explants were placed in Leighton tubes and incubated at 17°–18° in one of the three media A, B, and C described in Volume 1, Chapter 2.

Among the media tested only medium B (consisting basically of Eagle's medium in Hanks solution plus 20% ox serum) enabled tick tissues to survive and grow.

The adult donor animals (same number of males and females) were not fed, and their tissues revealed a reduced metabolism. The results of adult tick organ culture vary according to the number of days after metamorphosis. In general, the persistence of contractions and peristaltic movements during the first few days of culture (3–17 days) were noted. The tissues remained alive up to 27 and 46 days, and in one case contraction movements continued up to the 163rd day. Histological sections of this explant on the 170th day showed several salivary gland and intestinal epithelium cells in good health. Cell multiplication outside the explants was always slight and was inhibited at each renewal of the medium.

The explants taken from nymphs in the course of metamorphosis showed a state of intense activity and acquired the structure of adult organs. The culture of nymphal organs produced cell proliferations that were greater than those of adult organ cultures.

In 1958, Rehácek maintained organs of *Dermacentor marginatus* Sulzer (five-day-old nymphs) in hanging drops in order to study the viability of the tick encephalitis virus. The organs explanted into four different media continued to develop; in particular, chitinization continues and the damaged organs partly regenerate by migration of cells.

Later, Rehácek and Pesek (1960), Rehácek and Hana (1961), and Rehácek (1962, 1963) performed other tick organ cultures which enabled the propagation of Eastern Equine encephalomyelitis (E.E.E.), but the modifications in the cultured organs (salivary glands, gut, ovaries, malpighian tubes and fat body as well as the hypodermis) were not specified.

Organ culture in acarids deserves to be developed since the results obtained by Rehácek (1962), for instance, clearly show the influence of the nature of the organ on the proliferation of viruses.

b. Organ Culture for Various Morphological and Endocrinological Purposes. As in all other invertebrate phyla, organ culture would be an excellent means for studying embryonic or larval development and endocrinology. Streiff and Taberly (1964) maintained organs of oribatid acarids *Xenillus tegeocranus* Hermann and *Platynothrus peltifer* Koch in culture on an Ag gel medium mixed with seawater, glucose, and egg albumin.

The genital system, the central nervous system, and the intestine retained their anatomical structure for seven days in culture without reseeding.

More recently, Fowler and Goodnight (1966) succeeded in obtaining the culture of cephalic nerve tissues and intestine of the opilionid *Leiobunum longipes* on the B medium of Martin and Vidler (1962).

These organs survived well in culture for periods ranging from a few weeks to one year. The neurosecretory tissues retained their activity, but no experiment has yet been attempted to determine the precise part played by this neurosecretion.

D. Mollusks

In 1968 Bayne published a short review of works dealing with the culture of mollusk tissues and organs and summarized the most interesting physiological results. This chapter deals with three main successive aspects of mollusk organ cultures.

1. "Mixed" Culture of Organ Fragments

Mollusk organs gave rise to many attempts at culture at the time when vertebrate tissue cultures were successfully achieved through the utilization of embryo juice and blood plasmas.

If the hectocotylized arm of dibranch cephalopods is capable of surviving naturally during the transmission of the spermatophores after autotomy at the moment of reproduction, all mollusk organs do not possess the same aptitudes. In particular, the attempts by Dobrowolsky (1916) at *Octopus* tissue culture in the hemolymph of this animal were not successful.

Tissues of the eulamellibranch *Anodonta* were later explanted by several research workers. Krontowski and Rumianzew (1922) kept *Anodonta* tissues alive for a fairly long period by the combined media technique. These authors never observed real growth in their cultures. Zweibaum (1925a,b) observed the survival of vibratile epithelium of *Anodonta* gills for a period of up to 63 days in Ringer solution diluted by half. Zweibaum showed the dominating influence of the osmotic pressure of the medium on viability, while the addition of peptone, glucose, or *Anodonta* foot extract did not help. At the optimum osmotic pressure ($\Delta = 0.15°$), at first the fragments move rapidly and then those possessing some connective tissue cover themselves with an epidermis by proliferation of epithelial cells. These spherules then gradually utilize the lipoproteic granulations of the leucocytes and the proteic grains of the epithelial cells.

Important experiments in pulmonate mollusk tissue cultures were made from 1928 to 1934 by several authors. Gatenby (1931) explanted fragments of the *Helix aspersa* Müll. snail's mantle in hanging drops of snail blood without sterile conditions. In such cultures pieces of muscle remain alive and present contraction movements for a week. Gatenby and Duthie (1932) and, later, Hill and Gatenby (1934) described the behavior of small fragments of the pulmonary cavity wall and the mantle of the *Helix aspersa* in hanging drops composed of blood and various physiological solutions. The cultures lasted only a few days (three to five) and were sometimes made without disinfection. The authors describe the migration of the epithelial cells, and since they did not observe any mitoses in the explants, they believed that the cells of the mantle in culture or in the course of normal regeneration multiplied by amitosis. Bohuslav (1933a) cultured several postembryonic organs of the *Helix pomatia* L. on a medium composed of a homologous or heterologous peptone solution. Among the organs explanted (alimentary canal, salivary glands, seminal receptacle), the author established that it is mainly the epithelial cells of the alimentary canal that participate in the formation of the structures, whereas the growth zones of the seminal receptacle provide the fibroblastic formations.

Konicek (1933) explanted fragments of the heart (atrium and ventricle), the lung cavity, and the pedious tegument of the *Helix pomatia* snail on a medium of 0.45% agar containing a nutrient composition of peptone solution, glucose, and Ringer solution. He cultured three or four small organ fragments in one drop of medium so as to facilitate the study of the growth potential of the tissues. This technique of culture on gel medium therefore differed from the one developed by Et. Wolff and Haffen (1952) in which the organs were put on the surface of the gel medium and could directly take in the ambient oxygen and utilize the nutrient substances of the medium by diffusion. Konicek replanted his cultures every five to seven days; under these culture conditions he rarely achieved the tissue culture that he wanted: Only the amoebocytes emigrated from the atrial explants and proliferated around them. The other tissues showed no signs of growth, but the pulsations continued for three weeks in the atrium fragments, whereas they lasted only a few hours in the ventricle fragments. After the healing of the section, the skin presented peristaltic movements which lasted an average of 10 days, during which the secretory function and the vibratile movements of the tegumentary epithelium continued. Thus, Konicek achieved a sort of organ culture; only the composition of his medium was responsible for his failure to obtain cellular multiplication, for in the same year Bohuslav (1933b) obtained the multiplication of the cells of the *Helix pomatia* atrium connective tissue cultured in a medium similar to the one he used for the culture of the alimentary canal (physiological solution with albuminous by-products of ox fibrine) with the pH adjusted to the optimum value of 6.8. In the drop of liquid in which they were transported the explants continued to pulsate for a few days (two to four) then ceased. In the meantime, the connective tissue cells, resembling fibroblasts, formed a zone of peripheral membranous and plasmodial growth, the edges of which attached themselves to the cover slip on which the drop had been deposited. At this time, towards the 11th day, the pulsations reappeared and might persist up to 22 or even 39 days. The evolution of these explants shows the importance of environment factors on their behavior after the beginning of cellular proliferation. The absence of replanting and the adhesion of the edges of the cardiac fragment to the slide enabled it to recover its physiological properties. In this way, the author showed the important part played by the tension of the cardiac tissues in the manifestation of automatic contractions which has since been proved experimentally in several vertebrate species (Rybak, 1964).

After this series of experiments, other mollusk tissues were not cultured until 1949. That year Bevelander and Martin undertook the culture of the mantle in marine mollusks to study the mechanism of calcification. Thus, fragments of *Pinctada radiata* mantle, explanted in mollusk blood plasma, show cellular migration after 24 hours, but they also produce fibrous material (conchin) and secrete mucus. After four to six days the cultures often present crystals identical in formation to those of the normal or regenerating shell. These crystals generally form some distance from the edge of the mantle, in a mucus exudate.

It is only recently, following extensive research on the needs of vertebrate cells particularly by the technique of culture on synthetic media, that media allowing either tissue culture [see results of Flandre and Vago (1963) in Volume 1, Chapter 12] or organ culture were developed especially for mollusks. However, some authors still compose media which allow cellular migration while preserving the structure of the organs. Thus, the recent results of mollusk organ explantations by grouping them according to the aim of the culture will be given.

2. Evolution of Organs in Liquid Medium for Physiological and Parasitological Research

One of the most difficult cases of experimental survival to distinguish from organ culture is undoubtedly that of perfusion in normal physiological conditions. The functioning of the alimentary system of the snail *Lymnaea stagnalis appressa* Say was maintained for more than 60 hours in a physiological saline by Carriker (1946). In invertebrates, the heart has frequently been studied by perfusion and immersion in physiological solution. This organ shows remarkable vitality, and among mollusks the snail's heart in particular was used for numerous experiments on cardiac automatism (Cardot, 1933). This explantation technique enabled the testing of the action of various substances which, added to the medium in which the heart is immersed, increase its life span, particularly when the temperature is kept at about 2° (Jullien *et al.*, 1955). Using this technique, Ripplinger and Joly (1961) showed that the addition of crushed nerves to the medium maintained heart beats for 48 to 72 hours. After fractioning the hemolymph by dialysis, the dialyzable fraction provoked contractions, whereas the protein fraction was inoperative. Besides the immediate physiological data, these results are of interest for making a more appropriate culture media for this organ.

Gastropods are often intermediary hosts to trematode parasites, so the culture of organs which shelter certain phases of the cycles of these parasites is a means of studying the unknown ways of their development cycles. Up to now such cultures have mainly supplied data on the development of the explanted structure.

Benex (1961) adapted the Thomas (1947) technique of culture in antiseptic and nutritive synthetic medium to the tentacle of the planorb *Australorbis glabratus* Say in order to study the penetration and development of the miracidiums of schistosomes in this organ. The explantations were made into liquid media (see Volume 1, Chapter 2). The cut surface healed rapidly. For a week the tentacles contracted spontaneously and swam by waving their vibratile cilia. Later, the contractions were produced only after stimulation. On about the 15th day a structural dedifferentiation began and led to dedifferentiated lacunary spheres which died during the fourth week.

Following the hypothesis proposed by Thomas (1941c) on the essential role of the nervous system in maintaining differentiation, Benex (1964) showed the influence of the nervous elements of the tentacle on the prolongation of survival and on the persistence of its differentiation. When the ganglion cells at the base of the tentacle and the eye are retained, the long shape of the explant persists for 12–15 days. In the absence of these nervous structures the spheriform organization begins as early as the 10th day. Similarly, the structural dedifferentiation of the tentacles begins about the 15th day with the presence of the eye and about the 10th without it.

Using the same culture technique Benex (1965) studied the influence of certain sulfamides (sulfadiazine, elcosine) on the duration of survival and on the persistence of tissue differentiation. These two substances delay the starting of tissue dedifferentiation in the internal tentacle structures of the planorb *Australorbis glabratus* Say.

Besides sulfamide by-products, Benex (1967a) tested the activity of several substances which improve cellular metabolism when added to a liquid medium composed of a balanced saline solution and glucose. Benex explanted the festooned border of the mantle and the densely ciliated gill filaments of *Mytilus*. Vitamin C allowed better tissue respiration but vitamins B_1, B_2, B_6, and B_{12} seemed to have no effect. The products which prolonged survival were the sulfamides (1162 F and sulfadiazine) and *para*-aminobenzoic acid (PABA). Continuing her experiments in mussel organ culture, Benex (1967b) defined the influence of the chemical configuration of the bodies liable to increase the survival of her explants. After trying various chemical structures she noted the necessity for a benzene nucleus and a free NH_2 in the *para* position.

In the United States, Chernin (1963, 1964), Burch and Cuadros (1965), and Burch (1968) have explanted organs of *Australorbis glabratus* Say. Chernin (1963, 1964) kept heart and fragments of ovotestis and digestive gland in culture in either a basic salt solution or a complex medium at pH 7.3–7.5; under these conditions Chernin did not observe cell divisions in the heart. In the following year Burch and Cuadros (1965) developed a culture medium suitable for mollusk cells and organs as a first step to establish an appropriate medium for the *in vitro* culture of trematode parasite larvae.

The medium used was a complex mixture of various partial saline solutions, medium 199 with peptone added, calf fetal serum, and mollusk extract. Numerous fragments of organs (foot, mantle, esophagus, oviduct, gonad) of *Australorbis glabratus* Say, *Helix pomatia* L., and *Pomatiopsis lapidaria* Say were enclosed in sterile plastic petri dishes containing 5 ml of medium. The gonad tissues of the three species were maintained in culture for 60 days at pH 7.0. Live spermatozoa were active for the 60 days. Stages of prophase I of the meiosis (the diakinesis in particular) were observed in abundance as well as some metaphases I. After 14 days of culture many cells in mitotic metaphase were seen in a primary explant of *Helix pomatia*, indicating cellular divisions and a quantitative growth of the cultures. Also, subcultures of gonads and oviduct were achieved by replanting. The pedious muscle was maintained in culture under good conditions for 60 days at pH 8.5 in two species (*H. pomatia* and *A. glabratus*) and for 45 days in another species. The three categories of explants then presented a peripheral outgrowth of cells that were able to be replanted. The mantle, the esophagus, and the oviduct were also successfully maintained in culture for 45 days.

In Eagle's diluted liquid medium and in the presence of antibiotics Hollande (1968) kept *Helix pomatia* multifide glands alive for 33 days and studied the evolution of the secretion granules.

3. Organ Culture on Gel Media for Metabolic, Endocrinological, and Embryological Studies

In 1961, P. Sengel explanted gill filaments and intestinal diverticulae of the marine bivalve *Barnea candida* L. on two culture media derived from the standard Et. Wolff and Haffen (1952) medium. One of them (medium M) is a mixture of 1% agar and seawater, the other (medium St M) is 1% agar, seawater, and chick embryo extract. On these media the organs survive for seven days without replanting. The gill filaments underwent no structural change and the vibratile activity of the ciliated epithelium was maintained. However, the structure of the alimentary diverticulae was radically modified. The dark cell crypts disappeared, whereas the high clear cells flatten so that the lumen of the tubules was increased. This technique can be compared with the one used by Konicek (1933) for the culture of snail organs. However, as the tissues were not covered by nutritious liquid they were able to directly take in the ambient oxygen. In this way it is not necessary to replant so frequently, and replanting can often be avoided since it impedes the development of the organ.

This culture technique has since been applied to gastropod and cephalopod mollusks to obtain supplementary data on the sexual differentiation of these animals (Gomot, 1970). In fact, it has been known since the observation of cases of parasitic castration and experimental castration in pulmonate gastropods (Laviolette, 1954a,b) or in male cephalopods of the genus *Octopus* (Taki, 1944) that humoral correlations sometimes exist between the gonad and the genital tract. Other humoral correlations have been shown in *Octopus* by experimental excisions or sections between the differentiation of the female genital canals and the optic glands or between the gonads and the optic glands (Wells, 1960). Severing the optic tentacles and injecting crushed slug brains led Pelluet and Lane (1961) to assume that spermatogenesis is stimulated by a hormone from the tentacle, whereas the brain produces a feminizing hormone. Lane (1962, 1963) assumed that these hormones come from various types of neurosecreting cells which she described.

Since experiments in excision and reimplanting of organs are frequently very intricate in mollusks, the organ culture technique is a means of studying the correlations which may exist between various organs as they can be made to act on one another in the same medium independently of any other factor. The detailed results of this technique's contribution to hormone research are given in this volume, Chapter 5, and here we shall deal mainly with the behavior of isolated organs determined by the composition of the medium.

a. **Hermaphrodite Prosobranch Gastropods.** Among hermaphrodite prosobranch gastropods, research on sexuality was done in three protandrous marine species: *Calyptraea sinensis* L., *Crepidula fornicata* Phil. and *Patella vulgata* L.

In the *Calyptraea sinensis* L. the immature mollusks present a male genital apparatus which afterwards changes into a female one, undergoing the phenomena of both regression and morphogenesis. As the anatomy of the *C. sinensis* does not allow direct operations on the animal, Streiff and Peyre published in 1963 the composition of a hormone-free medium, A_6, which keeps embryo and adult organs of this mollusk alive. Derived from the media developed by Vakaet and Pintelon (1959) and by P. Sengel (1961), medium A_6 contains Gey solution with 1% agar, sterile seawater, and albumin. The first attempts at culture showed that the survival of the organs varies according to their nature, so Streiff (1963, 1964) modified medium A_6 in accordance with the organs to be cultured and developed four varieties, three of which, A_7, A_{7c} and A_{8c}, differ in their albumin content and by the presence of cysteine in some cases (A_{7c} and A_{8c}), whereas in the fourth, B_2, the albumin is replaced by ox serum.

On medium A_6 the ocular tentacles and the penis were perfectly healthy after 83–92 days in culture and presented excellent contraction reflexes at the approach of a source of heat or light. Media A_7, A_{7c}, and A_{8c} were too rich and caused cellular migration around the explant.

Medium A_7 was suitable for the culture of the ovaries, the testes, and the seminal vesicles for 83 days. In the gonads the distribution of the nucleic acids remained the same as in the test group animals. A histological study of the ovaries showed that the oocytes increased and entered the first stages of vitellogenesis. Examination of the testes showed that they functioned normally. The central nervous masses maintained a morphology comparable to that of intact animals during the whole period of culture on medium A_7 (81 days) and granules of neurosecretion could be colored inside the nerve cells.

The fragments of intestine survived well on medium B_2 and on medium A_7 where the ciliate cells and the mucus secretion remained normal, whereas media A_6 and A_7 provoked disorders of the cilia and mucus secretion ratio. Although the hepato-pancreas is a delicate organ it was cultured on medium A_{7c} or A_{8c} for 21 days without necrosis and the "ferment" cells still contained polyhedral inclusions.

Later, Streiff (1967a) mainly used a simpler medium, MA_7, consisting of agar, seawater with glucose, and albumin. (The Gey and Tyrode solutions of medium A_7 were replaced by filtered seawater containing 1.3‰ glucose.)

On the anhormonal medium thus developed Streiff was able to culture organs for seven months in darkness and at a temperature of 18°. This enabled him to discover the hormonal factors conditioning the development of the genital tract and the gonad (Streiff, 1966, 1967b,c). Concerning the genital tract, the results obtained in the *Calyptraea* are apparently contrary to those described in the pulmonates and male cephalopods because, for the first time in a mollusk, "this experimental study brings proof of the independence of the genital tract from the gonad" (Streiff, 1967b).

This conclusion is based on the observations of the three groups of cultures condensed in Table I.

Experiments associating male or female genital tract with the nervous system (the set of ganglions of the circumesophageal ring) or with the ocular tentacle as well as cultures on media based on hemolymph show that the differentiation of the genital tract is determined by hormonal substances which pass into the hemolymph. The evolution of the male genital tract depends on two antagonistic hormonal substances: "One, secreted by the male ocular tentacle, is responsible for the development and maintenance of the male morphology and physiology of the tract; the other, coming from the central nervous system, ensures the regression of the penis and of the deferent canal" (Streiff, 1966).

The female genital tract, on the other hand, seems to differentiate under the influence of a substance secreted at the level of the central nervous system during a very short period (Streiff, 1967b).

The determination of the gonad cycle was studied in the same man-

TABLE I

CULTURE OF *Calyptraea* GENITAL TRACT[a]

Organ in culture	"Donor" animal's stadium	Number of days of culture	Results	Conclusion
Genital tract (male) + presumed (female) tract area	Male (5 and 8 mm in diameter)	212	No change	Male and female genital tract equal a neutral organ
Complete ♀ genital tract	Female (15 mm)	212	No modification	
♀ Genital tract in the course of differentiation	Females in the process of differentiation	212	No modification	

[a] From Streiff (1967b).

ner. Culture on a hormone-free medium revealed that the undifferentiated goniae of immature gonads or gonads in the male phase undergo an oogonial autodifferentiation.

Associations similar to those used for the study of the genital tract (presented in detail in this volume, Chapter 5) show that "male gameto-genesis can only take place in the permanent presence of a masculinizing hormone secreted by the male central nervous system" (Streiff, 1967c). The female phase is supposed to result mainly from the absence of this hormone and from the intervention of hormonal influences initiating the phases of vitellogenesis and the development of the female tractus.

A similar experimental study has just been made in another pro-tandrous hermaphrodite gastropod, the *Crepidula fornicata* Phil. by Lubet and Streiff (1970). Five series of organ cultures of the male genital tract cultured alone or in association with a pleural ganglion of a female *Crepidula* show that this ganglion secretes a hormone determining the dedifferentiation of the male genital tract. This same hormone also causes the regression of the male genital tract of another species (*Calyptraea sinensis*). Experiments in bilateral ablation of various ganglia of the nervous system linked with these culture experiments en-abled these authors to conclude that the cerebral ganglia play a deter-mining role through the intermediary of the tentacles in the development of the male genital tract.

Sex inversion in *Patella vulgata* L., which was long considered as a gonochoric species, is a phenomenon about which precise data is still lacking (Choquet, 1966). While studying the reproduction cycle Choquet (1964) simultaneously explanted genital glands of *P. vulgata* L. so as to follow their *in vitro* behavior.

He used a medium derived from that of Wolff and Haffen and com-posed by Durchon and Schaller (1963) for the culture of polychaete organs. The medium is basically composed of agar, seawater contain-ing glucose, albumin, horse serum, and chick embryo extract; this com-position is given in Volume 1, Chapter 2. The gonads are explanted in May at the beginning of genital revival, they heal very quickly, and survive for over three months. At the outset they present spontaneous contractions and are capable of moving about on the surface of the medium. Histological study shows that the germinal cells of both sexes undergo a normal gametogenetic differentiation.

Studying the problem of sexual differentiation of the limpet, Choquet (1965, 1969) also studied the behavior of gonads in the male phase put in culture in various periods of the genital cycle and cultured either alone or in association with cerebral ganglia or with tentacles or with the cerebral ganglion–tentacle complex. The fragments of *Patella vulgata* L.

gonads in the male phase cultured alone always showed a fresh spermatogenetic growth.

According to the results obtained by the associations quoted, it seems that male sexual differentiation in *P. vulgata* is controlled by endocrine factors of cerebral or tentacular origin. In periods of sexual recess the association of the fragment of testicle with a cerebral ganglion stimulates spermatogenesis, whereas the tentacle and the cerebral ganglion–tentacle complex inhibit spermatogenesis.

b. Hermaphrodite Pulmonate Gastropods. Another sort of hermaphrodism without sexual inversion particularly attracted the attention of Guyard and Gomot (1964): the pulmonate gastropod *Helix aspersa* Müll. The gonad of this snail is composed of follicles in which male and female gonocytes are to be found mixed together in all seasons. In the course of the differentiation of the hermaphrodite gland in young animals, first, spermatogonia and their nutrient cells with dense chromatin network and, immediately afterwards, ovocytes are observed to differentiate from the germinative epithelium. From then onwards the gonad is hermaphrodite and produces both categories of gametes, the ripening of which is slightly asynchronous during the whole life of the animal. In these animals it would seem that the male and female genetic potentialities are juxtaposed without either one dominating. However, the question to be asked is what causes the undifferentiated cells of the germinative epithelium to finally differentiate into either male or female? According to his histological observations, Ancel (1903) thought that the differentiation of the female lineage depended on the presence of the nutrient cells, whereas Pelluet and Lane (1961), after experiments by ablation, thought that the sexual differentiation is the result of the hormonal action of the brain and eye tentacles.

Elective ablation of nerve ganglia which are surrounded by a thick connective mass is difficult to execute in live snails and it is also followed by a disturbance of the metabolism of the animal which may interfere with sexual differentiation. Guyard and Gomot (1964) cultured juvenile gonads of *Helix aspersa* on various nutrient media and on neutral media in association with presumably endocrine organs.

The gonad fragments were explanted at the stage when the three categories of cells become distinct (spermatogonia, nutrient cells, and ovocytes). Under these conditions, if the sexual determination of the germinal cells mainly depends on the environment as Ancel (1903) thought, the modifications of the culture medium should make it possible to direct the differentiation of the gonocytes towards either sex.

The survival of gonad tubules was achieved on simple media (Gomot

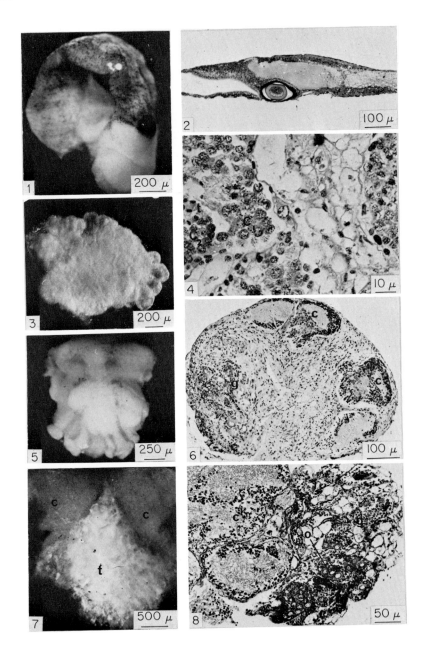

and Guyard, 1964). On medium T (mainly agar dissolved in Gey and Tyrode solutions), developed by Et. Wolff *et al.* (1953) to test the survival of vertebrate embryonic organs, the explants survive for about 20 days. Histological examination of the explants showed that a certain number of spermatogonia become spermatocytes during the first four days and that the male cells then disappeared. On the other hand, the ovocytes underwent a slight phase of growth and surrounded themselves with follicular cells. These female cells persisted up to the 20th day and then degenerated in turn so that the explant was reduced to connective tissue. On medium R, which is even simpler, containing only 1% agar in Ringer solution with glucose added, the sexual cells evolved in quite the same way and survived for no longer than 18 days.

When fragments of juvenile gonad at the same stage are explanted onto media enriched with an organic liquid (Guyard and Gomot, 1964) the life span of the explants is lengthened and the differentiation of the germinal cells is influenced by the nature of the organic adjuvant

---

PLATE I

ORGAN CULTURE OF MOLLUSKS

FIG. 1. Macroscopic view of an eye tentacle of a young *Helix aspersa* 12 mm in diameter after 23 days of isolated culture on G_2 medium. ($\times$32)

FIG. 2. Histological section of an eye tentacle of a young *Helix aspersa* 10 mm in diameter after 6 days of isolated culture on R medium. The explant is firmly fixed to the substrate. ($\times$90)

FIG. 3. Macroscopic view of a gonad of a young *Helix aspersa* 12 mm in diameter after 23 days of culture in G_2 medium. The glandular tubules are surrounding the gonadal mass with rounded papillae. ($\times$29)

FIG. 4. Histological section of a gonad of a young *Helix aspersa* 19 mm in diameter. After 24 days of culture on H medium, spermatogenesis is over. In addition to a large number of spermatocytes I in prophase of meiosis (C_1), spermatocytes II (C_2) and spermatids (t) may be observed at various stages of their development. ($\times$580)

FIG. 5. Macroscopic view of a cerebral ganglion of a young *Helix aspersa* 12 mm in diameter after 23 days of culture in G_2 medium. The connective sheath enclosing the adult ganglia is not yet opaque and the three cerebral stages may be distinguished. ($\times$22)

FIG. 6. Autologous association gonad–nerve collar of a young *Helix aspersa* 10 mm in diameter after 34 days of culture on T medium. The separate explants have united to form a single spherical mass. The histological section shows the gonad (g) on the left and the cerebral ganglia (c) on the right ($\times$90).

FIG. 7. Macroscopic view of an autologous testicule (t)–cerebral ganglia (c) association in a *Viviparus viviparus* adult after 21 days of culture on G_2 medium ($\times$14).

FIG. 8. Histological section after 6 days of culture on G_1 medium of an association between a gonad of *Helix aspersa* 11 mm in diameter and a cerebral ganglion (c) of an adult female *Viviparus viviparus*. Only ovocytes (o) are developing in the gonad ($\times$140).

(Photographs by A. Guyard and B. Griffond.)

(embryo juice or adult snail hemolymph). On the standard Et. Wolff and Haffen medium (1952) containing chick embryo juice the gonads easily survive for 50 days without replanting; they presented movements of superficial fibrillation and contraction. In the central cavity of the tubules no new spermatogonia appear but the existing ones multiply, change into spermatocytes I, and degenerate from the 15th day of culture. The cells of the germinative epithelium develop into ovocytes; those already existing increase in volume and surround themselves with follicles (see Fig. 5 in Volume 1, Chapter 2). This culture medium intended for embryonic vertebrate organs keeps the hermaphrodite gland of *H. aspersa* alive, but it is not a neutral medium, i.e., the chick embryo juice favors the female sexual lineage.

On medium H based on adult snail hemolymph the gonads survive for 50 days without replanting. As on the standard medium the interstitial tissue abundantly develops and the ovocytes increase in size. Moreover, after 20 days of culture spermatocytes in maturation division and spermatids, which become spermatozoa, are observed (Fig. 4, Plate I). So this medium allows normal and accelerated differentiation of male and female sexual cells.

These experiments in culture on media of different compositions show the undeniable influence of the nature of the medium on sexual differentiation of the cells. It may be supposed, from the origin of the organic liquids used, that chick embryo juice favors the female lineage by supplying nutritious substances (trophic role) while it is possible that the hemolymph contains one or more hormonal factors secreted by endocrine organs.

The search for the organs which may produce hormonal factors was undertaken by Guyard (1967, 1969a) using the technique of associated organs. To do this he developed two neutral complex media, G_1 and G_2 (see Volume 1, Chapter 2) basically made up of medium 199, peptone, and yeast extract which allow the hermaphrodite gland to survive for several weeks without the differentiation of either sexual lineage being disturbed (Fig. 3, Plate I). These media also permit the culture of the central nervous system (Fig. 5, Plate I). The associations achieved are succinctly presented in Table II together with the results concerning the behavior of the gonad.

According to these results it seems that the sexual differentiation of the germinal cells of the juvenile gonad depends upon factors in the medium probably determined by neurohumoral secretions coming from the cerebral ganglia, in the case of the female cells, and undetermined for the male cells.

Experiments on the culture of the optic tentacle were undertaken to determine the eventual role of secretory cells in sexual differentiation.

TABLE II

Action of Organs Associated with Snail Hermaphrodite Gland on Neutral Medium[a]

Type of association	Organs	Results
Autologous	Juvenile gonad of the snail + nerve collar (Fig. 6, Pl. I)	Lengthened life span, normal autodifferentiation
Homologous	Juvenile gonad of the snail + adult snail's cerebral ganglion	Feminization of ovotestis (presumed endocrine role)
Heterologous	Juvenile gonad of the snail + female adult *Paludina* cerebral ganglion (Fig. 8, Pl. I)	Feminization of ovotestis (presumed endocrine role)
Heterologous	Juvenile gonad of the snail + cerebral ganglion adult male of *Paludina*	Life span decreased
Heterologous	Juvenile gonad of the snail + *Paludina* testicle or ovary	No influence (no endocrine role)

[a] From Gomot and Guyard (1968).

Figures 1 and 2 of Plate I show that this tentacle survives well in culture. No endocrine influence of the tentacle has yet been revealed although Badino (1967) obtained a stimulation of spermatogenesis by this *Arion rufus* Drap organ.

Recently, Guyard (1969a,b, 1970, 1971) cultivated an undifferentiated *Helix aspersa* gonad primordium on a neutral culture medium G_2 and observed that the gonia autodifferentiated in ovocytes. This observation showed that there is *in vivo* inhibition of the autodifferentiation of the ovocytes during spermatogenetic differentiation which normally appears at first in the immature gonad.

The ovotestis and the optic tentacle of an adult *Australorbis glabratus* Say (gastropod pulmonate) were kept in culture by Vianey-Liaud and Lancastre (1968) on a medium consisting of a solid agar phase covered by a liquid phase. Only in a hormone-free liquid phase did Vianey-Liaud (1970) find that ovogenesis could continue for 36 days and that the spermatocytes and the spermatids degenerated 20 days after explantation. When the central nervous system is associated with the ovotestis, meiosis in the spermatocytes may continue for 30 days after explantation, but spermiogenesis does not take place.

c. Dioecious Prosobranch Gastropods. As research on the determination of sexual differentiation in gastropods has been mainly concerned with hermaphrodite species, Griffond (1969) undertook a comparable study in our laboratory on a species of separate sexes prosobranchs: *Viviparus viviparus* L. On medium G₂ composed by Guyard (see Volume 1, Chapter 2) *Viviparus* organs survived well. The testis was cultured for 25 days, and under these conditions there was no further spermatogenesis, but the spermatocytes evolved into spermatids and spermatozoa.

In cultures of juvenile ovaries (females with shells measuring 7 mm high) the ovocytes which had already begun their vitellogenesis before being put into culture perceptibly increased in volume within 15 days. In adult ovaries explanted during the rest period the ovocytes did not begin vitellogenesis during the culture.

The cerebral ganglia survived very well for 15 days in culture. The undeveloped genital tract in young males was also explanted and was easily maintained in culture for 20 to 30 days. The *in vitro* associations of gonads with various ganglia of the nervous system (Fig. 7, Plate I) should yield more precise data on the manner of sexual differentiation in this gonochoric prosobranch.

In the gonochoric marine prosobranch *Littorina littorea* L., Streiff and Le Breton (1970) showed by a comparative study that the presence or absence of the male genital tract is due to two hormonal factors as in morphogenesis and the dedifferentiation of the male genital tract of the protandre hermaphrodite species (*Crepidula fornicata* Phil.). A factor from the right ocular tentacle level of the male brings about the morphogenesis of the external male genital tract in the female. The other factor from the pediopleural complex of the female ensures the regression of the penis in the male.

d. Cephalopods. In cephalopods, following Wells' experiments (1960) on the endocrine role of the optic gland in *Octopus*, Durchon and Richard (1967) studied *in vitro* the role of the optic gland in the maturation of the *Sepia officinalis* L. ovary.

To achieve this the authors used cuttlefish of known age from the stock of Richard (1966). The organs were explanted by the Et. Wolff and Haffen technique (1952) onto a medium developed by Durchon and Schaller (1963) for the culture of polychaete annelids organs (see Volume 1, Chapter 2).

The optic glands of four- to six-month-old cuttlefish were cultured for a month at 21° and in darkness. Under these conditions, as early as the fourth day in culture, the secretion process began in the main cells with the appearance of cytoplasmic vacuoles with fine granular content. The nuclei of these cells increased regularly in size. The secretions of the main cells accumulated in the intercellular spaces and the gland took on the appearance of reticulate syncytium. The optic gland of *Sepia*, separated from the brain, began activity like that of the *Octopus* after *in vivo* suppression of cerebral inhibition (Wells and Wells, 1959).

Immature ovaries cultured alone survived very well: the ovocytes continued their growth but the germinative cord showed no resumption of activity.

From these data Durchon and Richard (1967) and Richard (1970) prepared associations of the optic gland and the ovary. The association of optic gland (juvenile or adult) and ovary of young animals (ovocytes with a diameter under 0.3 mm) provoked no change in the behavior of the ovocytes. But when the ovocytes were larger than 0.3 mm in diameter their size doubled in 20 days. In both cases after 20 days in culture mitotic activity persisted in the germinative cord. Thus, the endocrine influence of the optic gland under the control of the brain is necessary in the differentiation of the ovary from the previtellogenetic stage onwards.

The associations of optic glands and testes of young or adult cuttlefish prove that the secretion of that gland also ensures the multiplication of the spermatogonia. Also, the secretion of the optic gland is independent of the sex of the donor.

In the introduction to his research on experimental cephalopod embryology Marthy (1970) has recently shown that the behavior of the embryonic explants of *Loligo vulgaris* Lam. depends on the stage of development at the time of explantation. The explanted cellular complexes before organogenesis prefer to develop on histiotypic culture while those explanted after organogenesis show signs of an organotypic culture.

E. Annelids

The greater part of experiments of organ culture in annelids have been done with the aim of obtaining more precise knowledge of certain aspects of reproduction and regeneration in these animals. The following experiments show the main results obtained in the Polychaetes, Oligochaetes, and Hirudinea classes. The compositions of the media are set out in the chapter on techniques, and certain experiments on associations are reviewed in the chapter on hormonal research.

1. Polychaetes

The Et. Wolff and Haffen (1952) gel medium organ culture technique has been recently adapted by Durchon and Schaller (1963) for Polychaeta Annelida. The first experimental cultures (Durchon, 1963) dealt with various organs (parapodia, intestine, muscles, prostomium) belonging to four species of Nereidae (*Nereis pelagica* L., *Platynereis dumerilii* Aud. and M. Edw., *Perinereis cultrifera* Grübe, *Nereis diversicolor* O. F. Müller).

The medium used allowed the organs to survive well, in particular, those of the *N. diversicolor* which could be cultured for a month in a sealed embryological watch glass kept in the dark and at a constant temperature of 20°. Under these conditions several parapodia, associated in culture rapidly join together (Fig. 9, Plate II). Later, Durchon and others mainly dealt with the experiments concerning the endocrine activity of the brain and the action of the cerebral hormone at the cellular level during sexual reproduction.

a. **Determination of Sexual Differentiation.** The work of Durchon (1952) had shown that the nereid brain secretes a hormone inhibiting the maturation of the male genital products in all species and epitoquium in those presenting this phenomenon. In fact, ablation of the prostomium of an immature worm causes precocious genital maturity, whereas reimplantation of brains into the coelom of a decerebrated animal prevents this maturation. It was possible to obtain the same results by *in vitro* culture of parapodia explanted either alone or in association with a prostomium (Durchon and Schaller, 1963; Durchon and Dhainaut, 1964).

In the *N. diversicolor* (species without epitoky) when the parapodium is explanted by itself it heals rapidly. If the donor animal is a male the spermatogonia contained in the coelomic cavity of the parapodium undergo rapid maturation and after five to seven days in culture some spermatozoa are emitted by the nephridial pore. If the parapodia come from a female the ovocytes rapidly increase during the first 10 days in culture and then undergo radical changes before degenerating.

In the case of associated parapodium–prostomium, the joining of the organs is rapid (Fig. 10, Plate II). After a month the genital male and female products still retain the appearance they had when put into culture.

Subepitoke parapodia of species with epitoky (*P. cultrifera*, *N. pelagica*, *P. dumerilii*) cultured on the same hormone-free medium differentiate oar-shaped chaetae characteristics of the heteronereis stage. In these species it was impossible to obtain the development of the parapodial lamellae; the authors supposed that this absence of morphogenesis was due to the lack of indispensable substances in the medium. However, on the same medium Malecha (1967a) obtained rapid male gametogenesis as well as differentiation of heteronereid chaetae and of parapodial lamellae in portions of 1, 2, or 3 metameres in *Nereis succinea*

Leuckart cultured singly. The association of identical metameres with a prostomium of the same species or with a prostomium of a different species without epitoky (*N. diversicolor*) prevents somatic and germinal differentiation.

These experiments confirm the inhibiting effect of the brain on gametogenesis and on the appearance of epitoky (Durchon, 1952). They also allow the assumption that this action is direct and without the intervention of a "relay organ." Last, they prove once more, as was shown *in vivo* by Hauenschild (1956) and Boilly-Marer (1962), that the inhibiting hormone is not specific and that the brain of a species without epitoky is capable of blocking the somatic transformations of heteronereis stage.

Using the same technique, Durchon and Dhainaut-Courtois (1964) were able to add some precise data on the localization of the inhibitory center of sexual maturation in *N. diversicolor*. Parapodia were taken from male animals and cultured either singly (test group) or in association with anterior or posterior halves of the prostomium of immature animals. In the majority of cases only the posterior cerebral portion produces the hormone-inhibiting genital maturation.

Durchon (1967) added that it was not yet possible to explain the mechanism of the inhibitory hormone at the cell level in nereids. However, the first results obtained by Durchon *et al.* showed the important part played by this hormone in the regulation of synthesis of nucleic acids (Dhainaut, 1964; Durchon and Dhainaut, 1964; Durchon and Boilly, 1964). Using culture media with tritium-labeled thymidine it was possible to follow the synthesis of DNA in the course of spermatogenesis in *N. diversicolor*. In the spermatogonia the process is slow in the presence of hormone and more rapid without it. This DNA synthesis is followed by a phase of growth (one to two days after explantation) and then by meiosis. Once the synthesis of DNA is accomplished, the adding of prostomium to a parapodium in the course of culture is unable to interrupt spermatogenesis.

The observations made on the organ culture of ovocytes of the same species in the presence of hormone show that after 10–20 days their structure does not change (Figs. 11 and 12, Plate II). The ribonucleoproteins are regularly divided among the vitelline granules which are homogeneously distributed in the cytoplasm. In the absence of hormone the nucleus grows in size as early as the 10th day, the ribonucleoproteins condense around the nucleus while the vitellus is rejected to the periphery of the ovocyte (Figs. 13 and 14, Plate II). This disturbance in ovocyte metabolism increases up to the 20th day and leads to degeneration. Ultrastructural examination shows that experimental accelerated matu-

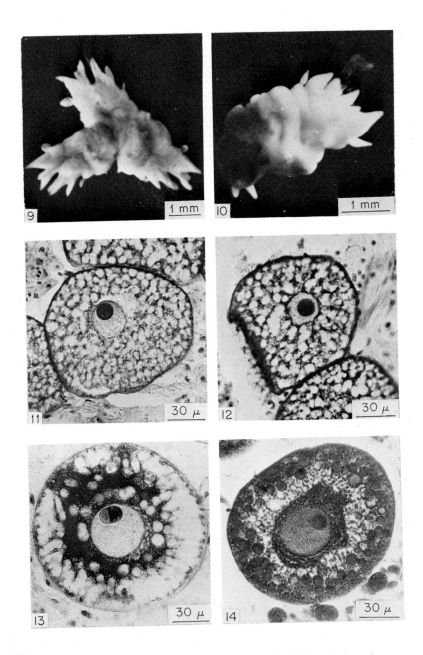

ration of the ovocytes in the absence of hormone is accompanied by modifications of the ovocytary membrane and by the disappearance of the Golgi material which takes place during normal maturation. But in the perinuclear zone enigmatic annulated lamellar structures form, possessing an affinity for bases which is considered by Durchon and Boilly (1964) to be in correlation with a high RNA content. It therefore seems that the cerebral hormone acts as a regulator of nucleic acid synthesis at the gonocyte level.

The study of the somatic modifications of epitoky, which include the phenomena of multiplication and dedifferentiation of cells previously differentiated, will be improved by studies using organ culture techniques incorporating in the media various substances that inhibit mitosis or protein synthesis. These have already been used in research on the cellular processes which are responsible for the phenomena of metamorphosis.

b. Culture of Regenerating Segments. The technique of *in vitro* culture has also been used in invertebrates for the study of regeneration in the groups where this phenomenon exists. In polychaete annelids most species possess the faculty of regenerating an amputated part of their body. In the course of regeneration the following three main stages are always recognizable: healing, formation of a blastema, and growth and differentiation of the regenerating part.

The formation of a blastema is particularly important. In general it seems to result either from the migration of cells with embryonic characteristics (neoblasts) towards the zone of the cut or from the structural dedifferentiation of certain cells belonging to the injured zone. These cells are capable of multiplying and of taking part in the restoration of

---

PLATE II

ORGAN CULTURE OF POLYCHAETES

FIG. 9. Association of three parapodia of *Nereis diversicolor* cultivated for 13 days. (×9)

FIG. 10. Parapodia associated with a prostomium after 13 days in culture (*Nereis diversicolor*). (×14)

FIG. 11. *Nereis diversicolor* ovocyte at the beginning of culture. (×720)

FIG. 12. *Nereis diversicolor* ovocyte after 20 days in culture with the presence of hormones. The ribonucleoproteins show a homogeneous spread similar to that when explanted. (×580)

FIG. 13. *Nereis diversicolor* ovocyte after 10 days in culture without cerebral hormone. The ribonucleoproteins have formed in the perinuclear region and the nucleolus is heterogeneous. (×450)

FIG. 14. *Nereis diversicolor* after 10 days in culture without cerebral hormone. The vitellus are at the periphery whereas the basophile substances are concentrated around the nucleus. (×450)

the different tissues. These two modes are not mutually exclusive, for the experiments of various research workers (in particular, of Stéphan-Dubois, 1954, in *Lumbriculus*) have shown that both may exist simultaneously, i.e., in oligochaetes, in which the neoblasts are not totipotent like those of planaria but can only regenerate the mesoderm.

In polychaetes, as in oligochaetes, the experiments on irradiation with x rays made by Stéphan-Dubois (1958) on the *Nereis diversicolor* O. F. Müller have proved the existence of free cells (amoebocytes) which migrate from the neighboring segments towards the scar zone. It seems that these amoebocytes are totipotent since the blastema grows, segments, and does not degenerate as in oligochaetes. However, Herlant-Meewis and Nokin (1962–1963) studying the histological appearance of normal regeneration in the *N. diversicolor* thought that the free cells of the coelom play the greater part in the constitution of the mesodermal blastema, whereas the other tissues (ectoderm and endoderm) regenerate independently. The experimental results of irradiation with x rays are sometimes contradictory from one polychaete species to another, probably due to differences in constitution.

In polychaete species, in which the primordial layers regenerate independently, the origins of the regeneration cells have to be determined. Among the hypotheses formulated, Thouveny (1967) suggested that it may be supposed that "histogenesis uses the same cellular elements as normal tissue renovation or physiological regeneration, each layer possessing its own stock of renovatory cells." However, Abeloos (1965) thought that certain cellular types are, in fact, "reversible pseudo-differentiation which can supply undifferentiated cell types."

To study the modalities of cell dedifferentiation and the potentialities of the dedifferentiated cells of the regenerating parts, Thouveny (1967) used various types of action on fragments capable of regenerating (x rays, synthetic culture media). Using Chandebois' suggestion (1963a) that a tissue culture medium causes dedifferentiation in adult worms when added whole or sectioned, Thouveny (1967) developed a medium suitable for tissue cultures of polychaete annelids. This liquid medium consists of seawater, amino acids, glucose, and extracts of whole animals. The experiments were done on the *Polydora flava* Clap. and *Owenia fusiformis* Delle Chiaje species. The fragments immersed in this medium undergo normal healing as early as the first day. Blastemas, which may appear at both ends, grow more rapidly than test fragments put in normal seawater. In the *P. flava*, the anterior blastema frequently attains the volume of a normal segment of the strain within three days. But the blastemas are incapable of differentiation and no trace of metamerization appears on the epidermal surface of these regenerating parts. Histological examination shows a progressive dedifferentiation of the ectodermal and mesodermal layers. The endoderm maintains its epithelial structure, often isolating itself in the form of epithelial vesicles with the ciliate apical part directed outwards. These vesicles survived in the synthetic medium for 14 days, and it may be supposed that the endoderm maintains its function of assimilating nutrients. The author of this work thinks that the inhibitory action of the medium on regeneration is linked with the presence of amino acids supplied to the fragments, because seawater containing antibiotics and glucose has no effect on regeneration.

Such explantations show the independence of the healing factors and the formation of the blastema, and of the factors of growth and metamerization. They also emphasize once more the importance of the composition of the medium in the behavior of explanted tissues. As with vertebrate embryo tissues a rich medium favors cell multiplication and dedifferentiation, whereas a meagre one suits organ culture. Such cultures in a medium intended for tissue culture throw little light on the potentialities of the blastema cells, whereas it is thought that a true organ culture of the blastema or of various associations of its elements would supply data on the respective roles of the amoebocytes (coelomocytes) and of the other tissue layers.

This method will doubtless allow the interactions between ectoderm and mesoderm to be brought to light in the future. The importance of this has recently been experimentally demonstrated by Boilly (1967, 1969) in his study of regeneration in the polychaete annelids *Syllis amica* Q., *Nereis diversicolor* O.F.M., and *Aricia foetida* Clap. after selective destruction of the mesoderm by x-ray treatment or by "Thorotrast" intracoelomic injection. In sabellids, ten-segment pieces of abdomen of *Sabella melanostigma* and *Branchiomma nigromaculata* were cultured *in vitro* by Fitzharris and Lesh (1969). Surgical manipulation of various organ systems also revealed a mutually dependent tissue interaction of gut, nerve cord, and body wall for normal regeneration to occur.

2. Oligochaetes

As early at 1922, Krontowski and Rumianzew explanted fragments of earthworm (regenerating heads and tails) in a culture medium composed of a mixture of agar and hydrolysates of earthworms. Cell proliferation was the main results.

Fragments of *Lumbricus terrestris* L. intestinal wall were cultured by Janda and Bohuslav (1934) in a medium mainly composed of peptones obtained by peptonic digestion of the fibrin of ox blood and of earthworm tissues. The explanted tissues were kept in darkness at a temperature of 15–26°. Whatever the shape (square, triangular, or rectangular) of the explants they formed spherical cysts with inversion of the layers of the intestinal wall. The movement of the vibratile cilia of the intestinal epithelium communicated a rapid rotative movement to the cysts which persisted for up to 37 days.

These intestinal cysts surrounded themselves with a film of mucus which constantly increased in thickness. Nevertheless, they showed pronounced peristaltic movements for quite some time. Some cysts retained their shape for 13 months but most of them disaggregated into large special cells.

Trial cultures of *Lumbricus herculeus* Sav. fragments on gel medium were made by Gay in 1963, who aimed at discovering what cellular transformations intervene in the formation of regeneration blastemas. The basic culture medium was a mixture of 1% agar in Tyrode solution and a solution of amino acids to which the author added worm extract in concentrations varying according to the types of explants.

Normal healing of the explants took place in the fragments from the wall of the clitellum and in the long fragments of immature worms. The ectoderm and the endoderm knitted together. A healing epidermis formed but the cells of this blastema were different from those of a normal one. Healing of the short fragments was defective. The medium used did not allow a more detailed study of the regeneration for it causes dedifferentiation of muscle and endoderm cells. The epidermis retained its epithelial structure and showed sensitivity to the composition of the medium. Within 34 days of culture the ectoderm of clitellum segments (32nd to 37th) transplanted onto the synthetic medium, gradually acquired the structure of a simple epithelium by involution of the glandular formations. When 10% mature worm extract is added to the medium the glandular activity of the clitellum epidermis of an immature worm increased over the first seven days of culture, but after 28 days of culture almost all the epidermal glands had disappeared. Gay thought that the mature worm extract supplied the fragment with the hormones needed for the differentiation of the clitellum but he did not do an experiment to verify this.

The experimental technique of organ culture of the brain of another earthworm, *Eisenia foetida* Sav., developed by Berjon *et al.* (1965) enabled one of them (Berjon, 1965) to obtain precise data on the action of the cerebral ganglia on the secretory cells of the tegument. The medium used was composed of agar, a buffered saline solution, an isotonic solution of pyrrolidone polyvinyl in a 3.5% solution, and the amino acids solution of Et. Wolff *et al.* C medium (1953). By means of cultures of clitellum singly isolated at various phases of their differentiation, and cultures of them in association with the brains of mature animals Berjon obtained proof of the fact that the hormonal action of the brain on the development of the secondary sexual characters in earthworms is a direct action (see this volume, Chapter 5).

Durchon (1963) pointed out that Schaller obtained good survival of various organs (tegument, gonads, prostomium) of *Eisenia foetida* Sav. on the medium used by C. Sengel (1960) for the culture of planarian blastemas. On this medium isolated segments present regeneration at both the cephalic and the caudal ends.

Using this organ culture technique Marcel (1966) made associations of cephalic regions (prostomium plus three segments), decerebrated or not, with segments of the esophageal region or the middle section of the body. In this way he showed that the brain exercises an inductive effect on the stomodeal ectodermal part of the alimentary canal, the only part capable of regenerating the pharynx.

Using the same technique, Marcel (1967a) showed that the anterior regeneration of anterior or posterior parts of the earthworm (8th–13th and 18th–23rd) is inhibited by homogenates of the segments in front of them. Homogenates of posterior segments have no effect on this regeneration. It can be agreed with Marcel that inhibitory substances controlling cephalic regeneration exist in the *E. foetida* as they do in the planaria (Lender, 1956; Et. Wolff *et al.*, 1964).

To analyze this phenomenon of inhibition Marcel (1967b) incorporated crushed nervous tissues of the first six segments (brain, subesophageal ganglia, and nerve cord), crushed "nerveless" cephalic regions, and crushed nerve cord of segments 30–36 into the culture medium. Pieces removed from between the 8th and the 13th segment underwent inhibition of regeneration of the head in 81% of cases in the presence of cephalic nerve homogenate, in 50% of cases in the presence of crushed "nerveless" cephalic region, and in 25% of cases in the presence of crushed nerve cord of segments 30–36 or in the test group. So, it seems that the inhibitory substance of the cephalic region is localized in the nervous tissue in front of the point of severing. The rather high rate of nonregeneration on media containing crushed "nerveless" cephalic region is likely to be due to the presence of a developed superficial nervous system, the cellular bodies of which are seated in the wall of the animal's body (Avel, 1959). After studying cephalic regeneration in *E. foetida*, Marcel (1970) thought that the regenerating morphogenesis could depend on the combined action of a trophic substance spread evenly along the ventral nerve cord and an inhibiting substance spread in a cephalocaudal gradient in the nervous system.

Vena *et al.* (1969) obtained the appearance of sexuality in *Enchytraeus fragmentosus* Bell (Oligochaeta Enchytraeidae) which reproduces solely by fragmentation. Serial cultivation of cephalic or caudal fragments were made in Petri dishes lined with moist filter paper, or better, in Petri dishes with a mat of glass wool in place of filter paper and kept at 25°.

Sexual maturity of these worms is macroscopically apparent although the worms are only about 1.5 cm when fully grown. Experiments are in progress to determine if the modified cultivation method enhances the reproductive physiology of *E. fragmentosus*.

3. Hirudinea

Busquet (1939) noticed the activity of the isolated penis of the leech which kept up its rhythmic movements when immersed in oxygenated Ringer solution maintained at 20–24°C. He also noted that various organ extracts extended penis functioning up to six hours and that testicular extract increased functional survival to 20 hours.

Hirudinea organ culture is quite recent and so far only organs of *Hirudo medicinalis* L. have been explanted (Malecha, 1965, 1967b). The culture method is that of Et. Wolff and Haffen (1952). The first cultures were made on a medium derived from that of C. Sengel (1960). It is a gel medium in which the physiological solution is that of Ammon and Kwiatkowski (1934) with glycogen and 12-day-old chick embryo juice added.

Fragments of tegument with the subjacent muscles heal in three days and remain alive for over two months. In the oviduct–albumen gland– vagina complex, the epithelium of the oviduct survived for 13 days while the albumen gland decreased in volume; the vaginal epithelium still persisted after 27 days of culture.

The male genital atrium–prostate gland–penis complex retained nor- mal structure after 13 days of culture. All these organs contracted at the slightest stimulation during the whole tissue in culture and histological examination confirmed the maintenance of the muscular structure.

The gonads were also cultured. The ovaries mainly containing ovo- gonia when explanted maintained their structure for 20 days; they did not mature. The behavior of testes in culture depended on their physio- logical state at the moment of explantation and on the culture medium. The spermatogonia showed no division on the media used. On the gel medium derived from that of Sengel, after seven days of culture, testes in spermiogenesis developed spermatocytes I. In liquid medium the testes containing spermatocytes I exhibited growth in some cases after seven days in Ammon and Kwiatkowski solution with 2‰ glycogen.

Following these first attempts at organ culture Malecha (1967a,b) began an *in vitro* study of male genital maturation in *Hirudo medicinalis* L. The organs were explanted onto a gel medium derived from that of Durchon and Schaller (1963), the seawater being replaced by a Ringer solution for leeches developed by Nicholis and Kuffler (1965). This medium allowed the survival of both testes and nervous tissue. The cultures were kept at 20° in darkness and replanted when the pH (7.5) varied.

In the testes cultured singly on this hormone-free medium some spermatogonia continued to divide, but the number of isogenic groups of spermatogonia and spermatocytes decreased. However, after three weeks of culture a few isogenic groups became spermatozoa.

When the testes were associated with the circumesophageal nerve mass with the first two ganglia of the nerve cord, half of their contents was represented by isogenic groups of spermatozoa and spermatids after three weeks of culture.

This technique allows the *in vivo* results obtained par Hagadorn (1966) with ablations and grafts of the brain in the Hirudinea, *Theromyzon rude,* to be verified *in vitro*. The endocrine influence of the circumesophageal nerve mass is indispensable for spermatogenesis.

4. Sipunculoidea

Experiments by Thomas (1932) in *Sipunculus nudus* dealt with the urns and enigmatic vesicles which exist in the coelomic fluid of the animal. These organs were cultured in Carrel flasks or in small tubes in homologous plasma. As early as the first planting there was multiplication of the vesicles and the urns. The organs were replanted after three weeks to a month. The urns continued their activity and moved about by the movements of their cilia for three months when they degenerated. The enigmatic vesicles persisted up to the sixth month.

F. Nemathelminths

In nemathelminths the "breeding" of nematode parasites as well as some organ cultures were attempted (see Lutz *et al.,* 1937). Certain techniques are similar to organ cultures.

Glaser (1932, 1940) and his collaborators cultured *Neoaplectana glaseri* Steiner an insect parasite, for several generations without loss of its infesting power on synthetic or organic media. In 1962, Jackson also obtained *in vitro* development of the nematode insect parasite *N. glaseri* by a culture technique on synthetic medium. The addition of liver extract might have allowed better growth.

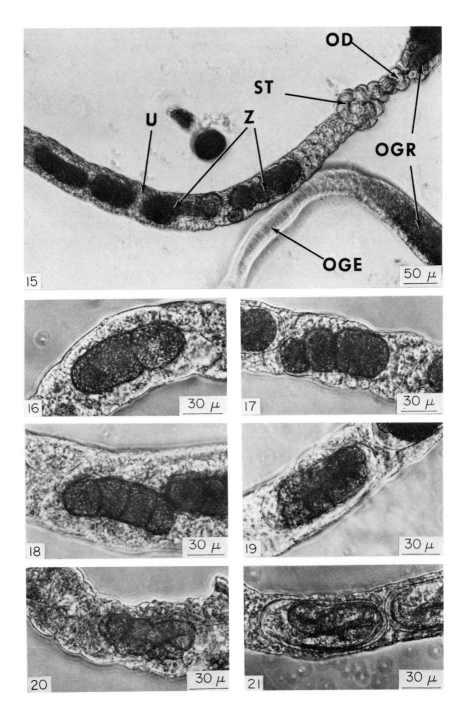

Studying nematode plant parasites, Hirumi *et al.* (1968) kept the female reproductive organs of *Meloidogyne incognita acrita* Goeldi alive. After sterilizing the surface of the galls, female genital organs (Fig. 15, Plate III) were put into a drop of culture medium consisting mainly of the Morgan *et al.* (1950) medium TC 199 and of calf fetal serum (see Volume 1, Chapter 2). Eggs developed inside the uterus and hatched after 14 to 15 days of culture. The larvae passed through the septum of the uterus and survived for another week in the culture medium (Figs. 16–21, Plate III).

Synthetic media were also used by Fatt (1967) for axenic cultivation of the free-living nematode *Caenorhabditis elegans* Bergerac to study the biochemical nature of temperature sensitivity and to attempt to identify some of the agents responsible for its reproduction at higher temperatures.

Recently, cultivation of *Ascaris suum* larvae in supplemented and unsupplemented chemically defined media was achieved by Levine and Silverman (1969). The second stage larva of the codworm *Terranova decipiens* was cultivated by McClelland and Ronald (1970) in Eagle's basic medium with 20% fetal calf serum over a period of four months.

G. Platyhelminths

Cultures of platyhelminths have two different aims. One is to understand the process of regeneration, particularly important in turbellarians, the other is to maintain adult trematodes or cestodes or their young larvae *in vitro* so as to study their development, their metabolism, or their reactions to antihelminth products. With these parasites there is a

---

PLATE III

ORGAN CULTURE OF *Meloidogyne incognita acrita*

FIG. 15. Female reproductive organ of *M. incognita acrita* in a culture, 3 hours after the initial cultivation. Phase contrast. OD, oviduct; OGE, germinal zone of ovary; OGR, growth zone of ovary; ST, spermatothica; U, uterus; Z, zygote. (×180)

FIGS. 16–21. Development of the embryos in organ cultures of the female reproductive organ *in vitro*. Phase contrast.

FIG. 16. A late two-cell stage embryo in an 18-hour culture. (×340)

FIG. 17. A three-cell stage embryo in a 30-hour culture. Note that one-half of the embryo developed more slowly than the other half. (×340)

FIG. 18. A four-cell stage embryo in a 40-hour culture. (×340)

FIG. 19. A 16-cell stage embryo in a 70-hour culture. (×340)

FIG. 20. A 7-day-old embryo in the culture. (×340)

FIG. 21. A first instar larva in the eggs of a 13th day culture. (×340)

kind of organism culture in which it is necessary to imitate the natural state in which the parasites live. Trial cultures of platyhelminth parasites can also be found in the work of Lutz et al. (1937).

1. Culture of Cestodes

On examining the attempts at cestode cultivation it seems that the survival and growth of adult tapeworms are less difficult to achieve than the development and differentiation of the various larval and juvenile stages.

Wardle and Green (1941) successfully maintained adult tapeworms of *Hymenolepis nana* Sieb. *in vitro* for 20 days, but to do so they had to use complicated sterilization techniques. Later, Raush and Jentoft (1957) achieved *in vitro* proliferation and survival for 30 days of *Echinococcus multilocularis* Vogel scolex. In preliminary experiments the scolex and the neck of *Hymenolepis diminuata* Rud. were cultivated by Schiller *et al.* (1959) in a medium mainly composed of horse serum and adult worm extract. The scolex regenerated strobila, and within 11 days growth resulted in a thirtyfold increase in size.

The greater part of research deals mostly with larval forms. Berntzen (1960) developed a technique of *in vitro* culture which enabled him to raise larvae of *H. diminuata* from infected insects to the adult state containing oncospheres in 15 days. Other attempts were made by Taylor (1961) to achieve the development of *H. diminuata* and *H. nana* larvae in anaerobic culture to sexual maturity. The media compared are too numerous to be reported here, but the majority of them were well tolerated by morula stages and young embryos which could survive for six to seven days in simple media. Whereas, in the most advanced embryos and cysticercoid or juvenile larvae only media made up of a mixture of horse serum, insect larva extract, and medium 199 (Morgan *et al.*, 1950) allowed a life span of nine days and conservation of the ability to infect mice.

The culture of cestode parasite larvae in natural media in anaerobiosis is interesting for it should allow better observation of the controversial aspects of normal development of the scolex in the cysticercoid larva. Using this technique Voge (1963) and Robinson *et al.* (1963) were able to obtain reproduction by budding in *Taenia crassiceps* Rud. larvae.

Several research workers had shown that carbohydrates represented the main source of energy of cestode parasites (see Read, 1961; Read and Simmons, 1963), but the *in vitro* experiments in this field were short term. So Taylor (1963) tried to maintain *Taenia crassiceps* larvae (easy to obtain in the laboratory) for as long as possible in synthetic media. The synthetic media on which he based his research were those developed for mammalian cells.

The Morgan et al. (1950) medium 199, the McQuilkin et al. (1957) medium NCTC 109, and the Waymouth (1959) medium are not as satisfactory as the Eagle medium (1955) for the culture of *Taenia crassiceps* Rud. By using the basic Eagle medium, Taylor et al. (1966) studied the metabolism of *T. crassiceps* larvae. They determined the respiratory ratio of the larvae, detected the excretion of lactic and pyruvic acids in the medium, and showed that glycogen is the main source of carbohydrates.

The culture conditions for these larvae are not yet optimal, but it is possible that modifications of this medium would allow a better growth and thus indicate the nutritional requirements of these animals. This may well lead in turn to the discovery of specific antimetabolites for the cestode adult or larva.

The influence of the external factors on the vitality of scolex of *Echinococcus granulosus* Batsch were shown by Benex (1968a). The scolex in *in vitro* culture showed two different aspects according to the explanted material (Benex, 1968b). First, to obtain true miniature hydatid cysts with laminated envelope, some fragments of germinal membrane must exist. Second, scolex without any germinal membrane tries to develop in to the strobilar stage. *In vitro* organotypic cultures of explanted pieces of germinal membrane from *E. granulosus* established that the formation of exogenous secondary cysts depends only on cells from the germinal membrane itself, without the scolex being conserved (Benex, 1968c).

2. Culture of Trematodes

The culture of trematodes has the same aims as those of cestodes. In 1962, Senft and Senft maintained adult *Schistosoma mansoni* in the NCTC 109 synthetic medium by Waymouth and the worms continued to lay their eggs during culture. This difference in behavior from the cestode larvae cultured in the same medium by Taylor (1963) perhaps reflects some difference in the metabolic needs between cestodes and trematodes or else it may be simply between larvae and adult platyhelminths.

Schistosoma mansoni Sambon adults or schistosomules, removed from the portal system of the hamster, were also cultured in total anaerobiosis by Bazin and Lancastre (1967) in a liquid medium with additional antibiotics.

Some *Dicrocoelium lanceolatum* Rud. were kept alive by Benex (1966) in a liquid culture medium composed of Hanks solution with colt serum, glucose, sheep red blood corpuscles, and antibiotics added. The criteria of survival are the deep muscular movements and the solidity of the fixation of the specimens on the walls of the recipient. This survival was used by Benex and Lamy (1967) for the *in vitro* experimental study of the therapeutic and, later, the prophylactic activity of various chemical substances.

3. A Study of Reconstitution and Regeneration in Turbellarians

a. **Cultures of Planarian Fragments and Reconstitution Experiments.** In the field of the extensive morphological and physiological research done by Child on the regeneration of planarians, Murray (1927, 1928, 1931) tried to develop a technique adapted for the *in vitro* culture of planarian tissues. Murray thought that this technique might supply some important data on the general problem of reconstitution, its mechanism, and the factors which governed it.

In the course of her research Murray earned credit for the discovery of an efficient technique for achieving asepsis in planarians by ultraviolet radiation. Small pieces of the body of *Planaria dorotocephala*, $\frac{1}{2}$ to $\frac{1}{4}$ mm square, were explanted. In spite of the ability for regeneration of this species, such small pieces of planarian put into water do not regenerate but die within a few hours.

In general, the hanging drop technique was used and the cells of the explant showed various degrees of migration and division according to the composition of the medium. In liquid medium mainly composed of diluted Locke solution either a cell culture (see Volume 1, Chapter 1) or a formation of organismic individuals was obtained from the explant or parts of the explant. These kinds of organisms, long or spherical, were surrounded with a cellular membrane often containing rhabdites. The spherical "individuals" survived for a short time, the long ones moved about in the medium and continued for 24 hours before disintegrating.

While studying the embryonic development of *Planaria torva* M. Schultz during *in vitro* culture on glass slides, Seilern-Aspang (1957) noticed that considerable flattening of the embryo caused a sort of polyembryony. In fact, after 10 to 12 days in culture under these conditions the embryo fragmented into peripherical offspring embryos, whereas the center continued to be formed of a syncytium of free blastomeres.

The study of the organization processes in the course of culture of pharynx fragments in the adult *Planaria gonocephala* A. Dugès suggested to Seilern-Aspang (1958) that culture on glass slides in only two dimensions allows cellular proliferation. On the contrary if the same fragments are placed at the end of a mica slide resting on the culture slide, i.e., in conditions where development is possible in three dimensions, a morphodynamic field appears which, after 21 days of organ culture, induces the formation of "individualities" composed of a parenchymatous mass surrounded by ciliate epithelial cells. Such "individualities" show regular cilia waving and in some cases move slightly. At best they survive for eight days.

Experiments in dissociation of adult planarian tissues were undertaken by Freisling and Reisinger (1958), who studied the conditions of reassociation of the various cells. When planaria of the species *Planaria polychroa* O. Schmidt are dissociated mechanically in Holtfreter solution, rapid reassociation of cells which form small spheres or "restitution bodies" is observed. The appearance of these spherules requires a contact of epithelium fragments which have retained their cilia with mesenchymatous elements in a homogenate medium which is sufficiently viscous. In order to observe these restitution bodies without being hindered by the homogenate, they are transported into a medium of adequate viscosity which is either gelatined Holtfreter solution or lobster abdominal sinus hemolymph, in which they survive for six days.

Fragments of epidermis without contact with mesenchyme cells are incapable of restitution. Fragments of ciliate gonoducts do not give any restitution formations either, and they die. Mesenchymatous and endodermal cells mixed together but without contact with ectodermal material are not capable of either restitution or autodifferentiation.

Histological examination of typical ciliate spherules shows that the cellular limits disappear. The muscular, connective, and glandular cells which compose the internal mass rapidly take on a syncytial aspect and the dense parts are replaced by hyalin or vacuolated parts. Amoeboid cells, similar to the neoblasts, migrate from the inside towards the restitution membrane.

Considerable internal vacuolization takes place at the end of the differentiation of the ciliate covering layer of the internal syncytium of the spherules. Afterwards, it is not rare for a spontaneous invagination of the superficial layer to appear, leading to a pseudogastrulation which shows a superficial similarity with the organization of acoelomate simple turbellarians. However, neither a nervous system nor a muscular layer exists in them. The authors think it to be a case of morphallaxis which finds no parallel either in the phenomenon of regeneration nor in embryonic development. The necessity for contact between the epidermis and the mesenchymatous cells in the reconstitution of these spherules brings out the importance of the tissular interaction mechanism in invertebrates which has since been brought to light in the processes of organogenesis in vertebrates. It may be that this interaction which ensures the continuity of the waving of the epidermis cilia of the spherules also plays an important part in the phenomena of regeneration.

Ansevin and Buchsbaum (1961) made several sorts of adult tissue cultures of the freshwater planarian *Dugesia tigrina* Gir. and studied the behavior of the cells according to the culture medium. They explanted either minced fragments of planarian or cells dissociated with 0.002% versene solution.

Liquid media inhibit cellular migration. The medium containing only planarian tissue extract allows the formation and survival of "restitution bodies" of Freisling and Reisinger (1958). Under these conditions if the explant is only formed of a group of a few cells, these cells only survive for 24 to 48 hours and then die. If the explant is a little larger, it becomes round and begins to regenerate. The epidermis forms a continuous covering layer while internal reorganization takes place. Such fragments begin moving after three or four days, and small planaria are reconstituted a few days later.

Addition of chicken plasma inhibits both the formation of "restitution bodies" and the survival and regeneration of larger explants.

Solid medium favors cell migration, but it also allows regeneration and morphallaxis inside the larger explants.

So planarian tissues cultured *in vitro* seem to show two antagonistic tendencies, one towards reorganization into a new organism, the other towards spreading on the substratum into an unorganized layer of cells. Each culture technique used favors one while suppressing the other.

In histiotypic culture of planarian fragments in a nutritive medium rich in amino acids, Chandebois (1963a,b, 1970) observed the disorganization of the explants, dedifferentiation, and cell proliferation. Under comparable conditions of cell cultures, Betchaku (1967) studied the behavior *"in vitro"* of planarian neoblasts in relation to some aspects of the mechanism of the formation of "restitution bodies" and regeneration blastemas.

The various experiments described show the very important part played by the size of the explants and the composition of the culture media in the phenomena of reconstitution from adult tissues, but apart from the necessity of ectodermal and parenchymatous tissue participation to produce a new organism, little new data on regeneration was obtained.

b. Cultures of Regeneration Blastemas. So C. Sengel (1959, 1963) left the culture of small fragments of planaria and turned to the culture of regeneration blastemas so as to study the factors liable to act on the morphogenesis of the regenerative material. Sengel, first of all, tested the survival of two- to three-day-old blastemas in various physiological solutions. Holtfreter solution, pure or diluted by half, is suitable for blastemas of *Dugesia lugubris* O. Schmidt, whereas this solution must be diluted 8 to 10 times for fragments of *D. tigrina* Gir. In these liquid media the blastemas heal rapidly but do not present any further differentiation.

To aid the development of explanted blastemas, Sengel adopted the Et. Wolff and Haffen (1952) culture medium in which the Tyrode and Gey solutions were replaced by Holtfreter solution with glucose added. The blastemas deposited on this solid medium healed, retained their flat shape and survived for at least four days in more than 80% of cases and for one to three weeks in 60% of cases. The chick embryo juice of this culture medium may be replaced by planarian extract, horse serum, or a solution of amino acids, but the use of these solutions does not give better results.

The still undifferentiated explanted blastemas have the shape of a small tongue from 0.2 to 0.3 mm long. They are made up of a mass of neoblasts surrounded by a very thin healed epidermis. The cephalic blastemas (future heads) and the caudal ones (presumed to be tails) present the same development during the first three to six days. Once healing is accomplished, pigment granules appear and the muscles differentiate and manifest themselves by contractions. On the other hand, the epidermal cells become ciliated and contain numerous rhabdites. Later on, each blastema differentiates according to its future destiny. A posterior bud only becomes a tail. An anterior bud becomes a head by differentiation of a brain and eyes. It may be concluded from these results that the 2½- and 3-day-old blastemas, although undifferentiated, are already determined fundamentally by the base. Younger blastemas could not be kept alive in organ culture.

After Sengel's first explantations of regeneration blastemas, two Italian workers, Manelli and Negri (1962) made similar experiments on the planarian *Planaria torva* O. F. Müller. The blastemas were removed five, seven, and eight days after the section. The Sengel medium not being suitable for the blastemas of this planarian, the research workers modified it by adding sheep plasma, but this technique favored cellular dissociation and migration. A standard medium with total planarian (*Dugesia lugubris*) extract was found to be the most efficient. The five-day-old blastemas explanted by these authors rapidly degenerated. The seven-day-old blastemas survived in culture for four days; the cephalic blastemas healed more easily than the caudal ones. In the course of culture, cells with rhabdites and pigmentary cells differentiate, but no mitoses are observed. The differences in development of the blastemas studied by these authors and by Sengel are doubtless due to differences between species of planaria and to differences in the age of the blastemas cultured.

Comparing the *in vitro* development of cephalic blastemas from sections at different levels of the body of *D. lugubris*, Ziller-Sengel (1967) noticed that the rapidity with which the eyes appear is greater in the blastemas from the most cephalic region and least in the blastemas from the most caudal part. This experiment verifies the existence of an anteroposterior gradient for the time of regeneration of the head as observed by Brønsted (1946).

This culture technique also enabled Ziller-Sengel (1964, 1967) to associate two or several regeneration buds. These rapidly joined and each element of the association differentiated according to its future destiny. In particular, the association of a cephalic with a caudal blastema tended to lengthen in a cephalocaudal direction and assumed the shape of a small planaria. In this case formation of a pharynx between the 7th and the 15th day in culture, and sometimes of digestive caeca, was to be seen. In the association of cephalic blastemas, each element tended to give a head with a brain and two eyes; often extra eyes differentiate. Caudal blastemas associated together only gave tails. The determination of the regenerated parts was therefore precocious and irreversible.

Explants maintained in culture on gel medium are incapable of growing. Ziller-Sengel tried to readapt them, after association, with a liquid medium. This readaptation did not succeed and the author thinks that this failure may be attributed to a lack of renewal of the reserves of neoblasts which divide very little in culture, doubtlessly due to faulty nutrition.

After Lender's experiments (1955, 1956) showing that planarian head homogenate added to water for the breeding of "decapitated" planarian inhibits the regeneration of the excised brain, Ziller-Sengel (1967) cultured 2½ to 3-day-old cephalic and caudal blastemas on media respectively containing head or tail homogenates. No specific inhibitory action was noted as to the differentiation of the blastemas in culture. It therefore seems that the inhibition hardly effects an already determined blastema. However, this determination is not completely irreversible for the same author obtained partial or total inhibition of cephalic differentiation by associating an older cephalic blastema (4½ days) with undifferentiated blastema (2½ days). The inhibition exercised by a live brain in the course of differentiation therefore seems more powerful than that of adult brain homogenate. This set of results (Ziller-Sengel, 1970) shows the importance of the inductive factors exercised very precociously and in a specific manner by the tissues of the stump in the course of regeneration.

It would be very interesting to be able to separate the epidermal from the "neoblastic" component in these blastemas and then to make associations of these components with complementary tissues of another region of the planarian or else from blastemas of a different age or with a different destiny. Such associations would doubtless allow the respective roles of the two tissues of the blastema in the course of regeneration to be brought to light, but the main difficulty lies in the material realization of such experiments with a small and particularly fragile material.

H. Coelenterates

Individuals from many of the coelenterate species are remarkable because they possess extensive morphogenetic potentialities that are preserved throughout their lives. These organisms are able to reestablish their structural and functional integrity under very varied natural and experimental conditions. So great can these powers of regeneration be, that an entirely new individual may be reconstituted from a single fragment containing a few cells isolated from the base of a sea anemone (*Sagartia*, for example). This natural "propagation by cuttings" is a particularly simple type of culture, and coelenterates offer an ideal experimental material for the study of the factors and mechanisms of regeneration and regulation. The fundamental aspects of the phenomena are the same as in the other phyla studied, and one of the main problems to be solved is the cause of the maintenance of postembryonic morphogenetic plasticity. Most of the research work on coelenterate reconstitution has been done on Hydrozoa. Some authors (Evlakhova, 1946) considered that the undifferentiated interstitial cells located at the base of the ectoderm were the sole source of regeneration material. Others considered that it was provided by the cells of the ectoderm and endoderm reassuming their proliferating activities (Beadle and Booth, 1938; Steinberg, 1954). Steinberg considered the interstitial cells to be simply undifferentiated nematoblasts.

It seems that the controversy between these extreme points of view may be resolved by a compromise in which the two sources prove to be mutually complementary. In fact, Brien and Reniers-Decoen (1955) showed that a hydra in which the interstitial cells had been destroyed by x-ray irradiation was able to bud and regenerate for 10 days, but afterwards the participation of the interstitial cells was essential for completing the final form of the young polyp. The histological observa-

tions of Tardent (1965) on *Tubularia* and *Hydra* also suggested that
the differentiated ectoderm and endoderm, as well as the undifferentiated
interstitial cells participated in the formation of the regeneration
primordium.

Coelenterate *in vitro* culture experimental research was particularly
interesting because of the following:

1. The simple structure of these diploblastic animals enabled the
destiny and morphogenetic potentialities of each of the isolated tissue
cell layers to be followed.

2. The amplitude of cell differentiation is very wide, ranging from
the embryonic type of interstitial cell to such highly specialized cells as
the nematocysts which are equivalent to organs. Achieving tissue dissociation and studying its development under culture should help to
elucidate tissue interdependence and its behavior in relation to the
composition of the media used.

Apart from the experiments on the survival in seawater of sea
anemone tentacles, acontia, and fragments of the margins of jellyfish
umbrellas as reported by Thomas (1963) and the culture trials of
Lewis (1916) with sea anemone septa in slightly diluted seawater to
which dextrose, peptone, and $NaHCO_3$ had been added, most of the
in vitro cultures had been carried out from the two points of interest
mentioned above, using either dissociation fragments of coelenterate
bodies or either of their two constituent tissue layers.

1. Culture of Bilayered Body Wall Fragments or of Dissociated Cells

From 1911 Wilson (1911a) carried out total cell dissociation with
the object of determining whether, once liberated from their anatomical
and physiological ties, hydroid cells would be able to recombine and
form masses of totipotent regeneration tissue as he had observed with
sponges (Wilson, 1907).

Hydranths and the coenosarc of the two hydroids, *Pennaria tiarella*
and *Eudendrium carneum,* were cut into pieces and pressed through
gauze (fine-mesh silk bolting cloth). In seawater the cells thus separated reaggregated in a few hours into masses 0.1–5 mm in diameter.
Despite a high mortality in a medium that was renewed two to three
times a day, certain nodules produced two-layered hydranths. Wilson
did not identify the cell types that participated in the formation of
the restitution masses, but he thought that hydra cells passed rapidly into
a simplified neutral state. He considered this to be a case of regressive

differentiation provoked by a change of the physiological state under the effects of isolation and shock.

Under similar conditions, the cells of the "sea feather" (*Leptogorgia virgulata*) which had a more complex organization lived for only a few days as 1-mm diameter spheres. Here, Wilson suggested, the medium inadequacy could have exerted an influence, but it was also possible that the isolated cells in which specialization was more emphasized than in the hydroids no longer possessed the genetic power to recombine and reconsitute the characteristic functional structure of the species.

Following Wilson's dissociation experiments with marine hydroids, fragmentation tests were carried out with *Hydra* (Issajew, 1926; Papenfuss, 1932, 1934; Weimer, 1934), during which the uniting of cell masses composed of both ectoderm and endoderm led to the reconstitution of new and complete animals. Culturing minced *Hydra* tissue, Chalkley (1945) noted that it could reorganize itself into one or more animals without producing any appreciable cell division. This *in vitro* study of reconstitution demonstrated that the phenomena of morphogenesis (the determination and maintenance of form) could be partly or entirely dissociated from the phenomena of proliferation and growth.

Li *et al.* (1963) noted that when *Hydra* cells were dissociated with trypsin and then cultured, they were very rapidly destroyed by microsporidia that contaminated the *Hydra,* hence the need for destroying the former in a bath of Fumidil B. Li *et al.* then observed that isolated and disinfected *Hydra* cells transplanted onto a rich culture medium (a mixture of Eagle medium, modified Earle solution, horse serum, and extract of *Hydra* buds) gave rise to a true culture of interstitial type cells.

Dissociated cells and fragments of coenosarc of *Tubularia* were placed in roller tubes and hanging drops by Tardent (1965). In hanging drop culture in a semisolid medium, a centrifugal cellular migration took place that culminated in the formation of a single layer of cells, as in the histiotypic culture of vertebrate embryonic tissue. With the roller tube technique, all the cell types disappeared after five to eight days of culture, except for the nematoblasts which survived and rapidly multiplied for several months provided that they were frequently transplanted. If the medium was not renewed for several days, the nematoblasts began to differentiate into nematocysts. This was an example of a change from a histiotypic to an organotypic culture caused by modifying the cultural conditions.

These interesting preliminary observations are worth completing with other experiments in which it would be very rewarding to succeed in isolating the various types of cells before culturing them, because it very rapidly becomes as difficult to follow what happens to them in a liquid medium (Wilson, 1911a; Li, 1962; Tardent, 1965) as in a semi-solid one (Tardent, 1965).

Up until now no one has succeeded in cloning particular types of cells in a chemically defined medium. To understand the factors controlling differentiation, Burnett *et al.* (1968) described a chemically defined physiological solution which supports growth and differentiation in stem explants of *Tubularia*. Cultures were observed over a period of six days and none of the differentiated cell types underwent a morphological change in culture.

However, Rannou (1968) found a transformation of the larva or colony cells of gorgonians of the species *Eunicella stricta* Bertoloni. In culture the cells differentiated into scleroblasts much earlier. Aggregates of cells, obtained by Frey (1968) after dissociation of fully differentiated medusae of *Podocoryne carnea* M. Sars, in a few cases produced stolos and structures of stolo-like shape. This means that at least part of the medusa cells have not irreversibly lost the ability to functional and structural metaplasia.

The explants of the gorgonian tissues *Eunicella stricta* and *Lophogorgia sarmentosa* survived for more than a month in seawater. The ectodermal culture technique of explants with the mesoglia facing mesoglia enabled Theodor (1969) to observe the effects of histotoxicity which were even more important when the species were well separated.

It may be concluded that in function of the species under consideration and according to the culture procedures used one of the following is obtained: (a) cellular reaggregation and the reconstitution of a new animal, (b) the survival of either explants or cells, or (c) cellular multiplication with or without apparent differentiation processes.

This simple technique could be adapted to help solve many of the cellular physiological problems.

2. The Fate of the Ectoderm and Endoderm when Cultured Independently

Up to now it is not really known whether one type of differentiated animal cell can transform into another type. The coelenterates, which possess one of the simplest of the metazoan structural organizations, have characteristics that have led to their being compared with plants and provide some most interesting research material.

Gilchrist (1937) fairly easily separated the ectoderm from the endoderm in the scyphistoma stage of *Aurelia* because the mesoglia between these two tissues was sufficiently thick to be broken. He stated that the fragments of ectoderm were capable of regenerating small complete polyps but that pieces of endoderm could not regenerate although they remained alive for several days. This finding was of considerable interest, but it still remained to be discovered how the regeneration of the endoderm was accomplished. Gilchrist simply suggested that it might be formed by the interstitial cells of the ectoderm but did not test his idea with an histological examination.

Thinking that the interstitial cells which were confined to the ectoderm of *Cordylophora* might produce the endoderm, Beadle and Booth (1938) separated the ectoderm from the endoderm in reconstitution masses derived from the coenosarc of this animal. They isolated four masses of supposedly pure ectoderm: One developed into a small hydranth, while sections of two others showed that an internal layer had begun to form. On so small a number of cases and despite other observations, the authors concluded that the endoderm layer redeveloped only when the ectoderm contained a few endoderm cells and that the interstitial cells of the ectoderm could not produce endodermal elements.

Papenfuss and Bokenham (1939) separated the cell layers of green and brown hydras and cultured the ectoderm and endoderm separately in small cups of filtered pond water in the presence of a control hydra. In no case did they obtain the regeneration of a hydra, and generally the tissue disintegrated within 24 hours. These results confirmed the absence of regeneration in hydra from ectoderm alone that had already been noted by Ischikawa (1890), who concluded that the interstitial cells were incapable of reforming the missing tissue layer.

Schulze (1918), who made a similar observation, suggested that the failure of the ectoderm tissue to regenerate was due to a lack of nourishment normally provided by the endoderm. At the time this hypothesis made little impact, and yet, to us it seems very important because recent experiments have demonstrated the significance of the ambient factors (nourishment, ion concentration, inductive substances) on the behavior of isolated ectoderm and endoderm cells.

For a number of years *in vitro* experiments with one or other of the tissue layers that make up the coelenterate body have been increasingly testing the different species and bringing to light interesting evidence concerning the differentiation phenomena.

Normandin (1963) separated enzymatically (by the action of pancreatine and amylase at pH 8.5) the endoderm and ectoderm of the body and peduncle of *Pelmatohydra*. The fragments of each were cultured separately at 15°–20° in a saturated solution of sulfadiazine in "pond" water sterilized on a Seitz filter. After a few hours, the ectoderm had disintegrated, but the endoderm tissue had reorganized itself into a solid ball in three hours. Eight to 24 hours after the operation this ball elongated and formed a central cavity. At the end of three days tentacles began to appear. From the third to the seventh day the tentacles differentiated around a hypostomal region which became functional after the seventh day. At the same time the author noted histological and histochemical changes which seemed to indicate that a certain dedifferentiation and redifferentiation had been taking place during the regeneration of the isolated endoderm of *Pelmatohydra*.

Gilchrist's experiments on the scyphistoma of *Aurelia aurita* Pér. and Les. were repeated by Steinberg (1963), who tried to reconstruct the origin of the endoderm from a histological study of a series of stages through which ectoderm fragments passed during regeneration. Hydranths were produced from the culture of small rectangles (0.5 × 1 mm) of ectoderm in filtered seawater. The author did not observe any interstitial cells either in the intact scyphistoma or at any stage during the reconstitution. On the other hand, he noted the appearance of amoeboid cells in the explants of ectoderm. The number of these cells increased without showing any mitotic activity. Gilchrist considered that the amoeboid cells which eventually formed the endoderm came from somatic ectoderm cells which showed considerable mitotic activity. According to this, somatic ectoderm can give rise to somatic endoderm by first losing its characteristics and becoming an "indifferent" amoeboid cell.

While awaiting the possibility of working with cell clones of known origin which would give the opportunity of discovering the potentialities of each type of cell, Zwilling (1965) repeated the ectoderm and endoderm dissociation experiments of Beadle and Booth (1938) with *Cordylophora lacustris* Allm. The author produced masses of pure reconstitution ectoderm from fragments of coenosarc tissue 0.1–0.2 mm in diameter. Although *Cordylophora* lived in fresh water, Zwilling noticed that the medium most favorable to the reconstitution of the coenosarc masses was a pasteurized mixture of 25–30% of filtered seawater in distilled water. During the first 10 hours fragments of *Cordylophora* ectoderm produced a hollow sphere whose external wall consisted of a single layer of small cells, while the cells of the interior were fragmented and disorganized. Fifty percent of the pure ectodermal masses reconstituted an internal layer with flagellae and formed a complete hydranth in five to nine days. When the author deliberately added endoderm cells, these always took up their position in the interior of the reconstitution sphere and the hydranths were more rapidly reconstituted. When the quantities of associated ectoderm and endoderm were equal, the hydranths reformed in two to three days. To make more certain that all the endoderm cells that might have contaminated the mass were eliminated Zwilling carried out the operations in two cycles. From 10-hour reconstitution masses he removed the thin external wall which he recultured. Here again he obtained the formation of a new endoderm layer.

Although these experiments may not have defined the precise roles of the ectodermal interstitial and somatic cells, they did show that the superficial ectoderm layer was able to form new endoderm.

Haynes and Burnett (1963) demonstrated that *Hydra viridis* L. could regenerate from a gastroderm explant consisting entirely of two types of cells, glandular and digestive cells, when it was cultured in an appropriate saline solution (0.1% NaCl, 0.2% $CaCl_2 \cdot 2H_2O$, 0.01% $KHCO_3$, 0.03% $MgSO_4 \cdot 7H_2O$). The new epidermis began to form on the explant within two to eight days, depending upon the temperature of the solution. When the explant was covered by the epidermis, the culture solution was gradually diluted with normal hydra culture medium. After the appearance of the mouth and tentacles, the animals were fed daily.

Histological and ultrastructural examination indicated that gland cells located at the periphery of the regenerating explant dedifferentiated into interstitial cells and subsequently redifferentiated into cnidoblasts. The digestive cells at the periphery of the explant transformed directly into epithelio-muscular cells (Davis *et al.*, 1966). During dedifferentiation and redifferentiation in the regenerating isolated gastrodermis of *H. viridis*, Davis (1970) observed the division of three cell types: (1) interstitial cells derived from dedifferentiated peripheral gland cells, (2) epithelio-muscular cells which come from the direct transformation of peripheral digestive cells, (3) cnidoblasts originating from interstitial cells. Hence, an explant that contained no interstitial cells was able to form some from the glandular cells, and this new-formed interstitial cell could differentiate into a cell type (cnidoblast) that was different from the type from which it arose (glandular cell). The authors claimed that this evidence refuted the idea that the interstitial cells represented a reserve stock of embryonic cells. Admitting that under the conditions of this experiment interstitial cells could be formed from differentiated glandular cells, it is still not known whether such a process occurs in the intact normal animal. It is, however, highly probable that in the presence of the epidermis the glandular cells will retain their original morphology.

Starting from this result, Burnett *et al.* (1966) isolated a number of animals regenerated from gastroderm and established clones from their asexual descendants. Tens of thousands of animals were obtained from a single explant. These populations were unique in that their interstitial cells were not produced by embryogenesis but by differentiated gastroderm cells. These authors made a histological and ultrastructural study of interstitial cells arising from glandular cells with the object of determining whether these interstitial cells arising from specialized somatic cells were capable of undergoing germinal differentiation. Sexual activity appeared after only seven weeks, and the electron microscope showed that the differentiation of the interstitial cells into spermatozoids and ovules was entirely normal. The authors concluded that there was no determined line of interstitial cells destined to form gametes. Their results showed that in hydra there was no separation into germinal and somatic cell lines during embryogenesis. This confirmed numerous earlier observations made by Brien (1941, 1961, 1966), who definitely established that "interstitial cells were, according to the particular circumstances, equally and reversibly somatic or germinal."

These endoderm culture experiments showed that the cells of this layer were able to undergo a metaplasia and take up morphogenetic activities again. Since the phenomena of dedifferentiation of glandular cells into neoblasts and of digestive cells into epitheliomuscular epidermal cells did not occur either in water culture or in an inorganic medium that differed from that of Haynes and Burnett (1963), Macklin and Burnett (1966) considered that the factor controlling these events might be a question of an adequate concentration of specific ions.

Macklin and Burnett (1966) also cultured specimens of entire *Hydra pseudoligactis* either with the gastric column opened longitudinally and spread out on a cork support, or the whole animal turned inside out, in ionic solutions that were stronger than normal. These ions were K^+, Na^+, Ca^{2+}, Mg^{2+}, SO_4^{2-}, Cl^- and HCO_3^-. The most striking changes at cell level were observed after treatment with strong Na^+ and Ca^{2+} concentrations. In particular, the combination of Ca^{2+} with Na^+ ions at the concentrations of 0.2% $NaHCO_3$ and 0.2% $CaCl_2$ provoked an obligatory differentiation of all the neoblasts into cnidoblasts. Treatment with this medium induced all the gland and basal reserve cells of the gastrodermis to transform into mucous cells. Furthermore, in a medium of high Na^+ concentration, the basophilic properties of the *Hydra* cells due to the cytoplasmic RNA disappeared, while at the same time the nucleus became metachromatic, probably as the result of a separation of the DNA–histone coupling.

The mechanism of the action was not yet known but it reproduced the cell transformations provoked by the application of a growth inducer (probably produced by the nerves of the hypostome) to a ring of hydra tissue taken from the gastric region (Lesh and Burnett, unpublished observations, according to Macklin and Burnett, 1966). This similarity of results suggested to these workers that the inducing agent might operate at the level of the cell membrane.

In a recent paper, Lowell and Burnett (1969) report the development of gastrodermis, and finally whole animals, from epidermal explants of *Hydra oligactis* and *Hydra pseudoligactis* demonstrating the totipotency of each cell layer in *Hydra*.

3. Development of Combinations of Ectoderm and Endoderm

The comparative study of the development in culture of fragments of bilayer coenosarc and of isolated and reassociated ectoderm and endoderm under different conditions permitted Diehl and Bouillon (1966a) to clarify the relationships existing between these two layers of the Cnidaria and their morphogenetic potentialities. These authors worked with the athecate hydroid *Cordylophora caspia* Pall. Their preliminary experiments first showed that eight days after irradiating the hydranths with x rays there was no sign of interstitial cells. The animals declined and finally disintegrated between the fourth and fifth week. Pieces of coenosarc extracted from the perisarc aggregated in a briny water culture (one-third seawater plus two-thirds fresh water) and in two or three days formed a spherical mass with outgrowths which reconstituted hydranths (in about four days) and stolons which became attached to the sides of the Petri dishes.

Fragments of coenosarc isolated one day after irradiation rapidly aggregated and reconstituted both stolons and hydranths in three days. In the following days no new hydranths appeared and at the end of a week the tissues disintegrated. This evidence agreed with the observations on the budding of irradiated hydras made by Brien and Reniers-Decoen (1955). Pieces of isolated ectoderm from normal stolons fused into a spherical mass after several hours and hydranths appeared five to eight days after the beginning of the experiment. These results confirmed those of Zwilling (1965). Fragments of ectoderm from animals irradiated one, three, and six days earlier always aggregated into spherical masses and remained alive for up to 15 days but no trace of morphogenesis was observed. Fragments of isolated endoderm also aggregated into spheres in three to five hours but after 12 to 24 hours these disintegrated, and under the conditions of the experiment, the endoderm alone never gave rise to complete animals. Endoderm irradiated with x rays behaved in the same way.

After these culture experiments with aggregates of normal or irradiated tissue, Diehl and Bouillon (1966a) used associations of normal with irradiated tissues. When normal ectoderm was associated with irradiated endoderm, there was a reconstitution of complete hydroids in a time similar to that necessary for ectoderm alone to reform a hydranth. This indicated that irradiated endoderm neither exerted an inhibitory effect nor played any morphogenetic role.

The opposite experiment revealed a great deal more evidence. Irradiated ectoderm combined with normal endoderm gave rise to a certain number of viable colonies. Granting that the irradiated ectodermal tissues had lost all morphogenetic activity and that every precaution had been taken to eliminate ectodermal cells, the authors concluded that the reconstituted animals were entirely reformed from normal endoderm elements which had dedifferentiated. These results suggested that the irradiated ectoderm cells acted as a protective medium in which the potentialities of the endoderm could be achieved.

The writer is of the opinion that these results may be compared with those of Burnett *et al.* (1966), who obtained the reconstitution of *Hydra* from endoderm alone provided that it was placed in a suitable saline solution. Thus, the irradiated ectoderm constituted the "medium" that permitted a metaplasia of the endoderm to take place and certain cells were capable of dedifferentiating into interstitial cells.

In the present case, it should be remembered that the ectodermal cells, even when irradiated, preserved their physiological activities during the days that followed. These activities with those of the endoderm normally determined the multiplication and differentiation of the interstitial cells. In the absence of any response from the latter, certain endoderm cells could recover their morphogenetic potentialities. It is to be hoped that current physiological and histological research will be able to define the gradual growth of the ectoderm cells as well as the influence that they exert. It would be interesting to know if the ectoderm from different regions of the colony had the same properties and to discover whether a relationship existed between its action on the endoderm and the inductive role played by the nerves of the proximal region of the hydra as proposed by Burnett (1966).

This method of tissue dissociation and reassociation was also applied (Diehl and Bouillon, 1966b) to the study of the factors which sometimes determined the formation of a hydranth and sometimes that of a stolon. Before the work of these authors there was little information on the relative importance of the endoderm and ectoderm in the morphogenesis of the hydranth and of the stolon. Moore (1952), however, showed that the grafting of oral hypostomal cones or of tissue taken from the polyp onto reaggregates of *Cordylophora caspia* Pall. tissues induced the appearance of hydranths. But in addition to the hydranths induced by this method, certain others appeared spontaneously in the reaggregation masses in the absence of grafting. The experimental method described by Diehl and Bouillon (1966b) enabled a control of the production of

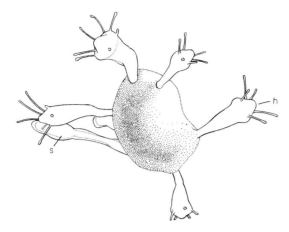

FIG. 22. A reaggregation mass formed from one-half ectoderm plus one endoderm of *Cordylophora* after 2 days of culture. Five hydranths (h) and two stolons (s) had appeared.

hydranths of *C. caspia* to be performed by varying the relative quantities of ectoderm and endoderm which are constituent parts of the wall of every hydroid.

By using only stolon tissue they were able to obtain, by dissociation,

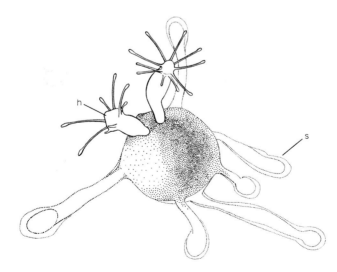

FIG. 23. A reaggregation mass formed from two ectoderm plus one endoderm of *Cordylophora* after 2 days of culture. Two hydranths (h) and six stolons (s) had appeared.

small masses of homogeneous tissues composed of either ectoderm or
endoderm. They then produced two categories of aggregation.

In one category they associated a quantity of ectoderm equal to one-
half of that which occurred in a stolon with each endoderm fragment of
0.2 to 0.5 mm (one-half ectoderm plus one endoderm). Two days after
preparing the aggregates, hydranths in particular had been formed
(Fig. 22).

In the opposite category (two ectoderm plus one endoderm) the emphasis was on stolon production with a few hydranths (Fig. 23).

Irradiated tissues showed the same behavior as normal tissues in
relation to the differentiation of hydranths and stolons (Fig. 24a and b)
but only for a limited period of life.

Since x rays destroyed the interstitial cells in *Cordylophora* (Diehl
and Bouillon, 1966a) in the same way as in *Hydra* (Brien, 1956), these
experiments on the culture of tissue associations proved that these cells
were no more responsible or necessary for the induction of a hydranth
(Diehl and Bouillon, 1966b) than for the budding of a *Hydra* polyp
(Brien, 1956).

These experiments demonstrate the importance of the cells of the
original tissues for maintaining and regulating the genetically determined
structural organization. They also reveal the persistence in these animals,
endowed with an astonishing plasticity, of tissue interaction mechanisms

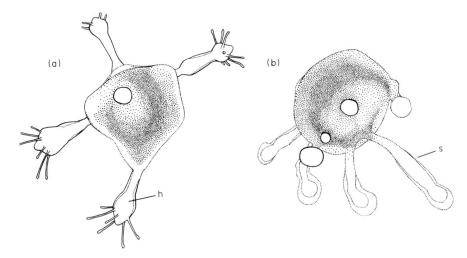

Fig. 24. Association culture of x-ray-irradiated ectoderm and endoderm in
Cordylophora. (a) One-half ectoderm plus one endoderm: normal hydranths (h)
with tentacles had appeared. (b) Two ectoderm plus one endoderm: stolons (s) differentiated, but no hydranth appeared.

characteristic of embryogenesis. There can be no doubt that the application to the coelenterates of the methods that were so successful in the study of embryo organogenesis in vertebrates will enable the essential phenomena of cell differentiation to be defined.

4. Studies of the Control Factors of Polyps Regeneration

The process of differentiation concerning the nature and the method of the transmission of the information that controls this phenomenon in the coelenterates was studied by Rose (1966, 1967) and Rose and Powers (1966). Using a method comparable with that employed by Et. Wolff *et al.* (1964) for the study of regeneration factors in planarians, these authors cultured *Tubularia* regeneration primordia. Under normal culture conditions, the reconstituting stalk fragments were able to preserve their structure and to regenerate the missing distal part of a hydranth in conformity with a well-defined polarity. In an attempt to obtain information on the nature of the control of this regeneration, Rose and Powers (1966) added aqueous extracts of homogenates of the distal and proximal regions of adult *Tubularia crocea* hydranths to the medium (a solution of chloramphenicol and pasteurized seawater) in which stalk fragments were reconstituting.

When extracts of the distal region were added to the culture medium in suitable concentrations, they maintained the proximal structures but inhibited the formation of the distal regions.

Extracts of proximal regions provoked the loss of proximal structures and suppressed all regeneration.

If extracts of differentiating primordium were added to the culture medium containing younger primordia, an inhibition of distal regeneration and a stabilization of proximal structures by extracts of distal regions were also observed. On the other hand, proximal extracts did not stabilize proximal structures.

Thus it was shown that aqueous extracts of regenerating *Tubularia* primordia contained substances that controlled differentiation.

Rose (1966) separated the various components of these extracts by electrophoresis and tested them on fragments of *Tubularia* in culture. Rose noticed particularly that inhibition persisted after acid extraction but that all regional specificity disappeared, whereas the inhibition was suppressed by treatment with trypsin which suggested to the author that the inhibitors might be proteins or polypeptides.

Finally, in an attempt to verify the hypothesis that the control of differentiation was determined by the migration from cell to cell in a polarized bioelectric field of specific inhibitors and repressors, Rose

(1966, 1967) submitted these repressors to an electric field. The author observed that they consisted of positively charged substances and that their progress could be changed by the application of an electric field. A review of the experimental work led Rose (1970a,b) to the conclusion that all levels of *Tubularia* hydranth have the genetic information to become any part of the hydranth or stem. By a series of repressions controlled by a bioelectric field of sufficient strength any level can become the most anterior part of the system for which they are not repressed.

The experimental study of the regeneration of polyp stem fragments of *Hydractinia echinata* was undertaken by Müller (1969a) to try and understand the control mechanisms of polarized regeneration. The segment cultures of the body column of the polyps showed that, at any axial level in the polyps, the apical regions could be regenerated whereas the potencies to regenerate basal parts were suppressed. If regenerative formations occurred at the basal end, they would become apical structures which lead to heteropolar development. The probability of such a heteropolar development is positively correlated to the distance between the apical pole and the basal cut. After removing the apical primordium the rate of heteropolarity rose within 3–12 hours after amputation of the hypostome which appeared to have a dominant action.

The polar pattern of potencies changed after the tissue dissociation. The reaggregated cell associations developed stolons.

Chemical analysis showed Müller (1969b) that not only did specific inducers exist, but that inhibitors existed which interfered with the dominant system. For the author this did not necessarily imply the existence of gradients of diffusible substances. Müller (1969c) performed similar experiments to analyze the biological and chemical factors in the sexual determination in *H. echinata* polyps. In contact with blastostyle bud tissue regenerating gasterozoids changed into blastostyles. Blastostyle extract applied in graduated dosages also increased the transformation rate. The causative factor was to be found in the range of fractions which also contained the tentacle-inducing substance.

The graft combination method (Burnett *et al.*, 1967) and the transplantation or parabiosis methods (Müller, 1969a) were also used to determine the cause of polymorphism of *H. echinata*. Müller (1969a) brought together the various hypotheses made by other research workers dealing with this subject.

Once the inducing factor of cell differentiation had been extracted from the tissues of *Hydra* by Lesh and Burnett (1966), *Hydra* regenerating annuli of 0.5 mm in apico-basal length were cultured in a control incubating medium or incubated in the material being tested for inductive activity for a four-hour period by Lesh (1970). At the end of this incubation period annuli were returned to the culture solution for 24 hours. These experiments demonstrated that hypostomal inducing materials, varying quantitatively can effect qualitative differences in the direction of differentiation of responding cell types.

Although the culture media in which coelenterate organic structures can be developed are in general very simple, the explantations offered varied a great deal. The plasticity of these animals offers a field of research particularly adapted to the study of cell differentiation due to the fact that these animals on reaching adulthood can still demonstrate various forms of morphogenesis comparable with those pertaining to embryonic life.

Future studies must try to precisely establish the nature, the localization of the inducing or stimulating factors, and the inhibiting factors whose ratio controls the regulating morphogenesis and differentiation processes.

I. Sponges

The sponges, like the coelenterates, are particularly interesting for organotypic culture because of the morphogenetic potentialities that most of them still possess in the adult state.

The study of reconstitution phenomena in sponges has been the object of a large amount of research work since the experiments carried out by Wilson (1907) on cell dissociation followed by the complete regeneration of these animals. In fact, the relatively restricted number of cell types and the less strictly determined morphology of the individuals than in the hydroids seem to offer simpler conditions for the analysis of the mechanisms of reconstitution. However, the difficulty of separating the cell categories has made the interpretation of the observations an arduous task and led to controversies similar to those that arose with the coelenterates and other invertebrates which can form restitution bodies.

Although the exact role of the different types of cells has not always been entirely clear, a renewed interest was shown in these experiments when it was realized that the cells of the embryonic organs of vertebrates possessed comparable properties. The study of sponge reconstitution provides excellent material for examining the histogenesis and morphogenesis of an organism; it should enable the cell interrelationships that are so important in organogenesis to be defined.

In this short review of the experiments carried out on the culture of sponge aggregates, the following two aspects of the experimental analysis of organogenesis will be separately considered: (1) the role of the main cell types and the recombination factors of the cells of the organized sponge, (2) the organogenetic potentialities of the larval tissues.

1. Reassociation Factors in Dissociated Adult Sponge Cells and the Roles Played by the Principal Cell Types during Reconstitution

The first sponge dissociations were achieved mechanically by Wilson in 1907. Lobes of *Microciona prolifera* Verr. were cut into small pieces and dissociated by being forced through a bag of fine bolting silk. The cells passed into the water in the receptacle, sank to the bottom, displayed an amoeboid activity, became attached to the substrate, and immediately began to fuse with one another. This cell culture in a medium of only filtered and renewed seawater gave a complete organogenesis and produced normal sponges. This differentiation took six to seven days. In this early work Wilson (1907) did not make a detailed study of the development of the various types of cells, he simply noticed that the regenerating masses consisted chiefly of archaeocytes and amoebocytes, but that it was possible that other types of cells also entered into their composition. He attributed an important role to the archaeocytes in sponge reconstitution. These archaeocytes might also arise partly from choanocytes and pinacocytes that had undergone regressive differentiation.

Culture trials with mixtures of dissociated cells of two very differently colored species of sponges (*Microciona* with *Lissodendoryx* and *Microciona* with *Stylotella*) showed Wilson that the cells reassociated to form pure regeneration masses containing only the cells of the same species.

The tendency for the cells to fuse vigorously together led Wilson to try to provoke a fusion of larvae. This he accomplished very easily by bringing them into mutual contact in a small cavity at the moment when the ciliated epithelium was being replaced by permanent epithelium. The masses composed of five to six larvae metamorphosed into perfect sponges, thus giving evidence of the presence of a very efficient regulation process.

Sponge dissociation was recommenced by Müller (1911) with the Spongillidae. The cells of *Spongilla lacustris* L. and *Ephydatia mülleri* Liebk. were dissociated by pressing between the fingers or by filtering through gauze. Müller found in the filtrate all the cell types, but three hours after the aggregation, the choanocytes has disappeared (probably phagocytosed); the masses consisted entirely of peripheral epithelium of totipotent pinacocytes and archaeocytes responsible for the neoformation of flagellated chambers. Like Wilson (1907, 1911b), this author considered reconstitution to be true regeneration from totipotent cells.

After filtration of the heterocoelous calcareous sponge *Sycon coronatum*, Huxley (1911, 1921a,b) also observed the formation of spherules within 24 hours and the reconstitution of an olynthus 17–28 days later. As Wilson (1911a) had observed in the hydroid restitution masses, Huxley also discovered that the size of these masses exerted a great influence upon their metamorphosis: Those closest in magnitude to the larvae showed the best differentiation into an organism. But Huxley's examination of *Sycon* did not permit him to come to the same conclusions as Wilson on the fate of the cells. Huxley considered that the reconstitution of the sponge was only a reorganizing of the cells in conformity with their original differentiation. Pipetting out the first cells to sediment, Huxley obtained cellular masses consisting almost entirely of choanocytes. These survived for 6–12 days but were incapable of forming a new sponge.

From 1925 to 1930, Wilson (1925), Wilson and Penney (1928, 1930), and Galtsoff (1925a,b, 1929) restudied the dissociation of *Microciona prolifera* Ver. Galtsoff made a point of trying to determine the respective roles of the different types of cells in the suspension by studying their changes during histogenesis of the aggregates. He identified the different types of cells by the structure of their nuclei and by the presence of characteristic cytoplasmic inclusions. In the first four days, a cellular rearrangement occurred without cell multiplication. At the end of this reassociation, the aggregates contained the outlines of flagellate chambers and mesenchyme canals, and later development consisted of the differentiation of these outlines. During this process the archaeocytes played the most important role in differentiating the various mesenchyme structures (gonocytes, archaeocytes, scleroblasts, collencytes, desmocytes). They never changed into choanocytes. The pinacocytes differentiated into choanocytes; they also formed the dermal membrane and the walls of the canals. His conclusions, therefore, did not agree with those of either Wilson or Huxley. Like Wilson, Galtsoff recognized that the archaeocytes could differentiate into several kinds of cells, but unlike Wilson, he did not consider that the dissociated cells could dedifferentiate and return to an embryonic state. Again, if he admitted the first part of Huxley's conclusions (that the destiny of the cells was not a function of their position), the second part (that the ultimate position of the cells was a function of the differences in their constitution) did not appear to him to be evident because the archaeocytes and their derivatives were able to occupy any position in the sponge. According to Galtsoff, a sponge could theoretically develop from a mass of coalescent cells when two sorts of cells, the archaeocytes and the pinacocytes, were present. Aggregates consisting entirely of archaeocytes obtained by centrifugation were incapable of restitution and died. The aggregates needed about 2000 cells for the regeneration of a sponge to be complete.

The conclusions of Galtsoff were questioned by Wilson and Penney (1928, 1930), who recommenced the early experiments of Wilson (1907). They still observed a sort of dedifferentiation due to the fact that the shock of dissociation provoked, as in the coelenterates (Wilson, 1911a), a change in the initial structure of the sponge cells. They confirmed, however, the opinion of Huxley concerning the choanocytes. These were probably derived from the filtered sponge and would be very abundant in the center of the aggregates. They also identified the presence of gray cells which were capable of regenerating the pinacocytes and formed the syncytial epidermal epithelium as well as the canals. The archaeocytes differentiated into glandular and rhabdoferous cells.

During the same period Fauré-Fremiet (1925) dissociated the marine sponges *Halichondria panicea* Pallas and *Hymeniacidon caruncula* Bowerbank and freshwater Spongillae. He studied in particular the formation of cell complexes and thought that the stability of the aggregates depended upon an equilibrium between the motility of the active cells and their tendency to agglutinate. This tendency could be modified by the medium. In a pure isotonic solution of NaCl the archaeocytes produced long, pointed pseudopodia while the contraction of the cell masses necessary for the formation of the complexes did not occur. In the presence of $CaCl_2$ the pseudopodia were lobulate and contraction was pronounced. Later, cultures of *Ficulina ficus* L. led Fauré-Fremiet (1932a) to conclusions identical with those arrived at with the cultures of *Microciona*. He emphasized that the aggregates could develop, like the parenchymula larvae, only after they had become fixed to some solid support. This attachment was due to the thigmotactic properties of the archaeocytes which spread over the support and formed a marginal membrane. During the development of the aggregates the regrouping of the cells which remained specifically distinct (verified by *intra-vitam* staining) was primarily, as Galtsoff thought, the work of the archaeocytes which directed the early stages of morphogenesis. The histiotypic development of the other cell types reacting mechanically and physiologically upon each other led to the formation of an organotypic structure. The dedifferentiated choanocytes regrouped together in hollow spheres on about the seventh day reconstituting the flagellated chambers. The archaeocytes produced the scleroblasts. The collencytes flattened to form pinacocytes.

Fauré-Fremiet (1932b) also showed that the choanocytes were essential for normal organogenesis.

Like Galtsoff (1925a,b), de Laubenfels (1932) analyzed the effects of the medium upon the reassociation of the cells of the marine sponge *Iotrochota birotulata* Higgin. From his investigation it was interesting to note that a solution of NaCl isosmotic with seawater as well as an aqueous extract of *Haliclona rubens* Pallas prevented the aggregation of *Iotrochota* cells. The aqueous extract, however, lost its effect when heated to 90°, but retained it at 60°. This suggested to the author that the agency was an organic substance, possibly an enzyme.

A colloidal hyalin substance was liberated into the seawater during dissociation, and this played an important part in the reassociation of *Iotrochota* cells. The reconstitution of new sponges was accomplished without intermediary dedifferentiation of respecialization.

The dissociated cells of five different sponge species, *Iotrochota birotulata, Haliclona rubens, Haliclona longleyi* n. sp., *Haliclona viridis* Duchassaing and Michelotti, and one species of *Halichondria*, were cultured as mixtures in pairs. The aggregates that formed contained only the cells of a single species, except in the case of the mixture of *Iotrochota* (dark purple) and *Haliclona longleyi* (pale yellow green) which produced bispecific masses that were obviously alive and healthy after 14 days.

Brien (1937) made a further study of the reconstitution of freshwater sponges using *Ephydatia fluviatilis* L. These were cut into fragments and pressed through the mesh of a fine bolt silk. The filtrate was cultured in a petri dish in freshwater that was kept very cool. The different cells of the viscous filtrate retained their individual characteristics. The choanocytes, in particular, kept their collars and flagellae and were often grouped into fragments of flagellated chambers. The free choanocytes showed little motility. However, the slow amoeboid movements of the archaeocytes and amoebocytes helped by their pseudopodia were clearly seen. Brien, however, considered that the amoeboid movement of the cells was not of great importance in reaggregation. The latter resulted, above all, from contacts between the protoplasmic extensions of the thigmotactic cells which were the archaeocytes and the amoebocytes. This protoplasmic adhesiveness probably brought about the coalescence of the cells which was indicated by the appearance of networks which thickened to develop into spherules.

So that organization would begin, the spherules had to be kept in well aerated running water which brought a supply of oxygen and removed the excess of bacteria. The spherules of 1 to 1.5 mm were the most liable to reorganize. Following the aggregation, according to Brien "the reorganization of the sponge is a phenomenon of morphallaxis, a setting of the different cells belonging to the original sponge which retain their distinctive characteristics without dedifferentiating." These conclusions agreed with those of Huxley. Brien insisted that, with the exception of the neoformation of scleroblasts, vacuolar, granular, and spherulate cells, histogenesis did not take place. Only morphogenesis and organogenesis were involved because of the nature of the cells in the culture. This was quite a different phenomenon from the organogenesis of the gemmule in which histogenesis was concomitant with and conditioned by organogenesis and morphogenesis.

During reorganization, organogenesis and morphogenesis arose out of an equilibrium between intercellular relationships. Hence, for sponge reorganization it was essential that the regenerates consisted of the three types of cells. This meant that here the archaeocytes did not possess the organogenetic totipotence that they acquired in the gemmule.

The morphological and histological observations of the reconstitution processes of dissociated cells were mainly concerned with the corneosiliceous sponges. However, Tuzet and Connes (1962) were able to prove that the calcareous sponge cells *Sycon raphanus* O.S. dedifferentiated after dissociation, then redifferentiated during reconstitution.

Experiments on sponge dissociation and a study of the phenomena of self-aggregation have been carried out since by other authors either with different species or with the species already mentioned. The results concerning the details of the development of one or more cell types and the method of coalescence of the cells varied from one worker to another to such a degree that it became difficult to classify them. In all cases, the main preoccupation had been: What caused reaggregation?

Sivaramkrishnan (1951) noted in the marine sponge *Callyspongia diffusa* the intervention of a hyalin protoplasmic network like that described by Galtsoff (1925a,b), while Spiegel (1955) was unable to confirm the presence of this network in the seven marine sponges he studied. He did, however, draw attention to an increase in the size of the pseudopodia during cell aggregation and demonstrated that specific antigenic substances existed on the surface of the cells taking part in cell adhesion.

Ganguly (1960) thought that the main regeneration factors of the freshwater sponge *Ephydatia* lay in the necessary and probably selective contact between the cells during the first few hours of their touching one another. Mookerjee and Ganguly (1964) studied the role of the pseudopodia and showed that physical and chemical factors intervened during aggregation of the dissociated *Ephydatia* cells. In tap water containing penicillin, the dissociated cells aggregate but subsequently dedifferentiate (Mookerjee *et al.*, 1965). The results suggest that reaggregation and regeneration of sponges is controlled by a mechanism related to the synthesis of proteins.

The question of cell behavior always revolved around the possibility or not of certain cells dedifferentiating. Sivaramkrishnan (1951) observed in both a *Sycon* and a *Callyspongia* species a dedifferentiation of choanocytes followed by their redifferentiation. Ganguly (1960) also indicated with an *Ephydatia* species that all the cells began by dedifferentiating and assuming a general rounded form, while their nuclei became deeply stained. Then, after four to six hours of culture, the choanocytes, pinacocytes, and scleroblasts very rapidly redifferentiated exclusively into their original states. The regenerates obtained by these workers contained all the cell types, but they did not produce a normal sponge. These observations were contrary to those of Brien (1937) on *Ephydatia* and this phenomenon of the change in aspect of cells like the choanocytes seemed to be a little too rapid to correspond to a real dedifferentiation.

The method of sponge cell dissociation used in the experiments described up to now was the same as that employed by Wilson for fragmentation and filtration. Humphreys *et al.* (1960a) produced viable cell suspensions of reconstituting sponges by the chemical treatment of marine sponges in cold seawater free from both calcium and magnesium (CWF-SW). In order to eliminate certain variable factors, such as cell migration and contraction of the cellular deposit that occurred during the self aggregation of mechanically dissociated cells, Humphreys *et al.* (1960b) kept the chemically dissociated cell material in suspension by a rotary movement of 80 rpm. The dissociated cells of *Microciona prolifera* Ver. cultured in this way, in seawater at 25°, rapidly accumulated at the center of the erlenmeyer flasks and within minutes began to cohere. After one hour they formed small clusters that coalesced into aggregates of 0.12 mm diameter in 24 hours. These aggregates were then able to develop, and with this technique Humphreys (1962, 1963) was able to evaluate the influence of various culture conditions (speed of rotation, pH, cell concentration, temperature, ionic composition of the medium). Two important results were worthy of attention. In the first place, the chemically dissociated cells remained dispersed at 5°, at 10° they formed only small aggregates; while optimum aggregation occurred between 18° and 25°. In the second place, the use of numerous media based upon modified artificial seawater showed that cell adherence required the presence of bivalent cations of which the most effective were calcium and magnesium.

A comparison between the behavior in culture of mechanically and chemically dissociated cells of the *Microciona prolifera* and *Haliclona occulata* sponges enabled Moscona (1963) and Humphreys (1963) to demonstrate the activity of an organic factor of cell adhesion that was released by chemical dissociation. In fact, the mechanically dissociated cells adhered together at 5° as effectively as at 24°. This suggested that the mechanically dissociated cells had retained their adhesive mechanisms which were lost during the chemical treatment in seawater free from Ca and Mg and could not be regenerated by cells kept at a low temperature. When the supernatant liquid from the chemical dissociation was added to dissociated cells cultured at 5°, these formed compact masses. The authors managed to concentrate the active factor by differential centrifugation. It had been obtained only by dissociation in seawater free from Ca and Mg; the homogenates of sponge cells were inactive. The preparations of the adhesive factor showed a remarkably selective activity. The preparations derived from the cells of *Microciona prolifera* Ver. produced aggregations in cells of this species but were ineffective in dissociated cells of *Haliclona occulata*, and vice versa. Moscona (1963) considered that there undoubtedly existed a similarity between this material and the antigens previously detected by Spiegel (1955). The adhesive factor was destroyed by boiling and inhibited by periodate, but it was not destroyed by ribonuclease or by deoxyribonuclease. Moscona (1963) also considered the possibility of a role for carbohydrates. More recent work by this author, Margoliash *et al.* (1965), indicated that it consisted largely of glycoproteins whose particles measured from 20 to 25 Å in diameter. The aggregating factor released by dissociated cells of *Haliclona variabilis* kept in seawater without calcium nor magnesium was also identified by Galanti and Gasic (1967) as a glycoprotein. Its properties are those of a globulin. These results were in agreement with the working theory of Moscona that had been suggested by vertebrate embryonic cell aggregation and according to which functional extracellular constituents might play an important part in the selective cohesion and organization of the cells.

The study of the mechanisms of cell union and their selectivity was of capital importance for the understanding of the differentiation of multicellular systems and tissue interactions.

However, the conclusions of Humphreys (1963) and Moscona (1963) that cell adhesion could be divided into three components, i.e., the cell surface, divalent cations, and an organic factor (intercellular material), were not unanimously accepted. In fact, Curtis (1962) suggested that during the different stages of reaggregation the cells passed through a series of forms of behavior on a time basis; thus, two cells which showed the same behavior at the same time would associate, while the others would remain separate. To test this idea, he used the cells of the four species of sponges: *Halichondria panicea* Pallas, *Hymeniacidon perleve*,

Microciona sanguinea Grant, and *Ficulina (Suberites) ficus* L. The chemical dissociation of the sponge fragments was accomplished with 0.004 M EDTA (ethylenediaminetetracetic acid) in a 0.55 M solution of sodium chloride buffered with *tris*-hydrochloric acid. After 24 minutes of treatment the cells in suspension were placed in a culture of filtered seawater at 16° to 17°. Cultured separately, the cell suspension of *M. sanguinea* were the quickest to reaggregate (three hours). Those of *H. panicea* reached the same stage in eight hours, while those of *S. ficus* and *H. perleve* in 15 hours.

The experiments of Curtis suggested that if the behavior of two cell types could be synchronized they would remain associated. For example, when he placed the cells of two sponges chosen from *Microciona*, *Suberites*, and *Halichondria* into culture at the same time, he obtained separate aggregates for each species. If, on the other hand, he added dissociated cells of *Microciona* to cells of *Halichondria* that had been reaggregating for eight hours, or cells of *Halichondria* to cells of *Suberites* that had been reaggregating for six hours, he obtained an aggregate of mixed cells. The adhesion between cells of different species was also obtained by Curtis (1970) in suspensions of mixed cells from the *Halichondria panicea* and *Haliclona occulata* sponges.

The origin of the specificity of the behavior of cells in terms of time could be due to the fact that the cells possessed some physicochemical property which reacted with a constituent of the medium when reaggregation commenced: different tissues having different rapidity constants for these reactions. Curtis believed that the mechanism of cell aggregation was not due to an adhesion of the cells caused by the presence of specific cementing substances, for the separated sponge cells reaggregated in culture at 3° even after multiple washings.

Thus it is evident that the exact determination of adhesiveness and cell reaggregation remains unknown, for the results of chemical dissociations seem to be diametrically opposed. The facts may be that both phenomena occur and are mutually complementary, and that the dissociations produced by EDTA (Curtis) did not liberate into the medium those substances liberated by the artificial seawater free from Ca and Mg used by Humphreys *et al.* (1960a,b) and Moscona (1963).

A study of the potentialities of cell types was also carried out by Borojevic (1963), who separated the various cell types by differential centrifugation. Separation trials had already been conducted by Galtsoff (1925b) followed by Brønsted (1936) without producing conclusive results. Agrill (1951) who centrifuged the cells of *Halichondria panicea* Pallas in a variable density medium obtained fractions of almost pure (90%) choanocytes under culture, these cells remained at their original number and did not redifferentiate into other cells. On the other hand, counting the cells of the other cell populations, he discovered that the pinacocytes had arisen from amoeboid cells. By this method, Borojevic (1963) obtained a suspension of perfectly dissociated choanocytes free from archaeocytes and containing less than 1% of collencytes. The choanocytes of the noncalcareous sponges, *Ficulina ficus* L. and *Hymeniacidon sanguinea* Grant, cultured in running water in an aquarium assumed an amoeboid form and the aggregate attached itself lightly to the support. A few hours later cytolysis commenced and the aggregates disintegrated. This confirmed the results of Huxley (1911) and showed that the choanocytes in the absence of other kinds of cells were not capable of reforming a choanoderm.

When studying demosponges, Connes (1966, 1968, 1970) observed that all the cell types took part in the reconstitution after they had undergone a more or less important morphological dedifferentiation. The disproportion between the various cell types of choanosome prevented the reconstitution of a long-living specimen.

After making comparative studies of calcareous sponges (*Leucosolenia complicata* Montagu, *Sycon lingua* Haeckel) and corneosiliceous (*Halichondria panicea* Pallas), Korotkova (1970) showed the specific features of the behavior of isolated cells, of aggregate formation, and of the development of the canal system. The most profound changes in organization and behavior were observed in calcareous sponges. The phenomena of morphallaxis were seen in the development of conglomerates of *Halichondria panicea*. Korotkova (1970)* referring to previous work done on the development of dissociated sponge cells, elaborated a general hypothesis for the regulation of morphogenesis during reconstitution. This regulation differed according to the organization of the canal system.

From all the work carried out on the culture of dissociated sponge cells it would appear that the reconstitution of a sponge requires the presence of the three main types of cells. It is to be hoped that the methods of differential centrifugation will enable the different cell types to be obtained in the pure state. Thus, accurately known combinations may be made and more precise information about the participation of the chief cell types during sponge reconstitution may be forthcoming.

2. A Study of the Organogenetic Potentialities of Larval Tissues

The embryonic development of sponges generally results in the formation of an embryo with a ciliated surface layer. In the calcareous

* For further details, see Korotkova's bibliography.

sponges the gastrulation of the amphiblastula is accompanied at the moment of its fixation by the invagination of the flagellate layer which forms the mass of the endoderm cells. These cells lose their flagellae, multiply, and then differentiate into flagellated endoderm cells (choanocytes) during the metamorphosis. In the noncalcareous sponges of the type *Oscarella* similar phenomena occur at gastrulation with invagination of the fixation pole. In the Spongillae, segmentation produces a massive embryo and the metamorphosis of the ciliated larva corresponds mostly to a reorganization of the cells. According to Brien and Meewis (1938), the ciliated surface cells migrate towards the interior where they become phagocytosed while the choanocytes differentiate within the internal mass and are organized into vibratile chambers.

Cell lineage during differentiation and organogenesis in sponges has not always been very clearly established and the origin of the choanocytes was often the subject of disagreement. Some authors considered that the choanocytes came from the ciliated surface cells of the larva; others thought they proceeded from the differentiation of the archaeocytes. These differences of opinion were due to the fact that the analyses of the histogenic processes had almost always been made from histological preparations. Again, the experiments on the separation of the larval cell layers and their *in vitro* culture were a novel method of studying the problem of cell descent.

Borojevic and Levi (1964, 1965) first dissociated the parenchymula larval sponge cells in order to study their powers of aggregation and their potentiality for differentiation. The parenchymula larvae of *Mycale contarenii* Martens and *Adocia elegans* Bow. were dissociated in solutions of EDTA using the Curtis' method (1962) and cultured in seawater. The cell suspension sedimented then formed aggregates and spherules which reorganized into functional sponges and not into a larval form. This phenomenon seemed to be due to the acceleration of the development of the already differentiated collencyte-pinacocyte type cells conditioning the differentiation of the choanoblasts. During the morphogenetic reconstitution, the collencytes showed a tendency to spread out: They formed the basal and marginal membranes of the spherules. The vacuolar larval cells disappeared. The ciliated cells coming from the larval envelope that were not phagocytosed by the archaeocytes changed very gradually into choanocytes. The authors were able, with the electron microscope, to see clearly the appearance of the collar before the disappearance of the structures and characteristic inclusions of the ciliated larval cell. Active phagocytosis by the archaeocytes which absorbed a great many of the ciliated cells was reported by Meewis (1939) and Brien (1943) to take place during normal metamorphosis of the larvae of

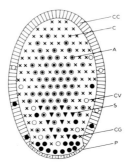

FIG. 25. Distribution of principal cell types in the free-swimming larva of *Mycale contarenii* Martens. A, archaeocytes; C, collencytes; CC, ciliated cells; CG, globiferous cells; CV, vacuolous cells; S, scleroblasts; P, pinacocytes.

the sponges *Oscarella* and *Haliclona*. In the case of the morphogenesis under consideration, archaeocyte phagocytosis reduced the number of ciliated cells until an equilibrium was established between the three principal types of cells. In fact, the spherules which contained relatively few ciliated cells developed rapidly with little phagocytosis.

Following these results, Borojevic (1966) microsurgically separated the external ciliated layer of the central part of the larva of *Mycale contarenii* Martens. He repeated this on several larvae in succession, culturing four or five cell groups of the same topographic origin on a glass slide in a drop of filtered seawater to which penicillin and streptomycin had been added. In an hour the united groups formed a reorganization spherule under culture at 18°–20°. The medium was changed every two days. Cultures of the central part of the larva mainly contained archaeocytes and collencytes (Figs. 25 and 26); the ciliated larval cells were entirely absent. The archaeocytes assembled in small groups which differentiated

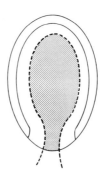

FIG. 26. Culture of the central part of the larva of *Mycale contarenii* containing a large number of archaeocytes and collencytes.

FIG. 27. Culture of the larval ciliary epithelium of *Mycale contarenii* which also contained a certain number of collencytes and some archaeocytes.

into choanoblasts and choanocytes. Within three to four days the culture had produced a small entire and functional sponge. In the culture of ciliated larval epithelium (Fig. 27) containing a few collencytes and archaeocytes, the ciliated cells rapidly formed a large number of vibratile chambers. The number of collencytes and archaeocytes then increased and normal equilibrium was established in two to four days, the culture producing a functional sponge. If the culture contained very few archaeocytes and collencytes, it did not spread out or develop but disaggregated after 7 to 10 days. Hence, the ciliated cells were not able to differentiate into other cell types. The culture of the posterior calotte (Figs. 25 and 28) covered with pinacocytes was incapable of forming a regular spherule.

These experiments indicated the possible double origin of the choanocytes: i.e., from the ciliated cells and from the archaeocytes (Figs. 29 and 30, Plate IV), and the regulator role played by the archaeocytes which represented a reserve of totipotent cells or which acted as phagocytes. They also clearly showed that the ciliated cells could not dediffer-

FIG. 28. Culture of posterior region of *Mycale contarenii* larva.

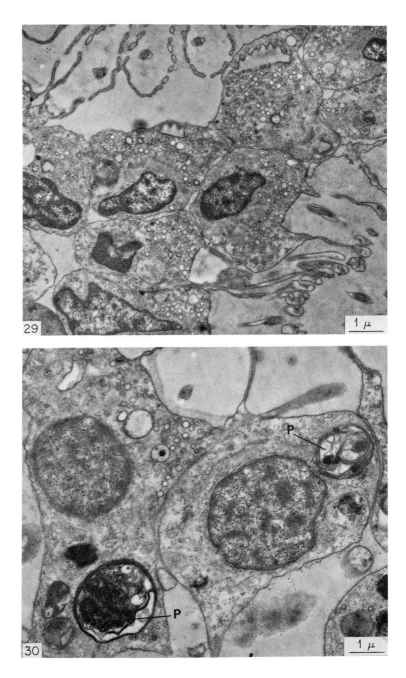

entiate in order to redifferentiate into other cells under the cultural conditions employed. Borojevic (1966) concluded that sponge development was equivalent to a succession of differentiations of embryonic cells in three main lines of sponge cells that did not correspond with the ecto-, endo-, and mesoderm of the other metazoans. The differentiation of cells during sponge development was equivalent to Wilmer's (1960) physiological conception of cell differentiation into epitheliocytes (ciliated cells), mechanocytes (collencytes–pinacocytes), and amoebocytes (archaeocytes).

III. GENERAL RESULTS AND DISCUSSION

In view of the diversity of structure of the cultured organs and the variety of reasons for their explantation, we shall simply try to review the main general headings under which the phenomena studied may be classified.

We preferred to begin by describing them in each phylum, because it is necessary at the outset to understand them in relation to the type of structural organization which produces them before comparing them with the similar phenomena observed in other phyla.

A. Culture Conditions

The nature and composition of the culture media are extremely varied. They differ according to the zoological group, since fragments of coelenterates or sponges may be cultured simply in fresh- or seawater, depending on the species, whereas the organs of mollusks or of nematodes are cultured on natural or synthetic media identical to those used for vertebrate embryo organ cultures.

In general, the culture media are devised so as to imitate as closely as possible the natural environmental conditions of the explanted organ. They also depend on the nature of the organ and on the purpose of the research; detailed analyses are given in Volume 1, Chapter 2.

We should like to emphasize the very remarkable fact that the culture

PLATE IV

Differentiation of Sponge Larval Cells of *Mycale contarenii* in Culture

Fig. 29. The choanocytes in two flagellate chambers in direct contact in a culture of the larval ciliary epithelium, 48 hours after operation.

Fig. 30. Young choanocytes differentiated in a culture of the internal part of the larva from archaeocytes. P, phagosomes in resorption.

of adult organs is as easy as that of embryonic or larval organs. Among invertebrates, other than insects, it is mainly adult or juvenile organs which have been cultured, and this has enabled the direct study of the correlations between organs. This is doubtless due to their structure, which is less complex than that of vertebrate organs, and to their small size.

Sterilization is an important condition which, for a long while, limited the culture of invertebrate organs or tissues other than that of sponges or coelenterates. Indeed providing that they are cultured in filtered natural running water (or frequently renewed water) the aggregates or fragments of sponges or coelenterates are not destroyed by bacteria or fungi. However, all organs requiring a culture medium containing tertiary or quaternary "nutriments" are liable to be destroyed by the proliferation of microorganisms present in the culture. The use of anti-biotics and ultraviolet rays and the sterile removal of organs (when they are themselves sterile), on the one hand, or the aseptic rearing of inverte-brates, on the other, have made invertebrate organ culture possible.

B. Main Applications

1. Keeping Organs Alive when Separated from the Organism

The term "culture" does not exactly apply to the explantations aimed at keeping the explanted organs alive. However, the techniques used for the survival of organs provide information for the culture of organs which must not only survive but also continue their development and/or their activity. In fact, methods used for organ survival are not funda-mentally different from culture techniques, the essentials of these tech-niques depend on the organ considered. The same medium may be used for the culture of organs with few requirements as well as for simply keeping alive organs which are more sensitive to explantation.

Moreover, although the distinction between survival and culture is clear in the case of the explantation of embryo organs, where culture is accompanied by differentiation and morphogenesis, it is much more subtle in the case of completely differentiated adult organs. Here the differ-ence between "survival" and "culture" depends on how long the structure and function of the organ are maintained. In general, when the ana-tomical integrity and functioning of the organ are maintained without modification for several days or weeks, it is considered to be "culture," whereas "survival" lasts only a few hours to a few days. The criterion of "time" is not at all absolute, as survival or culture are more or less easy to obtain depending on the zoological phyla concerned. Experimental survival in nonsterile liquid medium, as studied by Thomas (1941, 1963), enabled him to distinguish zoological groups in which separated organs survive badly and become infected (coelenterates, annelids, crustaceans), whereas others groups have organs which are resistant to some extent, such as vermidians, mollusks, and chiefly echinoderms.

In the case of experiments in comparative physiology, survival is sometimes maintained by simple perfusion with physiological saline solutions over the period that the observations are made.

In the field of parasitology, survival experiments have been carried out either to keep alive organs containing parasites at some stage of their development or to keep the parasites themselves alive separately from their host.

Planorb tentacles "survive" without structural modifications for 15 days in a complex liquid medium devised by Benex (1961) according to the principle of Thomas' techniques. As from the 15th day a "structural dedifferentiation" begins, mainly affecting the internal muscles and leading to a progressive decrease in the volume of the fragments which die during the fourth week. The survival of internal parasites is more difficult to obtain: The composition of the infested organs and the metabolic requirements of the parasite organisms must be taken into account; the technique is similar to the methods of "rearing." Very similar also to these techniques are those devised for the development of viviparous invertebrate embryos (nematodes, Hirumi et al., 1968) or for the development of eggs laid in cocoons (lumbricidae, André, 1962). In these extreme cases, resembling rearing techniques but where there really is "culture" because there is embryo development, the term "typical organism culture" given by Et. Wolff (1965) to the culture of mammal or bird embryos, may be applied.

2. Study of the Phenomena of Differentiation during Ontogenesis

The technique of organ culture has enabled embryologists to determine the potentialities of the presumptive primordia of the organs of vertebrate embryos at various stages of their development. In invertebrates, this technique offers the same possibilities, but up until now it has mostly been used in insects where a number of authors have studied the differentiation of the organs of larvae (imaginal discs, gonads). These experiments, which throw light on the factors necessary for the normal development of embryonic or imaginal rudiments are presented and discussed in this volume, Chapter 1.

In the other zoological phyla the culture of embryonic larval or juvenile organs is rarer. However, explantation of primordia or of organs of the genital system undergoing differentiation has been achieved in several mollusks. This technique has made it possible to show interesting aspects of the sexual differentiation of gonads.

Thus, the orientation and development of the germinal cells of the hermaphrodite juvenile gonad in the snail *Helix aspersa* may be determined by the composition of the medium. Gomot and Guyard (1968) obtained the differentiation of both sexual strains of the hermaphrodite gland of the snail *Helix aspersa* in a complex synthetic medium. By modifying the culture medium one can promote the differentiation of one or another strain. For example, in a medium consisting basically of embryo extract, only the formation of ovocytes is observed, whereas in the presence of hemolymph of the adult snail the germinal cells develop into spermatocytes and spermatozoa. In order to determine the influence of either trophic or possible hormonal factors, associations of gonads at the undifferentiated stage and of brains show that normal sexual differentiation is favored by the presence of juvenile brains. The association with adult mollusk brains in separate sexes shows the elective influence of the female brain on the differentiation of ovocytes whereas the male brain has no effect. These experiments prove the complexity of the determinism of sexual differentiation which results from the interference of trophic and hormonal factors (Guyard, 1971).

In a prosobranch mollusk of protandrous hermaphrodism, Streiff (1967a) notes that the young gonad in the male phase, cultured *in vitro* in a medium in the absence of hormone, differentiates to form an ovary. This observation, in addition to the lack of evidence proving hormonal control of female differentiation in the insect *Lampyris noctiluca* (Naisse, 1966) and spontaneous completion of female organogenesis in the absence of androgenic hormone in crustaceans (Charniaux-Cotton, 1957, 1967) led Charniaux-Cotton (1967) to suppose that female differentiation in invertebrates is an autodifferentiation. This hypothesis deserves to be confirmed by new experiments and the culture of gonads at the undifferentiated stage should be undertaken for this purpose in most of the zoological phyla. Organ culture also provided Streiff (1967b) with experimental proof that the differentiation of the organs of the genital tractus of the mollusk *Calyptraea* is independent of that of the gonads and depends only on the functioning of the cephalic hormonal system.

Lastly, the separation of cellular groups of sponge larvae and their culture *in vitro* enabled Borojevic (1966) to confirm the supposition made by Borojevic and Levi (1965) that the peripheral flagellated cells of the *Mycale contarenii* sponge larva change into choanocytes in the course of metamorphosis. At this early stage the flagellated cells are no longer capable of producing other types of cells. Cultures of internal cells of the larva show that the archeocytes of siliceous sponges can become differentiated into choanocytes, in the species where this is not the case in the course of normal development. This quality is a potentiality and appears when there is no longer a normal equilibrium between the cells; the archeocytes thus act as regulators of equilibrium by phagocytizing the excess cells and by developing into the type of cells that are missing. Organ culture is of precious assistance for all the methods of morphological investigation used up until now in the subtle study of the origin of different cellular strains in the sponges, and also serves in revealing certain important phenomena of regulation at the larval stage, since the central mass and superficial layer are independently both able to reconstitute a small functional sponge.

3. Humoral Correlations between Tissues and Organs

The main humoral correlations studied in culture *"in vitro"* in invertebrates other than insects are those involved in the determinism of sexual reproduction. Some are particularly marked. This is the case for the direct endocrine action of the polychaetes brain on the maturation of the genital products. Other examples of direct or indirect endocrine influences have been described in Oligochaetea, Hirudinea, Crustacea, and Mollusca and are analyzed in this volume, Chapter 5.

The enormous contribution of organ culture to this field is due to two main causes. First, adult invertebrate organs, small in size and simple in structure, nourished very often by soaking in the internal liquid, are much easier to culture than adult vertebrate organs. Secondly, experiments in removing organs, very often made difficult by their small size and their position, produce secondary metabolic disorders which impede the primary effects of the organ being removed.

These two techniques, i.e., surgical ablation and organ culture, are complementary to one another and should be practiced at the same time. These means do not always suffice to determine precisely the endocrine function of the organs under consideration and this is the case for the complex neural gland-nerve ganglion-vibratile organ of the *Molgula manhattensis* cultured by P. Sengel and Kieny (1962).

4. Analysis of the Process of Morphogenesis and Differentiation during Regeneration

Organ culture has been used in various ways to study the process of regeneration and we have seen that the data provided by this technique varies according to the type of organization of the animals considered. The fundamental questions, common to all cases of regeneration concern, firstly, the origin of the cells which take part in the formation of the regeneration blastema and, secondly, the process of morphogenesis in the course of the reconstitution of an individual.

In annelids and in turbellarians where regeneration was studied in *in vitro* culture this technique did not yield any results with regards to the origin of the blastema cells, but it provided detailed data on the factors involved in morphogenesis and differentiation. Trial cultures of fragments of oligochaetes in a gel medium enabled Gay (1963) to observe that healing is a function of the length of the segments and of the region of the section in the body of *Lumbricus herculeus*. Segments of *Eisenia foetida* also cultured in a gel medium formed anterior and posterior blastemas (Durchon, 1963).

Thouveny's experiments (1967) on the culture of polychaete fragments in a liquid medium rich in amino acids show that the phenomena of healing and of blastema formation are independent of growth or segmentation. His histological observations seem to be in favor of the hypothesis that dedifferentiated cells of each tissue of the section take part in the formation of the blastema. But they provide no data on the respective parts played by coelomocytes (amoebocytes) whose existence has been proved by the experiments on irradiation by x rays of Stephan-Dubois (1958) and by the ectodermic and mesodermic cells which seem to be maintained in a state of structural dedifferentiation by the culture medium.

Planarian blastemas cultured by C. Ziller-Sengel (1967) on a gel medium show potentialities far superior to those of annelids since they autodifferentiate themselves, depending on their origin, into an anterior or posterior region of the planarian. These experiments also show that the determining influence of the trunk tissues is a function of the anterior–posterior regeneration gradient. These blastemas, however, are incapable of regulation, and the reconstitution of a small planarian (up to now nonviable) requires the association of an anterior blastema with a posterior blastema.

These experiments in the culture of fragments of oligochaetes or of planarians gave rise to the discovery of regulating effects (induction or inhibition) on regeneration. Marcel (1966, 1970) showed the inducing function of the cerebral ganglia in the regeneration of the pharynx of the *Eisenia foetida*, whereas the regeneration of the head is inhibited by homogenized nerve tissue anterior to the section zone, as was observed by Lender (1955, 1956) who bred "decapitated" planarians in the presence of planarian head extract. The differentiation of planarian cephalic blastemas is not disturbed by the head extract but is inhibited by an older brain undergoing differentiation.

Coelenterates provide the most spectacular results in culture *in vitro*. In fact, like planarians, these animals possess a well-developed faculty for regeneration and, what is more, their body is composed of two layers of tissue easily separable from each other. In these animals the question of whether an already differentiated cell can dedifferentiate and re-differentiate itself into another type has caused similar controversy to that existing on the subject of the origin of blastema cells in the regeneration of worms. Without going through the whole set of material facts presented in the above paragraph it does seem evident that in a certain number of cases the research worker agrees in admitting that one of the layers, in some instances the ectoderm and in others the endoderm, is capable of regenerating a complete individual. In the case of the ecto-derm (*Aurelia, Cordylophora*) where the interstitial cells are localized, the regeneration of the endoderm seems to derive from the differentiation of the interstitial cells and at first sight does not give rise to a problem of previous dedifferentiation of certain cells. In the case of the endoderm (*Pelmatohydra, Hydra*) the authors who obtained regeneration of the hydranth describe phenomena of cellular transformation. According to Macklin and Burnett (1966) the phenomena of the dedifferentiation of glandular cells into neoblasts and of digestive cells into epidermal cells observed by Davis *et al.* (1966) is caused by the ionic composition of the culture medium.

The importance of the factors of culture on the differentiation of cells is also confirmed by the experiments on the separation of tissues and their reassociation after irradiation of one of them as performed by Diehl and Bouillon (1966a). Indeed, these research workers induced regeneration of the endoderm of *Cordylophora* by associating it with the ectoderm rendered incapable of regeneration by irradiation with x rays. The results obtained by Burnett's team in modifying the medium or by Diehl and Bouillon in associating it with an "inductor" tissue are comparable with the phenomena of metaplasia determined in vertebrate embryo tissues by Fell (1960) by adding vitamin A to the culture medium or by ourselves (Gomot, 1961) by submitting the endodermic epithelium to the inducing influence of a dermic fragment.

The experiments in the culture of fragments of coelenterates are very similar to vertebrate embryo organ culture. These experiments recently enabled Diehl and Bouillon (1966b) to explain the process of stoloni-zation and budding of hydranths by an ecto–endodermic interaction in which a predominance of the ectoderm produces the formation of stolons, whereas an association consisting principally of endoderm differentiates mainly the hydranths.

5. Research on the Causes of Cellular Cohesion and Organic Reconstitution after Dissociation

The dissociation of adult coelenterate or sponge tissues followed by spontaneous aggregation of the cells and by the reconstitution of an organism is a surprising phenomenon which presents biologists with a twofold problem. As early as the first experiments on mechanical dissociation carried out by Wilson (1907) the two aspects of this problem were considered as two successive stages in which the first is necessary for the second to occur. During the stage of aggregation of the cells, the amoeboid movements of the cells and the tigmotactic properties of some of them, observed by many research workers, certainly intervene in the autoaggregation.

Since the research work done by Herbst (1900) on the cohesion of the first blastomeres of the sea urchin egg in the course of division, the essential importance of calcium ions in the medium for maintaining cell combination is well known. The influence of ionic composition has never been contested. Fauré-Fremiet (1925) observed a difference in the shape of the pseudopods of the archeocytes in a solution of NaCl and in the presence of $CaCl_2$. The former inhibits cellular retraction while the latter favors it. Loeb (1922, 1927) also notes that the osmotic pressure and the pH of various physiological solutions affect the agglutination of the amoebocytes of *Limulus* in the same way as that of the archeocytes of dissociated sponges. Loeb is of the opinion that these combination phenomena result from a modification in the consistency of the external film of protoplasm of the amoebocytes or of the corresponding cells.

This influence of the ionic composition on cellular adhesiveness is now used to obtain chemical dissociations. Humphreys *et al.* (1960a) chemically dissociated sponge tissues by keeping them in artificial seawater without Ca or Mg. This technique is also used for dissociating vertebrate embryonic tissues. But after this type of dissociation the sponge cells do not behave in the same way as after mechanical dissociation. Mechanically dissociated sponge cells adhere together at 5°, whereas chemically separated ones are no longer capable of doing so. Moscona (1963) and Humphreys (1963) reestablished cellular cohesion at 5° by adding the supernatant of the liquid which made them suspect the existence of an organic factor of adhesion. This adhesive factor shows a selective activity and the authors were able to concentrate it.

Up until now it has not been possible to isolate this organic factor by other means of chemical dissociations. Curtis (1962), who achieved chemical dissociations by EDTA, observed cellular reaggregation at 3° even after frequent washing. Moreover, he showed differences in behavior of the cellular types and experimentally proved that the structure of the aggregates depends on the constants of reactivity of the types of cells present. The causes of cellular adhesiveness and aggregation are not very well known since apparently contradictory mechanisms have been described. In fact, since the experimental conditions are not the same, it may be supposed that this reveals two different types of physiological properties of the cell cytoplasm which is capable of adapting itself in several different ways to the artificial conditions imposed.

The organic reconstitution which occurs after the phase of aggregation also lends itself to controversy as to the respective parts played by the various cell types. The extreme positions may be summed up as follows: On the one hand, the reconstitution is thought to be essentially a phenomenon of reorganization of the cells which keep their distinctive characters without dedifferentiation; on the other, reconstitution is thought to be simply regeneration from totipotent cells which are either archeocytes or else cells issued from the dedifferentiation of one or several cell types. Discussion is often concerned with the behavior of the choanocytes after dissociation, certain authors interpret the change of shape which they undergo and their loss of flagella as a true dedifferentiation. It seems that this is not the case and that, at any rate, the choanocytes, which it has been possible to isolate by the technique of differential centrifugation, are not capable of reforming a choanoderm (Borojevic, 1963). The reconstitution of a sponge requires the presence of three main types of cells of this animal and it certainly consists mainly of morphogenetic movements rather than cellular transformations.

Techniques for separating the various cell types will make it possible, we hope, to show the cellular interactions which lead to the reconstruction of a type of organization which normally results from the progressive differentiation of the embryo and the larva.

These experiments on dissociation and reconstitution in sponges and in coelenterates provide us with an example of animal organization in which adult cells have retained the genetic power to recombine together to produce the functional architecture characteristic of their species. The behavior of these adult cells is comparable to that of vertebrate embryo cells, but in the latter the original totipotent genome of the egg cell is gradually restrained so that in a differentiated tissue the possible activities of a cell are governed by a small portion of the genome. It seems that the original genome of the cells of a metazoan is all the more reduced when the number of anatomical structures increases. In sponges, the genes which govern the organization of the individual are submitted to a certain delayed effect, since the larval cells reorganize themselves much more easily than the cells of the adult tissues (Borojevic and Levi, 1964), but this is not sufficient to prevent them from functioning in the course of adult differentiation of these animals. This restriction of the total genetic information seems to be gradual and is a function of the animals' complexity; in the planarians there is an onset of reorganization of dissociated cells, but it does not manage to reach a functional condition.

IV. CONCLUSIONS

With its recent applications to invertebrate organs, *in vitro* culture is a very general means of investigation applied to the study of living organisms. The ways of utilizing this technique vary depending on the organs and zoological phyla concerned. The use of this technique has proved to be very fruitful in the study of hormonal actions in the course of sexual differentiation and regeneration. Its contributions with regard to morphogenesis and differentiation are also spectacular in the phyla of the sponges and coelenterates where the simultaneous application of tissue separation and culture techniques makes it possible to demonstrate the larval regulation processes which are so important in the sponges.

The cellular transformations, produced in coelenterates due to the influence of the composition of the medium or by the inducing action of a tissue irradiated with x rays maintained in culture, are also rich sources of information. If the observations of Burnett *et al.* (1966) in *Hydra* are confirmed, they prove that previously differentiated somatic cells may be induced to change into cells capable of entering into meiosis and forming gametes.

These phenomena of dedifferentiation are perhaps impossible from somatic cells of metazoans of more complex organization since experiments on cell dissociation prove that the more complex the general organization which results in regional specialization of the cells, the more difficult the reorganization.

In organisms of simple architecture, comparable with that of some stage of vertebrate embryonic development, reorganization is easier. In this case it seems that the genetic determinism of reorganization is compatible with that of embryonic development, whereas in more highly developed metazoans it is not since the regional influences of the soma have probably modified it.

The blastic transformations obtained by the culture of blood lymphocytes of primates in the presence of phytohemagglutinin show the influence which the environment may have on cells in culture for revealing some of their lost capacities. Hence, this process of *in vitro* culture should provide us, in the future, with much information on the factors of normal morphogenesis and on the stage of dedifferentiation to which specialized cells may return.

Much work has been done in the field of invertebrate organ culture, but much remains to be done in the field of experimental embryology where few problems of organogenesis have been studied other than in insects.

How many captivating questions, however, remain to be answered in phyla where few, if any, experiments have been undertaken, i.e., protochordates, bryozoans, dicyemids?

We hope this review of the results obtained so far and of a few problems left in suspense will stimulate new and original research.

We should like to thank all the people who have assisted us in the completion of this chapter, by their advice and by their participation in the bibliography and illustrations.

REFERENCES

Abeloos, M. (1965). *Bull. Soc. Zool. Fr.* **80**, 228.
Agrell, I. (1951). *Ark. Zool.* **2**, 519.
Ammon, R., and Kwiatkowski, H. (1934). *Pflüegers Arch. Gesamte Physiol. Manschen Tiere* **234**, 269.
Ancel, P. (1903). *Arch. Biol.* **19**, 389.
André, F. (1962). *Bull. Soc. Zool. Fr.* **87**, 153.
Ansevin, K. D., and Buchsbaum, R. (1961). *J. Exp. Zool.* **146**, 153.
Arvy, L., Echalier, G., and Gabe, M. (1954). *C. R. Acad. Sci.* **239**, 1853.
Avel, M. (1959). *In* "Traité de Zoologie" (P.-P. Grassé, ed.), Vol. 5, p. 380. Masson, Paris.
Badino, G. (1967). *Arch. Zool. Ital.* **52**, 271.
Barrington, E. J. W. (1957). *J. Mar. Biol. Ass. U. K.* **36**, 1.
Barrington, E. J. W., and Franchi, L. L. (1956a). *Nature (London)* **177**, 432.
Barrington, E. J. W., and Franchi, L. L. (1956b). *Quart. J. Microsc. Sci.* **97**, 393.
Bayne, C. J. (1968). *Malacol. Rev.* **1**, 125.
Bazin, J. C., and Lancastre, F. (1967). *C. R. Acad. Sci.* **264**, 2907.
Beadle, L. C., and Booth, F. A. (1938). *J. Exp. Zool.* **15**, 303.
Benex, J. (1961). *C. R. Acad. Sci.* **253**, 734.
Benex, J. (1964). *C. R. Acad. Sci.* **258**, 2193.
Benex, J. (1965). *C. R. Acad. Sci.* **261**, 5233.
Benex, J. (1966). *Bull. Soc. Pathol. Exot.* **59**, 1, 99.
Benex, J. (1967a). *C. R. Acad. Sci.* **265**, 571.
Benex, J. (1967b). *C. R. Acad. Sci.* **265**, 631.

Benex, J. (1968a). *Bull. Soc. Pathol. Exot.* **61**, 66.

Benex, J. (1968b). *Ann. Parasitol. Hum. Comp.* **43**, 561.

Benex, J. (1968c). *Ann. Parasitol. Hum. Comp.* **43**, 573.

Benex, J., and Lamy, L. (1967). *C. R. Soc. Biol.* **161**, 592.

Berjon, J. J. (1965). *C. R. Acad. Sci.* **260**, 6212.

Berjon, J. J., André, F., and Meunier, J. M. (1965). *C. R. Soc. Biol.* **159**, 139.

Berntzen, A. K. (1960). *J. Parasitol.* **46**, 47.

Berreur-Bonnenfant, J. (1962). *Bull. Soc. Zool. Fr.* **87**, 377.

Berreur-Bonnenfant, J. (1963a). *C. R. Acad. Sci.* **256**, 2244.

Berreur-Bonnenfant, J. (1963b). *Bull. Soc. Zool. Fr.* **79**, 59.

Berreur-Bonnenfant, J. (1964). *Ann. Endocrinol.* **25**, Suppl. 14.

Berreur-Bonnenfant, J. (1966). *Bull. Soc. Zool. Fr.* **91**, 327.

Berreur-Bonnenfant, J. (1967). *C. R. Soc. Biol.* **161**, 9.

Betchaku, T. (1967). *J. Exp. Zool.* **164**, 407.

Bevelander, G., and Martin, J. (1949). *Anat. Rec.* **105**, 614.

Bohuslav, P. (1933a). *Arch. Exp. Zellforsch. Besonders Gewebezuecht.* **13**, 673.

Bohuslav, P. (1933b). *Arch. Exp. Zellforsch. Besonders Gewebezuecht.* **14**, 139.

Boilly, B. (1967). *Bull. Soc. Zool. Fr.* **92**, 331.

Boilly, B. (1969). *Arch. Zool. Exp. Gen.* **110**, 125.

Boilly-Marer, Y. (1962). *C. R. Acad. Sci.* **254**, 2830.

Borojevic, R. (1963). *C. R. Acad. Sci.* **257**, 961.

Borojevic, R. (1966). *Develop. Biol.* **14**, 130.

Borojevic, R., and Lévi, C. (1964). *C. R. Acad. Sci.* **259**, 4364.

Borojevic, R., and Lévi, C. (1965). *Z. Zellforsch. Mikrosk. Anat.* **68**, 57.

Brien, P. (1930). *Ann. Soc. Roy. Zool. Belg.* **61**, 19.

Brien, P. (1932). *Bull. Cl. Sci., Acad. Roy. Belg.* [5] **28**, 975.

Brien, P. (1937). *Arch. Biol.* **48**, 185.

Brien, P. (1941). *Ann. Soc. Roy. Zool. Belg.* **72**, 37.

Brien, P. (1943). *Bull. Mus. Roy. Hist. Natur. Belg.* **19**, 1.

Brien, P. (1956). *Bull. Biol. Fr. Belg.* **89**, 258.

Brien, P. (1961). *Bull. Biol. Fr. Belg.* **95**, 301.

Brien, P. (1966). *C. R. Acad. Sci.* **263**, 649.

Brien, P. (1969). Personal communication.

Brien, P., and Meewis, H. (1938). *Arch. Biol.* **49**, 177.

Brien, P., and Reniers-Decoen, M. (1955). *Bull. Biol. Fr. Belg.* **89**, 259.

Brønsted, H. V. (1936). *Acta Zool. (Stockholm)* **17**, 75.

Brønsted, H. V. (1946). *Biol. Medd. Dan-Vid. Selsk. (Copenhagen)* **20**, 3.

Bruslé, J. (1966). *C. R. Acad. Sci.* **263**, 1514.

Bruslé, J. (1967). *C. R. Acad. Sci.* **264**, 963.

Bruslé, J. (1968). *Proc. Int. Colloq. Invertebr. Tissue Cult., 2nd, 1967*, p. 63.

Burch, J. B. (1968). *Jap. J. Malacol.* **27**, 20.

Burch, J. B., and Cuadros, C. (1965). *Nature (London)* **206**, 637.

Burnett, A. L. (1966). *Amer. Natur.* **100**, 165.

Burnett, A. L., Davis, L. E., and Ruffing, F. E. (1966). *J. Morphol.* **120**, 1.

Burnett, A. L., Sindelar, W., and Diehl, N. (1967). *J. Mar. Biol. Ass. U. K.* **47**, 645.

Burnett, A. L., Ruffing, F. E., Zongker, J., and Necco, A. (1968). *J. Embryol. Exp. Morphol.* **20**, 73.

Busquet, M. (1939). *C. R. Soc. Biol.* **131**, 1024.

Cardot, H. (1933). *Ann. Physiol. Physicochim. Biol.* **9**, 630.

Carleton, H. M. (1923). *Brit. J. Exp. Biol.* **1**, 131.

Carrel, A., and Burrows, M. T. (1910). *C. R. Soc. Biol.* **69**, 293, 298, 299, 328, and 365.

Carriker, M. R. (1946). *Biol. Bull.* **91**, 88.

Chaet, A. B. (1964). *Amer. Zool.* **4**, 409.

Chalkley, H. W. (1945). *J. Nat. Cancer Inst.* **6**, 191.

Chandebois, R. (1963a). *C. R. Acad. Sci.* **256**, 1378.

Chandebois, R. (1963b). *Ann. Epiphyt.* [2] **14**, 141.

Chandebois, R. (1970). *Ann. Biol.* **9**, 543.

Charniaux-Cotton, H. (1954). *C. R. Acad. Sci.* **239**, 780.

Charniaux-Cotton, H. (1955). *C. R. Acad. Sci.* **240**, 1487.

Charniaux-Cotton, H. (1957). *Ann. Sci. Nat. Zool. Biol. Anim.* [11] **19**, 411.

Charniaux-Cotton, H. (1967). *C. R. Soc. Biol.* **161**, 6.

Chen, J. M. (1954). *Exp. Cell Res.* **7**, 518.

Chernin, E. (1963). *J. Parasitol.* **49**, 353.

Chernin, E. (1964). *J. Parasitol.* **50**, 531.

Choquet, M. (1964). *C. R. Acad. Sci.* **258**, 1089.

Choquet, M. (1965). *C. R. Acad. Sci.* **261**, 4521.

Choquet, M. (1966). *Cah. Biol. Mar.* **7**, 1.

Choquet, M. (1969). Thesis, No. 185. Nat. Sci. Doct., Lille.

Connes, R. (1966). *Bull. Soc. Zool. Fr.* **91**, 639.

Connes, R. (1968). *Bull. Soc. Zool. Fr.* **93**, 257.

Connes, R. (1970). *Bull. Soc. Zool. Fr.* **95**, 211.

Curtis, A. S. G. (1962). *Nature (London)* **196**, 245.

Curtis, A. S. G. (1970). *Nature (London)* **226**, 260.

Davis, L. E. (1970). *Exp. Cell Res.* **60**, 127.

Davis, L. E., Burnett, A. L., Haynes, J. F., and Mumaw, V. R. (1966). *Develop. Biol.* **14**, 307.

de Laubenfels, M. W. (1932). *Tortugas Lab. Pap. Carnegie Inst., Washington* **28**, 38.

Delavault, R., and Bruslé, J. (1965). *Bull. Soc. Zool. Fr.* **90**, 361.

Dhainaut, A. (1964). *C. R. Acad. Sci.* **259**, 461.

Diehl, F. A., and Bouillon, J. (1966a). *Bull. Cl. Sci. Acad. Roy. Belg.* [5] **52**, 138.

Diehl, F. A., and Bouillon, J. (1966b). *Bull. Cl. Sci., Acad. Roy. Belg.* [5] **52**, 1010.

Dobrowolsky, N. A. (1916). *C. R. Soc. Biol.* **79**, 789.

Driesch, H. (1902). *Wilhelm Roux' Arch. Entwicklungsmech. Organismen* **14**, 247.

Durchon, M. (1952). *Ann. Sci. Nat. Zool. Biol. Anim.* [11] **14**, 119.

Durchon, M. (1963). *Bull. Soc. Zool. Fr.* **89**, 45.

Durchon, M. (1967). "L'endocrinologie des Vers et des Mollusques." Masson, Paris.

Durchon, M., and Boilly, B. (1964). *C. R. Acad. Sci.* **259**, 1245.

Durchon, M., and Dhainaut, A. (1964). *C. R. Acad. Sci.* **259**, 917.

Durchon, M., and Dhainaut-Courtois, N. (1964). *C. R. Soc. Biol.* **158**, 550.

Durchon, M., and Richard, A. (1967). *C. R. Acad. Sci.* **264**, 1497.

Durchon, M., and Schaller, F. (1963). *C. R. Acad. Sci.* **256**, 5615.

Eagle, H. (1955). *Science* **122**, 501.

Evlakhova, V. F. (1946). *Dokl. Akad. Nauk SSSR* **53**, 369.

Fatt, H. V. (1967). *Proc. Soc. Exp. Biol. Med.* **124**, 897.

Fauré-Fremiet, E. (1925). *C. R. Soc. Biol.* **93**, 618.

Fauré-Fremiet, E. (1932a). *Arch. Anat. Microsc. Morphol. Exp.* **28**, 1.

Fauré-Fremiet, E. (1932b). *Arch. Anat. Microsc. Morphol. Exp.* **28**, 121.

Fell, H. B. (1960). *Nature (London)* **185**, 882.

Ferguson, J. C. (1964). *Biol. Bull.* **126**, 391.

Ferguson, J. C. (1968). *Comp. Biochem. Physiol.* **24**, 921.
Fischer-Piette, E. (1929). *C. R. Soc. Biol.* **102**, 764.
Fischer-Piette, E. (1933). *Arch. Zool. Exp. Gen.* **74**, 33.
Fitzharris, T. P., and Lesh, G. E. (1969). *J. Embryol. Exp. Morphol.* **22**, 279.
Flandre, O., and Vago, C. (1963). *Ann. Epiphyt.* [2] **14**, 161.
Fowler, D. J., and Goodnight, C. J. (1966). *Science* **152**, 1078.
Freisling, M., and Reisinger, E. (1958). *Wilhelm Roux' Arch. Entwicklungsmech. Organismen* **150**, 581.
Frey, J. (1968). *Wilhelm Roux' Arch. Entwicklungsmech. Organismen* **160**, 428.
Galanti, N., and Gasic, G. (1967). *Biologica* **11**, 28.
Galtsoff, P. S. (1925a). *J. Exp. Zool.* **42**, 183.
Galtsoff, P. S. (1925b). *J. Exp. Zool.* **42**, 223.
Galtsoff, P. S. (1929). *Biol. Bull.* **57**, 250.
Ganguly, B. (1960). *Wilhelm Roux' Arch. Entwicklungsmech. Organismen* **152**, 22.
Gatenby, J. B. (1931). *Nature (London)* **128**, 1002.
Gatenby, J. B., and Duthie, E. S. (1932). *J. Roy. Microsc. Soc.* **52**, 395.
Gay, R. (1963). *Ann. Epiphyt.* [2] **14**, 61.
Ghiani, P., Orsi, L., and Relini, G. (1964). *Atti Accad. Ligure Sci. Lett., Genova* **20**, 93.
Gilchrist, F. (1937). *Biol. Bull.* **72**, 99.
Glaser, R. W. (1932). *N. J. Dep. Agr., Circ.* **211**, 3.
Glaser, R. W. (1940). *J. Exp. Zool.* **84**, 1.
Gomot, L. (1961). *Colloq. Int. C. N. R. S.* Paris No. 125.
Gomot, L. (1970). *In* "Invertebrate Organ Cultures," *Doc. Biol.* **2**, p. 105. Gordon & Breach, New York.
Gomot, L., and Guyard, A. (1964). *C. R. Acad. Sci.* **258**, 2902.
Gomot, L., and Guyard, A. (1968). *Proc. Int. Colloq. Invertebr. Tissue Cult., 2nd, 1967,* p. 22.
Griffond, B. (1969). *C. R. Acad. Sci.* **268**, 963.
Guyard, A. (1967). *C. R. Acad. Sci.* **265**, 147.
Guyard, A. (1969a). *C. R. Acad. Sci.* **268**, 162.
Guyard, A. (1969b). *C. R. Acad. Sci.* **268**, 966.
Guyard, A. (1970). *Ann. Biol.* **9**, 401.
Guyard, A. (1971). Thesis, No. 56. Nat. Sci. Doct. Besancon.
Guyard, A., and Gomot, L. (1964). *Bull. Soc. Zool. Fr.* **89**, 48.
Haberlandt, G. (1902). *Sitzungsber, Akad. Wiss. Wien, Math. Nat. Classe,* **111**, Abt. 1, 69.
Hagadorn, I. R. (1966). *Amer. Zool.* **6**, 251.
Harrison, R. G. (1907). *Science* **26**, 415.
Hauenschild, C. (1956). *Z. Naturforsch. B* **11**, 610.
Haynes, J. F., and Burnett, A. L. (1963). *Science* **142**, 1481.
Herbst, C. (1900). *Wilhelm Roux' Arch. Entwicklungsmech. Organismen* **9**, 424.
Herlant-Meewis, H., and Nokin, A. (1962–1963). *Ann. Soc. Roy. Zool. Belg.* **93**, 137.
Hill, J. C., and Gatenby, J. B. (1934). *Arch. Exp. Zellforsch. Besonders Gewebezuecht.* **18**, 195.
Hirumi, H., Chen, T. A., and Maramorosch, K. (1968). *Proc. Intern. Colloq. Invertebr. Tissue Cult., 2nd, 1967,* p. 147.
Hollande, E. (1968). *C. R. Acad. Sci.* **267**, 1054.
Humphreys, T. (1962). Thesis, University of Chicago.
Humphreys, T. (1963). *Develop. Biol.* **8**, 27.

Humphreys, T., Humphreys, S., and Moscona, A. A. (1960a). *Biol. Bull.* **119**, 294.
Humphreys, T., Humphreys, S., and Moscona, A. A. (1960b). *Biol. Bull.* **119**, 295.
Huxley, J. S. (1911). *Phil. Trans. Roy. Soc. London* **202**, 165.
Huxley, J. S. (1921a). *Quart. J. Microsc. Sci.* **65**, 293.
Huxley, J. S. (1921b). *Biol. Bull.* **40**, 127.
Ikegami, S., Tamura, S., and Kanatani, H. (1967). *Science* **158**, 1052.
Ischikawa, C. (1890). *Z. Wiss. Zool.* **49**, 433.
Issajew, W. (1926). *Wilhelm Roux' Arch. Entwicklungsmech. Organismen* **108**, 1.
Jackson, G. J. (1962). *Exp. Parasitol.* **12**, 25.
Janda, V., and Bohuslav, P. (1934). *Publ. Fac. Sci. Univ. Charles, Prague* **133**, 3.
Jullien, A., Ripplinger, J., Cardot, J., and Rérat, D. (1955). *C. R. Soc. Biol.* **149**, 726.
Kanatani, H. (1964). *Science* **146**, 1177.
Kanatani, H. (1967). *Gunma Symp. Endocrinol.* [*Proc.*] **4**, 65.
Kanatani, H., and Ohguri, M. (1966). *Biol. Bull.* **131**, 104.
Kanatani, H., and Shirai, H. (1967). *Nature (London)* **216**, 284.
Kanatani, H., and Shirai, H. (1969). *Biol. Bull.* **137**, 297.
Kanatani, H., Kurokawa, T., and Nakanishi, K. (1969). *Biol. Bull.* **137**, 384.
Konicek, H. (1933). *Arch. Exp. Zellforsch. Besonders Gewebezuecht.* **13**, 709.
Korotkova, G. P. (1970). *Cah. Biol. Mar.* **11**, 325.
Krontowski, A., and Rumianzew, A. (1922). *Pflüegers Arch. Gesamte Physiol. Menschen Tiere* **195**, 291.
Lane, N. J. (1962). *Quart. J. Microsc. Sci.* **103**, 211.
Lane, N. J. (1963). *Gen. Comp. Endocrinol.* **3**, 6.
Laviolette, P. (1954a). *Ann. Sci. Nat. Zool. Biol. Anim.* [11] **16**, 427.
Laviolette, P. (1954b). *Bull. Biol.* **88**, 310.
Lender, T. (1955). *C. R. Acad. Sci.* **241**, 1863.
Lender, T. (1956). *Bull. Soc. Zool. Fr.* **81**, 192.
Lesh, G. E. (1970). *J. Exp. Zool.* **173**, 371.
Lesh, G. E., and Burnett, A. L. (1966). Unpublished observations. See Macklin and Burnett, 1966.
Lesh, G. E., and Burnett, A. L. (1966). *J. Exp. Zool.* **163**, 55.
Levine, H. S., and Silverman, P. H. (1969). *J. Parasitol.* **55**, 17.
Lewis, M. R. (1916). *Anat. Rec.* **10**, 287.
Li, Y. Y. F. (1962). *Anat. Rec.* **142**, 252.
Li, Y. Y. F., Baker, F. D., and Andrew, W. (1963). *Proc. Soc. Exp. Biol. Med.* **113**, 259.
Loeb, L. (1897). "Über die Entstehung von Bindegewebe, Leucocyten und roten Blutkörperchen aus Epithel und über eine Methode, isolierte Gewebsteile zu züchten." M. Stern & Co., Chicago, Illinois.
Loeb, L. (1922). *Science* **56**, 237.
Loeb, L. (1927). *Protoplasma* **2**, 512.
Lowell, R. D., and Burnett, A. L. (1969). *Biol. Bull.* **137**, 312.
Lubet, P., and Streiff, W. (1970). *In* "Invertebrate Organ Cultures," *Doc. Biol.* **2**, p. 135. Gordon & Breach, New York.
Lutz, F. E., Welch, P. S., Galtsoff, P. S., and Needham, J. G. (1937). "Culture Methods for Invertebrate Animals." Dover, New York.
McClelland, G., and Ronald, K. (1970). *Can. J. Zool.* **48**, 198.
Macklin, M., and Burnett, A. L. (1966). *Exp. Cell Res.* **44**, 665.

McQuilkin, W. T., Evans, V. J., and Earle, W. R. (1957). *J. Nat. Cancer Inst.* **19**, 885.

Malecha, J. (1965). *C. R. Soc. Biol.* **159**, 1674.

Malecha, J. (1967a). *C. R. Acad. Sci.* **265**, 613.

Malecha, J. (1967b). *C. R. Acad. Sci.* **265**, 1806.

Manelli, H., and Negri, A. (1962). *Boll. Zool.* **29**, 787.

Marcel, R. (1966). *C. R. Acad. Sci.* **262**, 2470.

Marcel, R. (1967a). *Bull. Soc. Zool. Fr.* **92**, 345.

Marcel, R. (1967b). *C. R. Acad. Sci.* **265**, 693.

Marcel, R. (1970). *Ann. Biol.* **9**, 527.

Margoliash, E., Schenck, J. R., Hargie, M. P., Burokas, S., Richter, W. R., Barlow, G. H., and Moscona, A. A. (1965). *Biochem. Biophys. Res. Commun.* **20**, 383.

Marthy, H. J. (1970). *C. R. Acad. Sci.* **271**, 2396.

Martin, H. M., and Vidler, B. O. (1962). *Exp. Parasitol.* **12**, 192.

Maximow, A. (1925). *Contrib. Embryol. Carnegie Inst.* **16**, 49.

Meewis, H. (1939). *Ann. Soc. Roy. Zool. Belg.* **70**, 201.

Mookerjee, S., and Ganguly, B. (1964). *Wilhelm Roux' Arch. Entwicklungsmech. Organismen* **155**, 525.

Mookerjee, S., Ganguly, B., and Gouri, C. V. (1965). *Indian J. Exp. Biol.* **3**, 1.

Moore, J. (1952). *J. Exp. Biol.* **29**, 72.

Morgan, J. F., Morton, H. J., and Parker, R. C. (1950). *Proc. Soc. Exp. Biol. Med.* **73**, 1.

Moscona, A. A. (1963). *Proc. Nat. Acad. Sci. U. S.* **49**, 742.

Müller, K. (1911). *Wilhelm Roux' Arch. Entwicklungsmech. Organismen* **32**, 397.

Müller, W. A. (1969a). *Wilhelm Roux' Arch. Entwicklungsmech. Organismen* **163**, 334.

Müller, W. A. (1969b). *Wilhelm Roux' Arch. Entwicklungsmech. Organismen* **163**, 357.

Müller, W. A. (1969c). *Wilhelm Roux' Arch. Entwicklungsmech. Organismen* **164**, 37.

Murray, M. R. (1927). *J. Exp. Zool.* **47**, 467.

Murray, M. R. (1928). *Physiol. Zool.* **1**, 137.

Murray, M. R. (1931). *Arch. Exp. Zellforsch. Besonders Gewebezuecht.* **11**, 656.

Naisse, J. (1966). *Arch. Biol.* **77**, 139.

Nicholis, J. G., and Kuffler, S. W. (1965). *J. Neurophysiol.* **28**, 519.

Normandin, D. K. (1963). *Diss. Abstr.* **23**, 3045.

Oyama, S. N., and Kamemoto, F. I. (1970). *Crustaceana* **18**, 309.

Pantin, C. F. A. (1934). *J. Exp. Biol.* **11**, 11.

Papenfuss, E. J. (1932). *Anat. Rec.* **54**, Suppl. 54.

Papenfuss, E. J. (1934). *Biol. Bull.* **67**, 223.

Papenfuss, E. J., and Bokenham, N. A. H. (1939). *Biol. Bull.* **76**, 1.

Pelluet, D., and Lane, N. J. (1961). *Can. J. Zool.* **39**, 789.

Petrovic, A. (1960). *C. R. Soc. Biol.* **154**, 1622.

Prop, F. J. A. (1959). *Nature (London)* **184**, 379.

Rannou, M. (1968). *Vie Milieu, Ser. A*, **19**, 53.

Raush, R. L., and Jentoft, V. L. (1957). *J. Parasitol.* **43**, 1.

Read, C. P. (1961). *Comp. Physiol. Carbohyd. Metab. Heterothermic Anim.* p. 3.

Read, C. P., and Simmons, J. E., Jr. (1963). *Physiol. Rev.* **43**, 263.

Reháček, J. (1958). *Acta Virol. (Prague), Engl. Ed.* **2**, 253.

Reháček, J. (1962). *Acta Virol. (Prague), Eng. Ed.* **6**, 188.

Reháček, J. (1963). *Ann. Epiphyt.* [2] **14**, 199.

Reháček, J., and Hana, L. (1961). *Acta Virol. (Prague), Engl. Ed.* **5**, 57.

Reháček, J., and Pesek, J. (1960). *Acta Virol. (Prague), Engl. Ed.* **4**, 241.

Richard, A. (1966). *C. R. Acad. Sci.* **263**, 1138.

Richard, A. (1970). *Ann. Biol.* **9**, 409.

Ripplinger, J., and Joly, M. (1961). *C. R. Soc. Biol.* **155**, 4, 825.

Robinson, D. L. H., Silverman, P. H., and Pearce, A. R. (1963). *Trans. Roy. Soc. Trop. Med. Hyg.* **57**, 238.

Rose, S. M. (1966). *Growth* **30**, 429.

Rose, S. M. (1967). *Growth* **31**, 149.

Rose, S. M. (1970a). *Biol. Bull.* **138**, 344.

Rose, S. M. (1970b). *Amer. Zool.* **10**, 91.

Rose, S. M., and Powers, J. A. (1966). *Growth* **30**, 419.

Rybak, B. (1964). *Vie Milieu*, Suppl. **17**, 25.

Schiller, E. L., Read, C. P., and Rothman, A. H. (1959). *J. Parasitol.* **45**, 45.

Schultz, E. (1907). *Wilhelm Roux' Arch. Entwicklungsmech. Organismen* **24**, 503.

Schulze, P. (1918). *Sitzungsber. Ges. Naturforsch. Freunde Berlin* **7**, 252.

Seilern-Aspang, F. (1957). *Zool. Anz.* **159**, 193.

Seilern-Aspang, F. (1958). *Zool. Anz.* **160**, I.

Senft, A. W., and Senft, D. G. (1962). *J. Parasitol.* **48**, 551.

Sengel, C. (1959). *C. R. Acad. Sci.* **249**, 2854.

Sengel, C. (1960). *J. Embryol. Exp. Morphol.* **8**, 468.

Sengel, C. (1963). *Ann. Epiphyt.* [2] **14**, 173.

Sengel, P. (1961). *C. R. Acad. Sci.* **252**, 3666.

Sengel, P., and Kieny, M. (1962). *C. R. Acad. Sci.* **254**, 1682.

Sengel, P., and Kieny, M. (1963). *Ann. Epiphyt.* [2] **14**, 95.

Sivaramkrishnan, V. R. (1951). *Proc. Indian Acad. Sci.* **34**, 273.

Spiegel, M. (1955). *Ann. N. Y. Acad. Sci.* **60**, 1056.

Steinberg, M. S. (1954). *J. Exp. Zool.* **127**, 1.

Steinberg, S. N. (1963). *Biol. Bull.* **124**, 337.

Stéphan-Dubois, F. (1954). *Bull. Biol. Fr. Belg.* **38**, 181.

Stéphan-Dubois, F. (1958). *Arch. Anat. Microsc. Morphol. Exp.* **47**, 605.

Streiff, W. (1963). *Gen. Comp. Endocrinol.* **3**, 98.

Streiff, W. (1964). *Bull. Soc. Zool. Fr.* **89**, 56.

Streiff, W. (1966). *Ann. Endocrinol.* **27**, 385.

Streiff, W. (1967a). Thesis No. 294. Nat. Sci. Doct., Toulouse.

Streiff, W. (1967b). *Ann. Endocrinol.* **28**, 461.

Streiff, W. (1967c). *Ann. Endocrinol.* **28**, 641.

Streiff, W., and Le Breton, J. (1970). *C. R. Acad. Sci.* **270**, 632.

Streiff, W., and Peyre, A. (1963). *C. R. Acad. Sci.* **256**, 292.

Streiff, W., and Taberly, G. (1964). *Bull. Soc. Zool. Fr.* **89**, 65.

Taki, J. (1944). *Jap. J. Malacol.* **13**, 267.

Tardent, P. (1965). *Regeneration Anim. Related Probl., Int. Symp., 1964*, p. 71.

Taylor, A. E. R. (1961). *Exp. Parasitol.* **11**, 176.

Taylor, A. E. R. (1963). *Exp. Parasitol.* **14**, 304.

Taylor, A. E. R., McCabe, M., and Longmuir, I. S. (1966). *Exp. Parasitol.* **19**, 269.

Theodor, J. (1969). *C. R. Acad. Sci.* **268**, 2534.

Thomas, J. A. (1932). *Arch. Zool. Exp. Gen.* **73**, 22.

Thomas, J. A. (1941a). Unpublished observations. See Thomas, J. A. (1963), p. 31.

Thomas, J. A. (1941b). *C. R. Acad. Sci.* **213**, 85.

Thomas, J. A. (1941c). *C. R. Acad. Sci.* **213**, 252.

Thomas, J. A. (1947). *C. R. Acad. Sci.* **225**, 148.

Thomas, J. A. (1963). "Survie et conservation biologique. Exposés actuels de biologie cellulaire." Masson, Paris.

Thomas, J. A. (1965). "Les cultures organotypiques. Exposés actuels de biologie cellulaire." Masson, Paris.

Thomas, J. A., and Borderioux, I. (1948). *Arch. Anat. Microsc. Morphol. Exp.* **37**, 263.

Thouveny, Y. (1967). Thesis, Nat. Sci. Doct., Aix-Marseille.

Trowell, O. A. (1952). *Exp. Cell. Res.* **3**, 79.

Trowell, O. A. (1961). *Colloq. Int. C. N. R. S.,* Paris **101**, 237.

Tuzet, O., and Connes, R. (1962). *Vie Milieu* **13**, 703.

Vakaet, L., and Pintelon, L. (1959). *C. R. Soc. Biol.* **153**, 174.

Vena, J. A., Hess, R. T., and Gotthold, M. L. (1969). *Experientia* **25**, 761.

Vianey-Liaud, M. (1970). *Bull. Soc. Zool. Fr.* **95**, 249.

Vianey-Liaud, M., and Lancastre, F. (1968). *C. R. Acad. Sci.* **266**, 1317.

Voge, M. (1963). *J. Parasitol.* **49**, 59.

Wardle, R. A., and Green, N. K. (1941). *Can. J. Res., Sect. D* **19**, 240.

Waymouth, C. (1959). *J. Nat. Cancer Inst.* **22**, 1003.

Weimer, B. R. (1934). *Physiol. Zool.* **7**, 212.

Wells, M. J. (1960). *Symp. Zool. Soc. London* **2**, 87.

Wells, M. J., and Wells, J. (1959). *J. Exp. Biol.* **36**, 1.

Weltner, W. (1893). *Arch. Naturgesch.* **59**, 209.

Wilmer, E. N., ed. (1960). "Cytology and Evolution," 1st ed. Academic Press, New York.

Wilmer, E. N., ed. (1965). "Cells and Tissues in Culture," Vols. 1 and 2. Academic Press, New York.

Wilmer, E. N., ed. (1966). "Cells and Tissues in Culture," Vol. 3. Academic Press, New York.

Wilson, H. V. (1907). *J. Exp. Zool.* **5**, 245.

Wilson, H. V. (1911a). *J. Exp. Zool.* **11**, 281.

Wilson, H. V. (1911b). *Bull. U. S. Bur. Fish.* **30**, 1.

Wilson, H. V. (1925). *Carnegie Inst. Wash., Yearbo.* **24**, 242.

Wilson, H. V., and Penney, J. T. (1928). *J. Elisha Mitchell Sci. Soc.* **44**, 79.

Wilson, H. V., and Penney, J. T. (1930). *J. Exp. Zool.* **56**, 73.

Wolff, Em. (1962). *Bull. Soc. Zool. Fr.* **87**, 120.

Wolff, Em. (1963). *Ann. Epiphyt.* [2] **14**, 113.

Wolff, Et. (1965). *In* "Les cultures organotypiques. Exposés actuels de Biologie cellulaire" (J. A. Thomas, ed.). Masson, Paris.

Wolff, Et., and Haffen, K. (1952). *Tex. Rep. Biol. Med.* **10**, 463.

Wolff, Et., Haffen, K., Kieny, M., and Wolff, Em. (1953). *J. Embryol. Exp. Morphol.* **1**, 55.

Wolff, Et., Lender, T., and Ziller-Sengel, C. (1964). *Rev. Suisse Zool.* **71**, 75.

Zerbib, C. (1966a). *Bull. Soc. Zool. Fr.* **91**, 203.

Zerbib, C. (1966b). *Bull. Soc. Zool. Fr.* **91**, 344.

Ziller-Sengel, C. (1964). *Bull. Soc. Zool. Fr.* **89**, 41.

Ziller-Sengel, C. (1967). Thesis, Nat. Sci. Doct., Paris.

Ziller-Sengel, C. (1970). *In* "Invertebrate Organ Cultures," *Doc. Biol.* **2**, p. 89. Gordon & Breach, New York.

Zweibaum, J. (1925a). *C. R. Soc. Biol.* **93**, 782.

Zweibaum, J. (1925b). *C. R. Soc. Biol.* **93**, 785.

Zwilling, E. (1965). *Biol. Bull.* **114**, 368.

3

In Vitro DEVELOPMENT OF INSECT EMBRYOS

Takeo Takami

I. INTRODUCTION

Tissue culture started as a technique for investigating differentiation, development, and physiological functions of tissues *in vitro*. This is evi-

137

dent in the early works of R. G. Harisson and R. Goldschmidt, who are well-known pioneers of tissue culture in vertebrates and in insects, respectively. More attention of researchers in this field, however, was soon attracted to proliferation of cells from explanted tissue fragments, and to subculture of these proliferated cells. And accumulation of efforts in this direction has now resulted in the establishment of many cell strains which are well adapted to cultural conditions in glass wares. In a sense, however, most of these proliferated cells are runaways from the organization of explanted tissues, losing the ability of organizing their original tissues or organs *in vitro*. These cells, therefore, are not suitable for use as materials for studying inherent natures of a tissue or organ as a whole, although they are very useful for analyzing biological or medical problems on the level of cells. Differentiation and development are usually hard to follow by means of tissue culture.

In insects, *in vitro* culture of embryonic tissues was accomplished far later than that of larval and pupal tissues. In the successful cultures of leafhopper embryonic tissues, explanted tissue fragments continued to contract and cells continued to proliferate for a long period (Hirumi and Maramorosch, 1964; Mitsuhashi, 1965a,b; Mitsuhashi and Maramorosch, 1964), but normal differentiation of the explanted tissues could not be seen even if they were brought into culture undamaged (Mitsuhashi and Maramorosch, 1964). In the silkworm, Grace's attempts to grow embryonic tissues have so far met with little success (Grace, 1958). The younger embryos failed to survive in culture more than 10 days. A large number of cells containing what appeared to be fat droplets were liberated into the medium, but no mitoses were observed. The gut and heart of the older embryos continued to contract for up to 14 days, but there was no growth of cells into the medium and no mitoses were observed. Larsen (1964, 1967) cultured fragments of the embryonic organs of the cockroach in TC 199 medium and observed that the fragments had continued to contract for a long time; for example, over $2\frac{1}{2}$ years in some heart fragments. He, however, said nothing about the differentiation of these explanted tissues. He only described the behavior of cells grown in the cultures.

The embryo is an organized mass of rudimentary tissues and organs that are destined to develop into larval or imaginal organs. It may therefore safely be said that the growth of embryos is a result of development and differentiation, which, in turn, can be observed by embryo culture. In this regard, it is worth making reference to some encouraging results of tissue culture which have recently been reported in the leafhopper (*Macrosteles fascifrons*) and the fruit fly (*Drosophila melanogaster*), that is, formation of canal-like structures from the developed cell sheet in the leafhopper (Mitsuhashi and Maramorosch, 1964) and formation of nonrandom aggregates, which looked like tubules or strings of cells, in the fruit fly (Horikawa *et al.*, 1966). Although the natures and functions of these structures have not yet been made clear, they may represent primitive organ formation or limited differentiation of cells in culture. Lesseps (1965) reported that when the cells of 6½ 16-hour-old *Drosophila* embryos (the embryonic development takes 22 hours at 25°) were dissociated, and cultured *in vitro*, aggregations of the cells took place and oenocytes, hypoderm, nerve, muscle, and tracheal cells took up characteristic positions in the aggregates.

With the above-mentioned facts for a background, the author will describe the present status of *in vitro* development of insect embryos, centering the presentation around his data obtained in the silkworm, *Bombyx mori* L., which is one of the popular materials for embryo culture in insects.

II. HISTORICAL SURVEY OF EMBRYO CULTURE IN INSECTS

In insects, while the maintenance of activities of embryonic cells in a physiological solution for a short time is familiar, as is shown by the works of Carlson *et al.* on the cell division in the grasshopper, there have been reported a rather small number of studies on the *in vitro* development of embryos or embryonic organs. The important reports on these themes are shown in Table I.

Slifer (1934) studied the behavior of grasshopper embryos which had been removed from the eggs and immersed in salt solutions of different concentrations. The formula of her normal solution used for the experi-

TABLE I
IMPORTANT WORKS ON INSECT EMBRYO CULTURE

Authors	Year	Materials
Slifer	1934	*Melanoplus differentialis*
Bucklin	1953	*Melanoplus differentialis*
Wada	1954	*Bombyx mori*
Jones	1956	*Locustana pardalina, Locusta migratoria*
Takami	1957	*Bombyx mori*
Koch	1961	*Carausius morosus*
Krause	1962	*Bombyx mori*
Mueller	1963	*Melanoplus differentialis*
Seidel and Koch	1964	*Carausius morosus*

TABLE II
SLIFER'S NORMAL SOLUTION[a]

Salt	Percent
NaCl	0.9
KCl	0.02
CaCl$_2$	0.02
NaHCO$_3$	0.02

[a] Bělař's formula recalculated by Slifer.

ments is mentioned in Table II. She used embryos of the common grasshopper (*Melanoplus differentialis*) at the age of about 12 days before hatching (at 25°) as materials. The embryos have just completed revolution at this stage, showing strong and distinct movements of the lateral body walls. The cessation or reappearance of these movements served as a useful endpoint for the detection of the effects of treatments in her experiments. In the most successful instance of her experiments to see how long development would continue in such a medium, an embryo survived for four weeks and, by the end of a month, had acquired most of the characteristics of an animal about to hatch. Since the salt solution contained no organic substances, it is certain that such a long-term survival and development of the embryo in it were attained by the supply of necessary nutrients from the yolk substances enclosed by the explanted embryo itself. And an addition of 0.25–0.5% dextrose to the solution had no noticeably beneficial effect on the survival or development of embryos cultured in it. It may be said, therefore, that in her case, the solution did not take the place of the yolk supplying nutrients to the embryo but only of the chorion or cuticle protecting the self-sustaining embryo. In the strict sense of the word, such a culture is possibly beyond the limits of embryo culture.

As a new technique, Bucklin (1953) applied the same method of embryo culture as did Slifer to the study of embryonic diapause in the grasshopper (*M. differentialis*). He dissected off the cuticle and the vitelline membrane and explanted the exposed embryo and yolk, which were enclosed by the serosal membrane, to hanging drops of Bělař's solution. Diapause was promptly terminated in the culture, and the embryo resumed movement and morphogenesis, underwent revolution, and developed at the normal rate to the stage of hatching. He (Bucklin, 1959) succeeded in hatching these explanted embryos by removing them from the drops, drying them off, and raising them through the first nymphal instar. In another series of experiments, Bucklin removed diapausing embryos, or pieces of embryos, from eggs divested of yolk and membranes, and then placed them in hanging drops. The yolk-free embryos, or pieces of embryos, also commenced post-diapause development in the culture. They could not undergo blastokinesis, but a high percentage of them showed the inception of mitotic activity. Cell division continued for a few days at approximately the same rate as in a normal embryo with yolk, after which time the rate of cell division decreased and the embryo finally died, presumably because of exhaustion of stored nutrients in the embryonic tissue. Bucklin says, on the basis of his embryo culture, that the explantation causes initiation of post-diapause development by acting directly on individual embryonic tissues, i.e., cells not only removed from the yolk but also isolated from influences of the brain and other known endocrine structures.

Wada (1954) dissected the chorion off from eggs of the silkworm (*Bombyx mori*) and cultured embryos with yolk, enclosed by the serosa, in Tyrode's solution containing glutathione and cysteine hydrochloride. The embryos explanted at the early reversal stage showed nearly normal development at first but could not swallow down the serosa. In the normal development, the embryo consumes the serosa at the blue head stage (Section III,A) which comes about two days before hatching at $25°$. When embryos were explanted at the diapause stage, they could not develop beyond the stage of bristle formation, which occurs about four days before hatching in an intact egg at $25°$. In these cultured eggs, it is possible that the metabolic interrelation between the embryo and yolk was nearly normal and the embryo depended on the enclosed yolk for nutrients, as was mentioned concerning the culture of grasshopper embryos by Slifer.

Jones (1956) applied embryo culture to the study of the relationship between the ventral head glands and the embryonic molt in the locusts *Locustana pardalina* and *Locusta migratoria*. Embryos which had been isolated in Ringer's solution (0.75 or 0.9% NaCl, 0.02% KCl, 0.02% $CaCl_2$, and 0.02% $NaHCO_3$) in nearly the same way as did Slifer were ligated either between the head and thorax or between the thorax and abdomen so that, in each, the embryo was divided into two blood-tight compartments. Experiments were performed on post-katatrepsis embryos which had attained either about two-thirds or about three-quarters of the length of the egg. The isolated embryos were placed in glass cells containing aerated sterile Ringer's solution. In this condition, the self-sustaining embryos remained alive for about two weeks, and during this time they showed signs of progress in development. He observed that, if the embryo was ligated between the thorax and abdomen before a critical period, the molt was limited to the thorax and, if ligated immediately behind the head, the body failed to molt.

The present author (Takami, 1957) adopted embryo culture as a technique for studying embryonic diapause in the silkworm (*Bombyx mori*). He explanted 20 30-hour-old yolk-free nondiapause embryos into an extract from nondiapause silkworm eggs and could culture them beyond the stage of appendage formation. Well-developed explants showed invaginations of the stomodaeum and proctodaeum, formation of bristles, pigmentation of the mandibles, and contraction of gut. The details of his work will be given in Section III.

Embryo culture was applied by Krause (1962) and Krause and Krause (1964) to the experimental study of early embryonic development of the silkworm (*B. mori*). They showed that germ bands younger than 26 hours at 26°, separated from the underlying yolk system, could develop *in vitro* until the stage of body segmentation. No embryos, however, could attain to the stage of appendage formation. They carried out their experiments by culturing embryos in the medium after Takami's formula, saying that the physiological solution recommended by Wyatt (1956) for culturing ovarian cells of the silkworm larva was not suitable for their embryo culture. Krause and Krause (1965) observed that a germ band which had been cut longitudinally through the median line into two halves often showed regulative development into twin-like embryos in *in vitro* culture.

Mueller (1963) studied the embryonic molt of the grasshopper (*Melanoplus differentialis*) by culturing whole embryos, or parts of embryos, after the method of Bucklin, in Bĕlař's isotonic salt solution made as recalculated by Slifer (Table II). On the basis of the results obtained, she came to the conclusion opposite to Jones', that is, the neuroendocrine system of the head and prothorax was not necessary for the control of the embryonic molts of *M. differentialis*. Especially interesting among her results, from the view point of *in vitro* development of embryos, is the development of parts of embryos isolated from the yolk system. The embryos of *M. differentialis* are staged in terms of the number of days of morphogenesis at 25°, disregarding the time normally spent in diapause after the 20th day. Accordingly, Stage 20 refers to any diapause embryo regardless of the time since oviposition. Stage 22 refers to an embryo which has been developing for two days after the cessation of diapause. Hatching occurs at approximately 38 days of development, i.e., Stage 38. Mueller cultured 235 metathoracic legs, isolated at Stage 20, in the medium with yolk from eggs of the same stage. Among these 235 legs, 83% completed the first embryonic molt (typical of Stage 27) and 53% became elongated (Stage 30). Embryonic pigment was observed in 18% (Stage 32), and 4% molted twice and developed nymphal pigment spots at the femoro–tibial joint and at the distal end of the leg (Stagle 36).

Among the works on embryo culture in insects, Koch's papers (1961,

1964) are notable. He made every effort to formulate a medium suitable for culturing embryos of *Carausius morosus,* and after a series of experiments, he obtained Normalmedium 1961, which was composed of salts, trypsinized extract from *Carausius* eggs, and heat-inactivated *Carausius* hemolymph. In this medium, for instance, an explanted 31-day-old embryo showed such distinct progress in development during the nine days of culture, as regeneration of a wounded antenna, segmentation of the antennae and appendages, increase in the number of eye facets, pigmentation of the eye, and progress in dorsal closure, etc. In this embryo rhythmical contractions of the rudimentary heart were observed to commence four days after the explantation (Seidel and Koch, 1964).

III. EMBRYO CULTURE IN THE SILKWORM, *Bombyx mori* L.

A. Normal Stages of the Development of the Nondiapause Embryo

There are abundant varieties of the silkworm, which are distinguished one from another by physiological characters, and they are not always the same in detail of the rate of embryonic development even in the same incubating conditions. The embryonic development, however, is most strikingly affected by diapause. In this paragraph, for the convenience of explaining the works in the silkworm, an outline of the normal development of the nondiapause silkworm embryo at 23° (Nakata, 1932) will be given. In the diapause egg, the embryo enters diapause at Stage 6 (mentioned below). A female of the silkworm lays about 500 eggs within half-a-day after copulation. The egg is a tiny disk of about 1.3 mm long, 1.0 mm wide, and 0.6 mm thick.

Stage 1	$1\frac{1}{2}$–2 hours	Fertilization takes place
Stage 2	10 hours	Cleavage nuclei enter the periplasm
Stage 3	12 hours	Formation of blastoderm begins
Stage 4	20 hours	Germ band differentiates
Stage 5	25 hours	Head lobes are formed. Primitive groove appears
Stage 6	30 hours	Germ band becomes slender to form spoon-shaped embryo
Stage 7	35 hours	Cells of the mesoderm are arranged in about 17 segmental groups
Stage 8	40 hours	Gnathal and thoracic appendages appear. Stomodaeum invaginates, and neural groove appears
Stage 9	2 days	Abdominal appendages appear. Proctodaeum invaginates. Maxillary and thoracic appendages are of about two segments, respectively. Labial appendages are smaller than maxillary ones

Stage 10 $2\frac{1}{2}$ days   Coelomic sacs begin to be formed. Labial appendages
                              are nearly the same as maxillary ones in length, and
                              the former are of two segments
Stage 11   3 days   Spiracles are formed. Silk glands invaginate. Ganglia
                              appear
Stage 12 $3\frac{1}{2}$ days   Antennae move forward to the line of stomodaeum.
                              Grouping of gnathal appendages begins, and head is
                              nearly demarcated from thorax. Anterior and poste-
                              rior midgut strands become continuous
Stage 13   4 days   Antennae come ahead to the line of labral lobes.
                              Mandibular appendages come to the line of stomo-
                              daeum. Both the labial appendages commence to
                              meet together. Both the silk glands meet together
                              at the anterior part. Abdominal part begins to move
                              inward the egg as the first step of embryonic reversal
Stage 14 $4\frac{1}{2}$ days   Posterior part of abdomen becomes ventral concave
Stage 15 $4\frac{3}{4}$ days   Dorsal closure is completed but an opening is left be-
                              hind the second thoracic segment
Stage 16   5 days   Embryonic reversal is completed. Gonads appear. Dor-
                              sal closure of midgut completes
Stage 17   6 days   Caudal horn appears. Caudal legs are doubly segmented
Stage 18 $6\frac{1}{2}$ days   Bristles appear. Pigment begins to deposit in mandibles
                              and ocelli. Claws appear on thoracic legs
Stage 19   7 days   Taenidia appear in tracheae
Stage 20 $7\frac{1}{2}$ days   Pigment begins to deposit in the head wall. Closure of
                              body wall is completed
Stage 21   8 days   Serosa is consumed. Head of embryo can be seen to be
                              bluish through chorion (blue head stage)
Stage 22   9 days   Pigment begins to deposit in the body wall, and embryo
                              can be seen to be bluish through chorion (blue body
                              stage)
Stage 23 $9\frac{1}{2}$ days   Pigmentation becomes darker. Embryo emerges from
                              the egg when placed in a light place

B. Culture of Eggs Removed from the Chorion

The chorion, which is the outermost hard envelope of the silkworm
egg, is a nearly complete barrier against the penetration of water or
culture media into the egg, and any sign of bursting or swelling of the
egg is hardly observed even after soaking a long time in water. In an egg
which is kept in water or culture media, the embryo cannot continue
to develop because of the inhibition of gas exchange through the respiratory canals piercing the chorion.

On the contrary, in a naked egg which has been removed from the
chorion and cultured in a suitable medium, the embryo can continue to
develop until it swallows down the serosal membrane (Stage 21 mentioned
above).

Our routines for culturing naked silkworm eggs *in vitro* are as follows. Prediapause eggs (before and at Stage 6) oviposited on water-proof kraft paper are convenient to handle. The eggs are usable for more than 10 weeks if they are stored at 5° on the second day after oviposition, whereas nondiapause eggs can only be stored within a week.

In order to disinfect the egg surface, the kraft paper carrying eggs is cut into small pieces with 10 or more eggs each, and the pieces are contained in a small wire basket (Fig. 1) to be soaked in a 3% formaldehyde solution for 30 minutes. Then they are rinsed five times in sterilized distilled water and soaked in 70% ethyl alcohol for 10 minutes before they are taken out from the basket under aseptic conditions. The pieces of paper are then dried under a germicidal lamp to prevent contamination.

To remove the chorion from eggs, treatment with sodium hypochlorite solution is not effective for the silkworm egg, although it is recommended for many kinds of insect eggs. At Stage 6, owing to the development of serosal cuticle, the chorion backed with the vitelline membrane becomes separable from the serosa which contains the embryo and yolk. The dissection of eggs is performed under a low-power binocular microscope. A piece of egg-carrying paper is held with a pincette in a small watch glass and a part of chorion is cut from each egg to make a sufficiently big opening for removing the naked egg. Since an embryo younger than Stage 6 occupies a wide area, spreading across the ventral side and over both the lateral sides on the egg surface, a slit is made along the median line of the dorsal side lest the embryo should be damaged. In an egg at Stages 6–13, a slender embryo is located on the ventral side, extending across the posterior pole to the dorsal side, so it is recommended that a lengthwise slit be made on the lateral side. A physiological solution con-

FIG. 1. A wire basket and silkworm eggs.

TABLE III

PHYSIOLOGICAL SOLUTIONS FOR DISSECTING EGGS AND PREPARING CULTURE MEDIA[a]

Substance	mg/100 ml		
	No. 1	No. 2	No. 3
NaCl	750	750	600
KCl	50	50	40
$NaH_2PO_4 \cdot 2H_2O$	5	5	20
$MgSO_4 \cdot 2H_2O$			20
$NaHCO_3$	5	5	20
Glucose		20	100
Lactalbumin hydrolysate (NBC)			400
Dried extract of yeast (Diago)			50
$CaCl_2 \cdot 2H_2O$	20	20	20
L-Arginine–HCl			10
L-Glutamic acid			10
L-Glutamine			10
L-Glycine			20
L-Histidine			50
L-Serine			20

[a] Preparation Nos. 1 and 2: $CaCl_2 \cdot 2H_2O$ and the others are dissolved in 250 ml and 750 ml of bidistilled water, respectively, sterilized by heating separately, and mixed together after cooling, or mixed together before sterilizing by filtration through a Millipore filter. pH is 6.7–7.0. No. 3: Dissolved in bidistilled water and sterilized by filtration through Millipore filter. pH is ca. 6.5.

Culture media for naked embryos, Nos. 1 and 2: Two parts solution plus one part egg juice. No. 3: Eight parts solution plus one part egg juice and centrifuged for 10 minutes at 1000–1500 rpm.

taining no egg juice (Table III) is then poured into the watch glass and a naked egg enclosed with the serosal membrane is taken out of the chorion in the solution by careful manipulation of a dissecting needle. The naked egg is picked up with a pipette of suitable size and placed on the underside of a coverslip or that of the cover of a small Petri dish to make a hanging drop culture. The size of the hanging drop should be sufficient to enclose the explanted egg. About 10 (if necessary, more than 20) hanging drop cultures are contained in one 40-mm Petri dish (Fig. 2). All these procedures must be carried out under aseptic conditions. The Petri dish is sealed with Parafilm as shown in Fig. 2. It is not necessary to change the culture medium in ordinary works.

The naked egg contains an embryo with a nearly complete yolk system, and such a self-sustaining embryo is good material for *in vitro* culture. It can be cultured well even in a salt solution containing no organic substances, such as No. 1 solution in Table III. According to Bucklin (1959), in the grasshopper *Melanoplus differentialis*, diapause

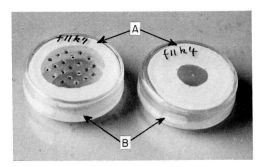

Fig. 2. Petri dishes containing 21 cultures of naked eggs (left) and one culture of a naked embryo (right). A. wet filter paper to keep moisture; B, Parafilm.

eggs removed from the eggshell can recommence development in a rather wide range of salt concentrations such as twice- and half-concentrated Ringers. The embryo of naked egg cultured *in vitro* often comes out into the surrounding medium through an opening of the serosa wounded by dissection, as if it were pushed out of the egg, and continues to develop there (Fig. 3). When this happens, the embryo is usually divested of the amnion on its ventral side. The amnion abnormally extends from the dorsal edge of the exposed embryo toward the mass of yolk from which the embryo has departed and encloses it. In such a culture, all the steps of embryonic development after the coming out can be traced under a microscope. The development proceeds at a nearly normal rate although the embryo has some inevitable defects in morphogenesis resulting from the abnormal developmental conditions. The midgut strand comes out of the body cavity before the dorsal closure is completed and reaches to the separated yolk mass along the abnormal extention of the amnion, resulting in abnormal formation of the midgut. This abnormality causes incomplete dorsal closure. There often occurs an evagination of the stomodaeum or proctodaeum.

Many of the embryos which have remained and continued to develop in the eggs can complete normal embryonic development to Stage 21. They come into the medium by swallowing the serosa and move actively (Fig. 4), often sliding on the surface of the glass with the surrounding culture medium. These embryos, however, cannot advance to hatch, being drowned in the medium, because they depend on the tracheal system for respiration after this stage. Bucklin (1959) said that he had succeeded not only in hatching explanted *Melanoplus* embryos by removing them from the drop but also raising them through the first nymphal instar. We have not yet succeeded in hatching cultured embryos in the silkworm.

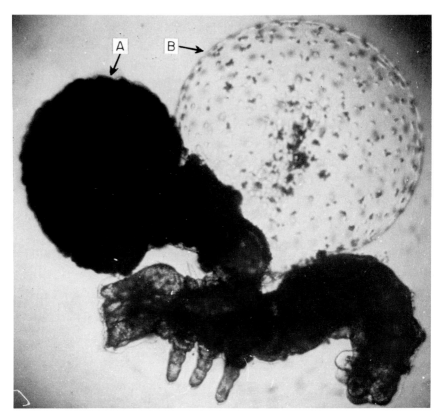

FIG. 3. An exposed embryo of a naked egg on the sixth day of culture. A, serosal membrane; B, amnion.

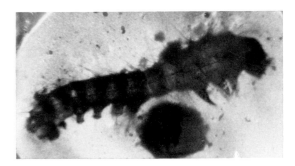

FIG. 4. A normally developed embryo of a naked egg cultured *in vitro*.

In the intact silkworm egg stored at 5° from the second day after oviposition, diapause tends to terminate with difficulty after about one month of chilling. In comparison, diapause is easily terminated in cultured naked eggs and post-diapause development is soon induced in them without chilling. Since the silkworm embryo enters diapause before appendage formation, the appearance of appendages can be taken as an unmistakable morphological criterion of the termination of diapause. In intact eggs the normal development of embryos from Stage 6 to the appendage formation (Stage 8) takes about five days at 23° after termination of diapause and about 10 hours in nondiapause eggs, while it takes about one day in naked eggs cultured at the age of two days without chilling.

In vitro culture of naked eggs is useful as a technique for studying various embryological, physiological, and pathological problems, but it is not suitable for studies on nutritional problems because of the self-sustaining development of embryos.

C. Culture of Eggs Divested of Serosa

In diapause eggs, yolk cells stick fast to the serosa, and the latter is hardly separable from the contents of eggs, that is, embryos and yolk cells. In nondiapause or pre-diapause eggs, the serosa is easily separable from the contents, and if an egg divested of the serosa is cultured *in vitro*, its embryo, which was at the beginning embedded in the mass of yolk, comes out of the mass as it grows. *In vitro* development of the embryo of such an egg proceeds in nearly the same way as that of the naked egg mentioned above.

D. Culture of Naked Embryos

From the above-mentioned development of embryos exposed in the medium, it is understood that the covering with serosa and amnion is not essential to the development of silkworm embryos *in vitro*. And it has been also observed that when a yolk-free naked embryo is explanted to a physiological solution and encircled with naked eggs at a distance to avoid contact, the encircled embryo grows pretty well without any contact with living yolk cells. This fact means that the direct contact with the yolk system is not always necessary for the development of embryos and gives us promise of success in *in vitro* culture of naked embryos.

1. Culture Media

In an intact egg, the embryo develops among the mass of yolk cells bathing in intervitelline fluid. This structure of the silkworm egg suggests that the intervitelline fluid must be the best medium for culturing or maintaining the embryo. The fluid, however, is difficult to separate because yolk cells are so fragile that the fluid sucked up with a pipette is inevitably mixed with their fragments or content, and the juice obtained from crushed eggs usually contains a large quantity of yolk materials pressed out of yolk cells. The author (Takami, 1957) made extracts from such egg juice and used it as a medium for culturing embryos. He picked eggs one by one with a dissecting needle and pushed them with a small pipette to suck up the juice. To make extract, one part of the juice was mixed with two parts of the salt solution No. 1 or No. 2 mentioned in Table III, and the mixture, contained in a small glass tube (about 3×30 mm), was centrifuged for 10 minutes at 1000–1500 rpm. The mixture was thereby fractionated into three layers, that is, a fatty upper, a sedimentary lower, and a rather clear middle fraction. The middle fraction, which had occupied a greater part of the tube and showed gradual change in clearness from the top to the bottom, was the extract used for culturing embryos. Heating of the juice was not necessary. In the extract from nondiapause or pre-diapause eggs, nondiapause or pre-diapause embryos explanted at Stage 6 developed well, being accompanied by cell divisions, as ectodermal segmentation, appendage formation, and stomodaeal and proctodaeal invaginations proceeded until the level of Stage 13. Bristle formation and pigmentation of mandibles and ocelli (Stage 18) were occasionally observed.

The development of explanted embryos is, however, affected by diapause of eggs which were used as materials for donors of either embryo or juice, and the deeper the diapause, the worse the result of culture *in vitro*. When the eggs are in a full diapause, the development of embryos, even if it started, is very little, scarcely going as far as Stage 8 (appendage formation). Thus it is desirable to prepare materials by storing two-day-old pre-diapause eggs at 5°. The eggs cold-stored at this stage can be used as good sources of juice for preparing extract or of embryos to be explanted for more than 10 weeks.

All the solutions mentioned in Table III are noticeable by their high ratio of sodium to potassium as compared with many other physiological solutions for tissue culture in lepidopterous insects. Seidel and Koch (1964) would be able to improve the culture medium for *Carausius* embryo by adjusting the ionic ratio Na:K on the basis of the quantitative analysis of the egg. They hold the opinion that the mole ratio Na:K changes with the development from egg to larva; that is, Na increases, while K decreases, in the ratio. They also calculated the mole ratio Na:K in Takami's No. 2 solution (Table III) as 19:1 and pointed out the necessity of the quantitative analysis of silkworm eggs to explain the good results of culturing embryos in the medium prepared with such a solution.

In case a medium contains no egg extract or hemolymph, it is possible that the ionic ratio produces a powerful effect on the explanted embryos. But the effect seems to be rather mild after mixture with egg extract or hemolymph. The physiological solution presented by Wyatt (1956) for the culture of the ovarian tissue of the silkworm *Bombyx mori* has a composition imitating the larval hemolymph and contains three sugars, four organic acids, and 22 amino acids, in addition to five inorganic salts. In a medium prepared by mixing this solution with egg juice (9:1) and centrifuging the mixture, silkworm embryos grow well although the mole ratio Na:K is very low (1:5) among the inorganic components. And in this medium, the inorganic salts could be replaced with those of Takami's No. 1 or No. 2 solution without any notable effect on the development of explanted embryos.

In the silkworm, at present, we have no exact knowledge about the chemical properties of egg juice necessary for the preparation of culture media. There are many data reported on the pH and free amino acids in the egg. These data, however, have been accumulated by the studies of egg juice or egg homogenate and give us nearly no information about the properties of the intervitelline fluid in which embryos bath during their development in intact eggs. Accordingly, the formulae of the author's solutions mentioned in Table III have been adopted only for the practical reason that they were of use for culturing embryos. Solutions No. 1 and 2 are now used for preliminary tests of the nutritional effect of organic substances to be mixed for making culture media, because they contain no nutritious organic components except glycogen disturbing the result of experiments. Solution No. 3 is commonly used for the preparation of culture medium by mixing with egg juice (10:1 or 20:1). It contains several kinds of amino acids which are possibly insufficient in lactalbumin hydrolysate. In the medium prepared with this solution, however, explanted embryos took a half-day longer than in that made of Grace's solution (1962) to develop from Stage 6 to Stage 9. The development of embryos in the medium made of Wyatt's solution was similar but somewhat inferior to that in the medium made of Grace's solution. The author tested a modified Grace's solution (Table IV), obtaining a little improvement in the development of explanted embryos (Fig. 5). The solution was prepared by changing the amounts of amino acids, especially of

TABLE IV
A Modified Grace's Solution for Silkworm Embryos[a]

Substances	mg/100 ml	Substances	mg/100 ml
Salts		L-Lysine–HCl	60
$NaH_2PO_4 \cdot 2H_2O$	114	L-Methionine	10
$NaHCO_3$	35	L-Phenylalanine	20
KCl	224	L-Proline	70
$CaCl_2 \cdot 2H_2O$	100	L-Serine	110
$MgCl_2 \cdot 6H_2O$	228	L-Threonine	30
$MgSO_4 \cdot 7H_2O$	278	L-Tryptophane	10
Amino acids		L-Tyrosine	10
L-Alanine	30	L-Valine	30
β-Alanine	20	Carbohydrates	
L-Arginine–HCl	100	Sucrose	2500
L-Asparagine	30	Fructose	40
L-Aspartic acid	30	Glucose	70
L-Cysteine–HCl	20	Glycogen	100
L-Glutamic acid	70	Organic acids	
L-Glutamine	70	Malic acid	67
L-Glycine	70	α-Ketoglutaric acid	37
L-Histidine	150	Succinic acid	6
L-Isoleucine	10	Fumaric acid	5.5
L-Leucine	20		

[a] Solution: Prepared following Grace's instruction (1962) and the pH is adjusted to 6.5 Culture medium for naked embryos: 9.5 parts of solution plus 0.5 parts of egg juice, or 9 parts of solution plus 1 part of heat-treated (60°, 5 minutes) and refrigerated ($-20°$) hemolymph from silkworm pupae. Centrifuged for 10 minutes at 1000–1500 rpm.

arginine, histidine, and proline on the basis of quantitative analysis of the silkworm egg (Sasaki *et al.*, 1957), decreasing the amount of sucrose, adding glycogen, and removing vitamins. The effect of vitamins on the *in vitro* development of embryos is not distinct in such a medium containing either an egg extract or hemolymph. In addition to these, the author has tested many other solutions including the one which contained phosphate buffer as a component (Takami, 1959b, 1963).

2. Method of Culture

Embryos are commonly cultured in hanging drops. For culture vessels, the author uses the same type of Petri dishes as mentioned above for the culture of naked eggs (Fig. 2) by reason of convenience in handling and in replacing culture medium. Depressed slides are also used.

From naked eggs which have been dissected out of the chorion in the same way as mentioned above (Section II, B), embryos are removed to the physiological solution by opening the serosal membrane with two dis-

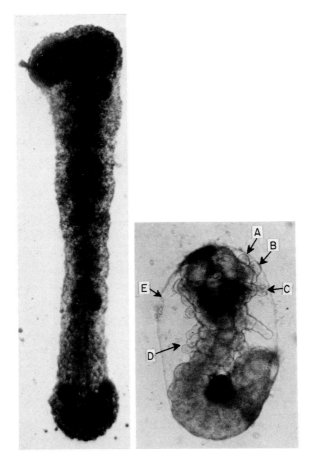

Fig. 5. *In vitro* development of a naked embryo. Left: a newly explanted embryo showing temporary swelling (about 1.3 times as long as the original length) caused by the explantation. Right: the same embryo on the ninth day of culture. A, antenna; B, mandible; C, maxilla; D, thoracic legs; E, amnion.

secting needles. The embryo removed from the egg is transferred onto the underside of a cover slip or that of the cover of a Petri dish, after roughly removing yolk cells by two or three operations of gentle sucking in and out of a small-bore pipette. The physiological solution which has accompanied the embryo is replaced with culture medium by repeating adding and sucking up of culture medium several times. During this replacement of medium, yolk cells stuck on the embryo are carefully removed. The cover slip is then sealed on a depression slide with melted paraffin, while the Petri dish is sealed with Parafilm (Fig. 2). The culture medium should be sufficient in amount to enclose the explanted embryo.

In ordinary work, culture is continued without renewal of the medium. The cultures are kept at 25° for incubation.

3. Development of Explanted Embryos

In vitro development of naked silkworm embryos progresses rather normally until the stage of appendage formation, and then it is compelled to be abnormal under the unnatural environmental conditions. The normal silkworm embryo in an intact egg clearly shows two cycles of growth in length. The first cycle of growth continues from the beginning of development to the appendage formation (Stage 8), being followed by shortening in length before the embryonic reversal (Stage 13). The second growth, which is remarkable in the abdominal part, begins together with the reversal in curvature of the embryo, continuing to the end of organogenesis (Stage 22) when the embryo becomes longer than the circumference of the egg.

There is every reason to believe that the first growth depends on surrounding yolk cells for necessary energy, while the second growth of embryos with nearly completed body wall depends on the yolk mass enclosed in the body for a considerable part of necessary energy. Mueller (1963) says, as to the development of explanted metathoracic legs of *Melanoplus* embryos, that isolated parts lack direct access to nutrients from the yolk because of rapid healing of the cut end of the femur and that a further barrier is imposed at the time of the first embryonic molt by the secretion of the relatively impermeable second embryonic cuticle around the explant. In our experiments, explanted silkworm embryos with little or no enclosed yolk grew well in the first cycle, but they showed almost no sign of growth in the second cycle, resulting in the formation of abnormally short-abdomened embryos (Figs. 5 and 6). In the culture of naked eggs, embryos which had come out of the egg into the medium at an early stage developed accordingly, with little of the enclosed yolk and were also inferior in the growth of abdomen in the second cycle.

There are many other factors inhibiting the development of embryos *in vitro*. The explanted naked embryos come in contact with the surface of glass and usually stick on it. In such a condition, it is natural that each part of the embryo cannot always continue to develop harmoniously with others. The embryonic reversal cannot take place in such a condition either. The midgut strands from the stomodaeal and proctodaeal invaginations, which are to extend on the surface of the embraced yolk system along the dorsal side of embryo, cannot extend normally in cultured embryos. The mass of yolk which is the source of nutrient for embryos has another important function in an intact egg. At the begin-

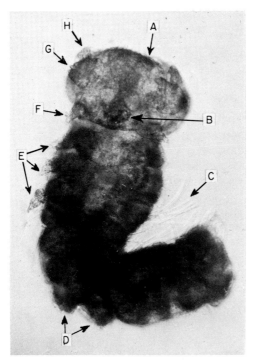

Fɪɢ. 6. A well-developed naked embryo on the 10th day of culture. A, labrum; B, labium; C, bristles; D, abdominal legs; E, thoracic legs; F, maxilla; G, mandible; H, antenna.

ning of development, it holds the embryo on its surface as a necessary foundation and later, being enclosed by the embryo, acts as an internal support of the embryo. Without such a foundation or support, an explanted naked embryo hardly keeps its normal form *in vitro*. Krause and Krause (1964) reported that morphogenesis of explanted embryos was disturbed by random behavior of yolk cells stuck on them.

Such being the case, if we expect embryo culture to obtain hatchable normal embryos, the results will be contrary to our hope. However, when we examine a cultured embryo not as a whole but in parts, we will find many interesting suggestions and problems from the viewpoint of development.

There is a critical stage in the post-diapause development of the silkworm embryo. At the beginning of post-diapause development in winter, it is possible for an embryo to increase in length until the critical stage accompanied with no cell divisions, while further development cannot take place without cell divisions. The embryo at the critical stage is nearly the same in shape as the nondiapause embryo at Stage **7**. It grows to the stage of appendage formation, accompanied by cell divisions, within a day or so at 25°. In nondiapause embryos, existence of such a critical stage cannot be evidenced. The appendage formation, therefore, can be taken as an accurate morphological criterion of the initiation of development accompanied with cell divisions in both diapause and nondiapause embryos. Suitability of materials, culture media, environmental conditions, etc. is sharply reflected in the formation of appendages *in vitro*.

Two halves of an embryo cut off longitudinally through the median line at Stage 5 did not reunite, even if they were cultured side by side and each of them developed into a unilateral embryo of Stage 8 with appendages of one side only.

IV. APPLICATION OF EMBRYO CULTURE

As was seen in the historical survey, diapause is one of the main problems which have been studied by means of embryo culture. The author (Takami, 1957, 1958) has compared the development of diapause embryos with that of nondiapause embryos in the extract from diapause eggs and in that from nondiapause eggs, and observed the following:

1. Nondiapause embryos at Stage 6 can be cultured well beyond the stage of appendage formation in the extract from nondiapause eggs, while fully diapause embryos (about 30-days old) are, for the most part, unable to grow in the same extract.

2. One- or two-day-old pre-diapause embryos still retain the ability to develop well *in vitro* as do nondiapause embryos.

3. The extract from one- or two-day-old pre-diapause eggs has nearly the same effect on the *in vitro* development of embryos as that from nondiapause embryos although the extract becomes less effective with increasing days after oviposition.

4. When pre-diapause eggs are stored at 5°, their embryos are prevented from becoming less active in *in vitro* development, and the extract from them is also prevented from becoming less effective on the *in vitro* development of embryos.

5. When diapause eggs are kept at 25° continuously, their extract possibly recovers its efficacy, which has fallen with the onset of diapause, on the explanted embryos after about 150 days, in advance of the termination of embryonic diapause.

Inability of naked diapause embryos to develop *in vitro* does not necessarily mean that the diapause of silkworm embryos cannot be terminated by explantation. Since the termination of diapause is promptly induced by explantation of naked eggs, it is probable that a longer survival and a higher percentage in the appendage formation of naked diapause embryos in side-by-side cultures with nondiapause embryos (Takami, 1959a) are due to the accelerating effect of explantation on the termination of diapause together with betterment of cultural conditions by the existence together of nondiapause embryos. Whether nondiapause embryos produce a diapause-terminating factor or not is another problem to be studied in future.

As to the study of embryonic molt, references were already made to Jones and Mueller's results in *Locustana pardalina*, *Locusta migratoria* and *Melanoplus differentialis*. The author (Takami, 1963) carried out similar experiments in the silkworm. In the silkworm, the prothoracic glands first appear as a pair of invaginations in the labial segment at about Stage 11. The author cultured partial embryos at Stage 6 *in vitro*, after cutting off the anterior part of embryos in various length, to see the effect of decapitation on the embryonic molt. He came to the conclusion that neither the brain nor the prothoracic gland seemed to be necessary for occurrence of the embryonic molt in the silkworm (*B. mori*).

Krause's works have introduced into embryology many problems to be solved by means of *in vitro* culture. The interaction between different tissues or organs in morphogenesis mentioned in the preceding section will also be an interesting problem in this connection. In normal development, the midgut strands from the end of the proctodaeal invagination extend forward on the surface of the embraced yolk mass along the dorsal side of embryo and become continuous with those from the stomodaeal invagination as the first step of midgut formation. In the development of embryo *in vitro*, the proctodaeal invagination often comes nakedly into the culture medium and the rudiments of midgut strands grown out from such an exposed invagination do not extend forward into the medium but remain as two lobes, or spread backward, on the surface of the invagination. And, in such a case, the Malpighian tubules originated on the anterior part of the exposed proctodaeal invagination also go back on the surface of the latter, although in normal development they elongate forward, then backward, and, after zigzagging, end on the rectum.

Application to virological studies is another big problem. In connection with the problem of the transmission of viruses through the egg, infections of cultured silkworm eggs with the nuclear- and cytoplasmic-polyhedrosis viruses have been studied (Takami *et al.*, 1966, 1967). Inoculated with the nuclear-polyhedrosis virus, polyhedron formation was confirmed to occur in the nuclei of various kinds of embryonic and

extra-embryonic tissues including those in which polyhedron formation
had never been reported to occur at the larval and pupal stages, i.e.,
serosa, yolk cell, amnion, epidermis, tracheal epithelium, ganglion, fore-
gut, midgut, hindgut, silk gland, muscle, fat body, gonad, and hemocyte.
On the contrary, in the case of cytoplasmic-polyhedrosis virus, polyhedron
formation was observed to occur only in the midguts of full-grown
embryos which had swallowed down the serosa and reached to the stage
about to hatch and could be found in neither rudimentary nor malformed
midguts exposed in the medium. These results suggest that there are
interesting relationships between differentiation of tissues and infection
of tissues with virus, which must be overlooked in the studies by cell
culture.

Incorporation of nucleic acid precursors and amino acids was recently
studied by Okada (1967, 1969) and Park and Yoshitake (1969) by using
the technique of embryo culture in the silkworm (B. mori L.).

REFERENCES

Bucklin, D. H. (1953). Anat. Rec. 117, 539.
Bucklin, D. H. (1959). In "Physiology of Insect Development" (F. L. Campbell,
    ed.), p. 94. Univ. of Chicago Press, Chicago, Illinois.
Grace, T. D. C. (1958). Aust. J. Biol. Sci. 3, 407.
Grace, T. D. C. (1962). Nature (London) 195, 788.
Hirumi, H., and Maramorosch, K. (1964). Exp. Cell Res. 36, 625.
Horikawa, M., Ling, L. N., and Fox, A. S. (1966). Nature (London) 210, 183.
Jones, B. M. (1956). J. Exp. Biol. 33, 174.
Koch, P. (1961). Verh. Deut. Zool. Ges., Saarbrücken p. 123.
Koch, P. (1964). Wilhelm Roux' Arch. Entwicklungsmech. Organismen 155, 549.
Krause, G. (1962). Verh. Deut. Zool. Ges., Wien p. 190.
Krause, G., and Krause, J. (1964). Wilhelm Roux' Arch. Entwicklungsmech. Organ-
    ismen 155, 451.
Krause, G., and Krause, J. (1965). Z. Naturforsch. B 20, 334.
Larsen, W. P. (1964). Life Sci. 3, 101.
Larsen, W. P. (1967). J. Insect Physiol. 13, 613.
Lesseps, R. J. (1965). Science 148, 502.
Mitsuhashi, J. (1965a). Jap. J. Appl. Entomol. Zool. 9, 107.
Mitsuhashi, J. (1965b). Jap. J. Appl. Entomol. Zool. 9, 137.
Mitsuhashi, J., and Maramorosch, K. (1964). Contrib. Boyce Thompson Inst. 22,
    435.
Mueller, N. S. (1963). Develop. Biol. 8, 222.
Nakata, T. (1932). Bull. Fukuoka Sericult. Exp. Sta. 1, No. 2, 1.
Okada, M. (1967). Jap. J. Exp. Morphol. 21, 495.
Okada, M. (1969). Jap. J. Devel. Biol. No. 23, 67.
Park, K. E., and Yoshitake, N. (1969). Appl. Entomol. Zool. 4, 171.
Sasaki, S., Watanabe, T., and Kondo, Y. (1957). J. Sericult. Sci. Jap. 26, 291.
Seidel, F., and Koch, P. (1964). Embryologia 8, 200.

Slifer, E. H. (1934). *J. Exp. Zool.* **67**, 137.

Takami, T. (1957). *Bull. Sericult. Exp. Sta. (Tokyo)* **14**, 577.

Takami, T. (1958). *J. Exp. Biol.* **35**, 286.

Takami, T. (1959a). *Science* **130**, 98.

Takami, T. (1959b). *Bull. Sericult. Exp. Sta. (Tokyo)* **15**, 477.

Takami, T. (1963). *J. Exp. Biol.* **40**, 735.

Takami, T., Sugiyama, H., Kitazawa, T., and Kanda, T. (1966). *Jap. J. Appl. Entomol. Zool.* **10**, 197.

Takami, T., Sugiyama, H., Kitazawa, T., and Kanda, T. (1967). *Jap. J. Appl. Entomol. Zool.* **11**, 182.

Wada, S. (1954). *Proc. 6th Annu. Meet., Kanto Branch, Sericult. Soc. Japan,* p. 28.

Wyatt, S. S. (1956). *J. Gen. Physiol.* **39**, 841.

II

Use of Invertebrate Tissue Cultures

4

INVERTEBRATE CELL CULTURE
IN GENETIC RESEARCH

Claudio Barigozzi

I. INTRODUCTION

The use of the cell culture *in vitro* has come to be considered, in recent years, one of most promising tools for the genetic analysis of the somatic or the embryonic cell. To obtain an idea of the achievements brought about in this new field, it suffices to refer to books such as "Somatic Cell Genetics" (edited by Krooth, 1964) and "Cytogenetics of Cells in Culture" (edited by Harris, 1964). The importance of the results obtained is, obviously, connected with the existence of suitable techniques for obtaining not only primary cultures but also established cell strains or lines of known genotype. Since such techniques are strongly dependent on the species used, the different zoological groups have made different contributions to the genetic analysis of the cultured cell. However, since the culture of invertebrate tissues is today still in its infancy compared with that of vertebrates and, especially, of mammals and birds, a logical consequence is that the results of genetical interest, too, are still in a very initial phase. Some difficulties in growing invertebrate cells *in vitro*, especially as regards insects, have however been partially overcome in recent years. Hence, it is to be hoped that there will as a new impetus in future years, especially in view of the fact that, among animals, several species of insects are gentically the best known. There is no doubt, in fact, that, as soon as convenient culture techniques are developed, exceptional facilities for investigation will be offered by species where the chromosome number is low, each pair unambiguously classifiable, and where both cytological and genetic markers exist.

II. GENETIC PROBLEMS

A. Outline of Main Problems

The main genetic problems that can be studied using *in vitro* cultures of cells and tissues can be outlined as follows:

1. Fine structural analysis at microscopic level of mitotic chromosomes can be done better *in vitro* than by utilizing *in situ* cells because of the increased resolving power offered by cultured cells. This practice is now the most generally used for mammals, including man.

2. When other techniques fail (for instance, with man), mapping of genes is brought about using cell hybridization techniques *in vitro* (Siniscalco, 1970).

3. The study of DNA replication, aiming also to discriminate between eu- and heterochromatin by means of autoradiography, achieves the clearest results when the silver grains can be assigned unambiguously to different chromosomes and to different chromosome sections. The cultured cells, when well spread, are the best material for examining autoradiographs.

4. One of the most striking phenomena in cultured cells is the change in chromosome number. This change occurs especially in established cell lines. Not only polyploid but also heteroploid cells appear, the mechanisms responsible being still imperfectly known in mammals, however.

5. Gene action can be studied in single cells wherever genetic markers are used, the phenotype of which manifests itself at cellular level (for instance, the enzymes). When hetero- and polyploidy give rise to different gene doses, the gene action can be studied.

6. Cellular differentiation can be studied *in vitro*, starting from definite developmental stages and by challenging cells with different media.

7. The complex relationships between viruses and cells are among the most difficult to study. Cultured cells can show different effects caused by viruses (chromosome breakages, cytocidal action, cell transformation) which cannot be detected otherwise.

B. Vertebrate versus Invertebrate Cultured Cells and Tissues: Advantages and Disadvantages for Studying Genetic Problems

As has already been pointed out, vertebrate cells *in vitro* are useful for the study of genetical problems but have, nonetheless, some serious limitations. We may emphasize again that, on one hand, vertebrate (and especially mammalian) cells can be cultured very easily *in vitro* and that clones can also be obtained easily. On the other hand, their chromosome numbers are high, and the possibility of classifying a genome by individual pairs is generally so slight that it is preferred to base caryograms on groups of pairs. This condition makes it very problematic to assign a given locus to a particular autosome, and in consequence, mapping is frequently impossible.

A second drawback of mammals is the limited number of known loci, except in a few species (mouse, rabbit). Finally, established mammalian cell lines show frequent changes in chromosome number. This makes studies on gene dosage frequently impracticable. If we consider invertebrates, and especially insects, we meet quite opposite conditions. In several orders the chromosome numbers are low and the morphological differences between pairs are clear cut. At the present stage of our knowledge, we can discard all invertebrate types except the arthropodes. Among these, insects are without doubt more promising than other classes; among insects, Diptera predominate because there are numerous genera where the chromosomal set is very easy to analyze and there is also an unparalleled amount of genetic information. This is the case of *Drosophila* species and, particularly, of *Drosophila melanogaster*. Finally, the interest offered especially from holometabolic species should be emphasized; the existence of the larval stage, where the early imaginal discs represent determined but not differentiated cell groups, permits the study *in vitro* of problems of development and differentiation.

III. CULTURING INSECT CELLS AND TISSUES

It is not the aim of the present contribution to report on the techniques now in use for culturing insect cells *in vitro*. It may however be useful to mention the basic lines which have proved the most successful.

The insect orders which are promising for genetic purposes are Lepidoptera, Orthoptera, and Diptera. The last, from the point of view of genetical research, is the most important.

The material utilized for culturing is obtained from (a) embryonic cells and imaginal discs, (b) pupal and larval tissues, and (c) hematocytes and ovarian tissues. The most recent information on the whole suggests the superiority of early or late embryonic cells.

The Diptera genera which have given the best results are *Aedes* (more than one species), *Culex* (different species), *Culiseta*, *Musca*, *Glossina*, *Anopheles*, and *Drosophila melanogaster*. Established cell lines are known for *Aedes aegypti* and *A. albopictus*, *Culex quinquefasciata*, *Anopheles stephensi*, and *Drosophila melanogaster*. The genera which offer the best perspectives for genetic work are *Aedes* and *Drosophila*. In both cases, the established lines reveal beautiful mitoses, the chromosomes of which are very easy to analyze. In *Aedes* the diploid chromosome number is six, and as is well known, it is eight in *Drosophila melanogaster*. Between *Aedes* and *Drosophila* there are considerable differences. Compared to *Drosophila*, *Aedes* is genetically practically unknown. This is the reason why, at this stage, *Drosophila* offers the most attractive possibilities.

Confining our interest to *Drosophila melanogaster*, we can reckon on two distinct techniques for obtaining established lines and on one for primary cultures. The former two were developed by Horikawa and Fox (1964), Echalier and Ohanessian (1968, 1969, 1970); Echalier *et al.* (1965), and Kakpakov *et al.* (1969). Primary (short-term) cultures can be easily obtained by the same technique employed by Horikawa and Fox. Both techniques start from embryos, i.e., from large collections of embryonate eggs. These are theoretically male and female in equal proportions, hence male and female cells are together in the same population.

At the present time, the Horikawa and Fox technique is employed only for short-term cultures, while that of Echalier and Ohanessian seems to have the more promising future. Cloning is not yet possible with *Drosophila* cells; this is probably the most forward step still to be achieved in established cell lines from the point of view of genetic problems.

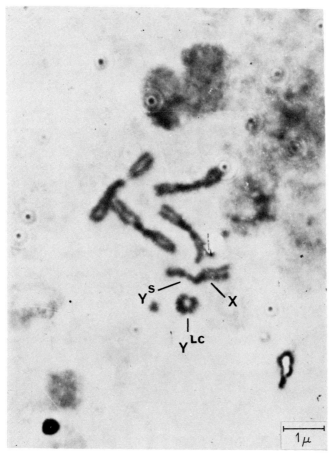

Fig. 1. Metaphase of a male embryonic cell of "Sterilizer." Note the ring chromosome and the short arm of the Y chromosome (Y^S) translocated to the X. Y^{CL}, long arm ring Y chromosome. (Unpublished. Courtesy of Miss C. Halfer; orcein.)

IV. GENETIC PROBLEMS STUDIED BY MEANS OF TISSUE CULTURE

As has been pointed out above, we shall deal mainly with *Drosophila*. The research lines regarding which some relevant results have been obtained are the following: (a) chromosome structure differentiation between eu- and heterochromatin and DNA replication, (b) chromosome mechanisms, (c) genome dynamics in continuous cell populations, (d) gene action and cell differentiation, and (e) genetic problems of cell–virus systems.

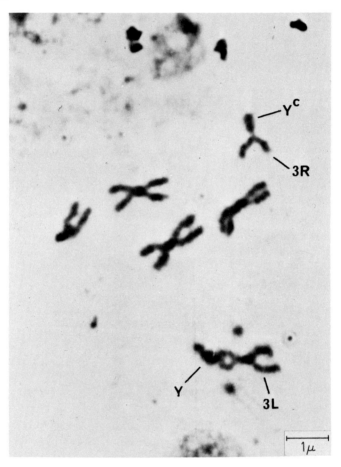

FIG. 2. Metaphase of a male embryonic cell of T (Y;3) P80. Note the centric portion of the Y chromosome (Y^c) translocated to the a portion of the right arm of the third pair and the acentric portion of the Y translocated to the main part of the third pair. 3R, right arm third chromosome; 3L, left arm third chromosome. (Unpublished. Courtesy of Miss C. Halfer; orcein.)

A. Chromosome Structure. Differentiation in Eu- and Heterochromatin and DNA Replication

Culturing *in vitro* of isolated cells (e.g., embryonic cells) is a powerful tool for determining the chromosome structure. In *Drosophila* species the main features of the chromosomal structure have been known since the beginning of cytogenetics. Very small structural details, however, are seldom detected with mitotic chromosomes. Evidence of this is the rather poor cytological knowledge of many aberrations (translocations, for instance) very frequently used in genetic research. Short-term cultures

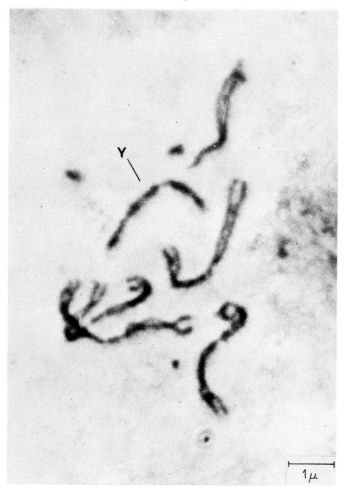

FIG. 3. Metaphase of a male embryonic cell of wild stock Varese. Y, Y chromosome. (Unpublished. Courtesy of Miss C. Halfer; orcein.)

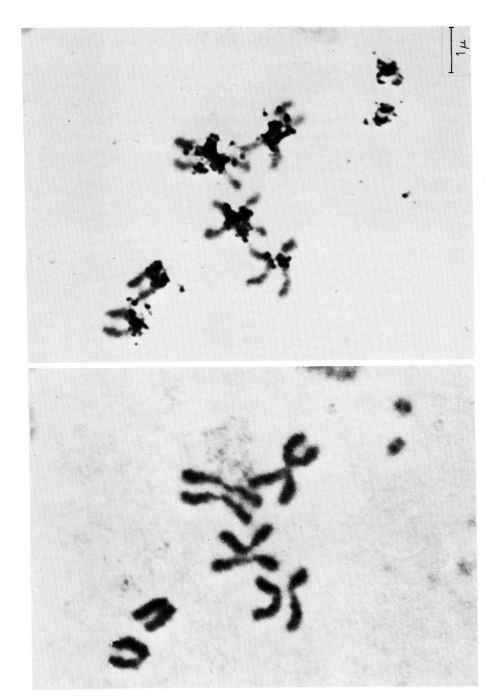

are sufficient to obtain a number of mitoses where chromosome rearrangements, ring chromosomes, etc., are unambiguously observable. A good example is provided by the work done by Halfer *et al.* (1969) on some translocations. Two of them are represented here ["Sterilizer" and T(Y;3)P/80] and were previously known through salivaries and genetic data only (Figs. 1 and 2). A second example (Halfer *et al.*, 1969) is that of the Y chromosome. This element was recognized by Cooper (1959) as composed by five blocks, two blocks belonging to the short arm and three to the long arm. Culturing of embryonic cells (Halfer *et al.*, 1969) shows these structures very clearly (Fig. 3).

Euchromatin and heterochromatin can be distinguished well in mitotic chromosomes of cultured cells, simply as a result of heteropycnosis. An additional criterion is provided by autoradiography after tritiated thymidine incorporation. In the normal chromosome set heterochromatin replicates later than euchromatin, as is generally the case (Lima-de-Faria and Jaworska, 1968). This fact is demonstrated in Fig. 4, which represents an early anaphase of a female cell. The silver grains (right) are not uniformly distributed along the chromosomal axes but restricted to the heterochromatic sections, i.e., to the primary constrictions and to the heterochromatic section of the X. Male cells show, additionally, that the whole Y is also late replicating (Barigozzi *et al.*, 1966, 1967). The possibilities, offered by *Drosophila*, of a deeper analysis of the DNA replication time are very considerable. The use of translocations involving heterochromatin permits one to attack the problem of how late replication is influenced by the position taken by the heterochromatin involved. Halfer *et al.* (1969) have shown (Figs. 5 and 6) that the translocated pieces behave in a manner interpretable as a position effect although heterochromatin remains later replicating than euchromatin. This line of work will lead to identification of the factor controlling DNA replication within a chromosome.

All the problems related to heterochromatin, it should be emphasized, find in mitotic chromosomes the best conditions to be studied in Diptera, since the polytene chromosomes fail to allow a fine analysis of the heterochromatin since this is clumped in the chromocenter.

B. Chromosomal Mechanisms *in Vitro*

In both primary cultures and established cell lines, mitoses are frequent. Therefore, it is easy to detect, by means of them, a variety of structural

FIG. 4. Early anaphase of a normal female embryonic cell of a stock carrying in males a translocation from Y to the second chromosome. Right: autoradiograph. (Unpublished. Courtesy of Miss C. Halfer; orcein.)

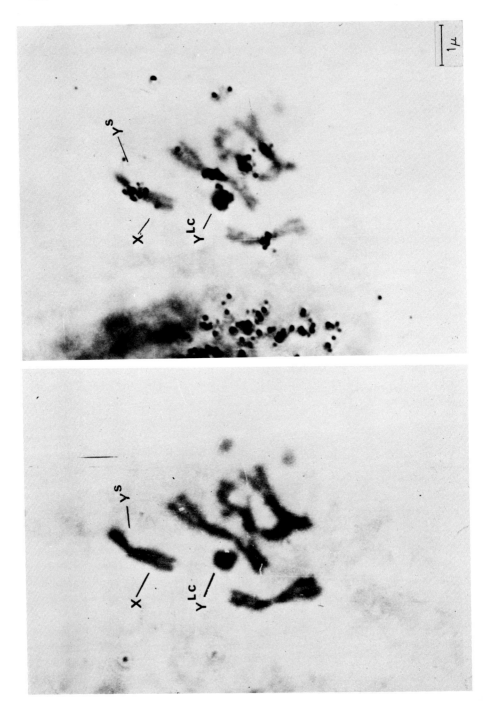

and numerical aberrations of the chromosome set. The mitosis has a typical average duration time, which depends on the culture conditions. In primary cultures of wild stock (Dolfini and Tiepolo, 1968) G_1 lasts, on the average, 2.5 hours, S lasts 21.25 hours, while G_2 has a very short duration (less than one hour). In a established cell line supplied by Echalier and Ohanessian, Dolfini et al. (1971) found different times: G_1, 1.8 hours; 1, up to 10 hours; and G_2, up to 7 hours. The differences indicate that the established line is in better condition than the primary cultures.

Even in primary cultures, chromosome aberrations appear very early. They consist mainly in losses or increases regarding the first and the fourth pair of chromosomes (Dolfini and Gottardi, 1966). Similar results have been found by Ottaviano Gottardi (1967) in cultures where cell concentration was very high. In this case chromosome losses predominate; hence, from 40 to 184 hours the modal number decreases from 8 to 7. The second and the third chromosome pairs are only exceptionally affected. The number of polyploid cells is very low, thus the hyperploid cells do not seem to derive from chromosome losses in polyploid cells. In short-term cultures it is possible that the aberrant chromosome numbers reflect conditions already present in the embryo. This explanation does not certainly hold for established lines.

An interesting problem in Diptera is that of the origin of polyteny, and of the relationship between polyploidy and polyteny. A number of polyploid cells capable of undergoing mitosis is already present in the six-hour-old embryo (Figs. 7 and 8). In the established cell line obtained by Echalier and Ohanessian and under investigation (Dolfini, personal communication) tetraploidy is present in 10–15% of the cells.

In the established cell line K_C a total loss of one of the fourth chromosomes has taken place, as well as a shortening of the long arm of the Y chromosome. Since the heterochromatic portion of the X seems to be longer than normal, a translocation between Y and X can be postulated. These and other possible cytological markers make this cell line very interesting for further work (Dolfini, personal communication). A particular form of polyploidization is cell fusion. This phenomenon, described as spontaneous and induced in several cell lines of different species, leads to the formation of allopolyploid cells where two genomes could indefinitely perpetuate or by Echalier and Ohanessian and under investigation (Dolfini, personal communication) has found a few cases

FIG. 5. Metaphase of "Sterilizer" (see Fig. 1). Right: the autoradiograph showing that the Y^S is earlier replicating than Y^{LC} (the ring-shaped portion of the Y). (Halfer et al., 1969; orcein.)

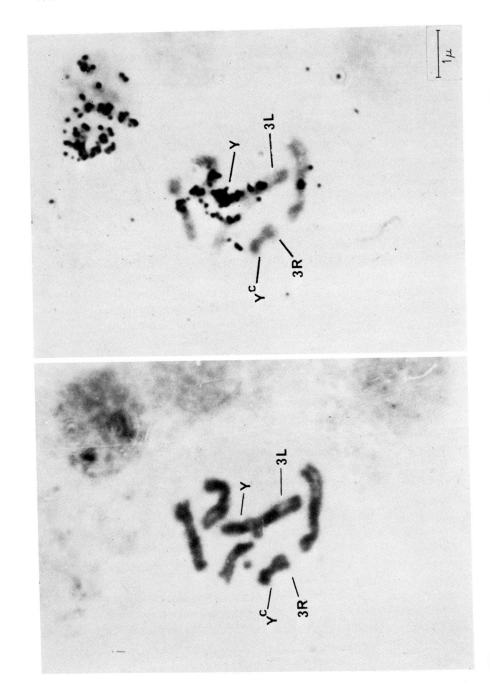

of cells which have, in the same nucleus, male and female chromosomes (XXXY) with the characteristics typical of the K_C strain cells. The possibility of spontaneous fusion is therefore demonstrated in *Drosophila*. Hybrid cells followed by decrease in chromosome number or by reduction to haploidy are of great interest for analyzing the genome in the somatic cell.

C. Genome Dynamics in Continuous Cell Populations

Every cell population, when genetically heterogeneous, is submitted to forces which tend to change or to preserve the original genetic composition. Established cell lines are an excellent material for the study of these phenomena in a model which is simplified by the fact that, by and large (disregarding fusion), cells behave as asexual individuals.

Cytological markers are an excellent means to check whether different genotypes are differently selected in the population and how strong is the importance of sampling through subculturing in respect with genotypic diversification.

In established lines of *Drosophila* originated from embryonic cells, a particular condition exists, irrespective of other chromosomal markers, i.e., that there is a theoretical 1:1 ratio of XX and XY cells. The situation already pointed out for K_C line is the only one, up to now, which gives us some indication of what can happen in a continuous culture with many subculturing steps. The different genotypes identifiable are:

XY (shortened Y)
XX (normal)
XX (one with probably and extra heterochromatic piece)
XO
(tetraploid cells)

The frequency variation from slide to slide, each being a sample of a subculture, are remarkably high; the range goes from cases having over 60% of XY cells to cases having over 60% of XX cells. Tetraploids are around 5% (Dolfini and Tiepolo, 1968). Clearly, this variation is largely due to sampling. Whether or not selection also plays a role, it is not yet possible to estimate.

Even though *Drosophila* has so far supplied the only data regarding these problems, it must be emphasized that much is to be hoped for from *Aedes*, where isoenzyme patterns have also been identified (Green, 1971).

FIG. 6. Metaphase of T (Y;3) P80 (see Fig. 2). Right: autoradiograph showing that Y^c is earlier replicating than the acentric portion of the Y. 3R, right arm third chromosome. 3L, left arm third chromosome. (Unpublished. Courtesy of Miss C. Halfer; orcein.)

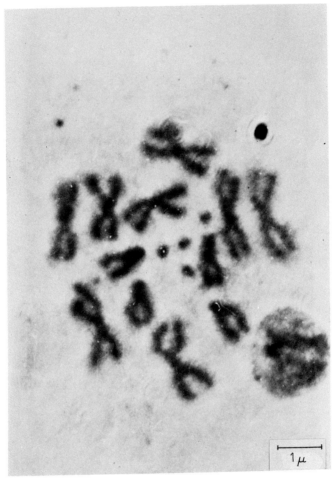

Fig. 7. Tetraploid female metaphase of a short-term culture of T (Y;2) B/b stock. (Unpublished. Courtesy of Miss C. Halfer; orcein.)

D. Gene Action. Differentiation and Multiplication *in Vitro*

The utilization of biochemical genetic characters for the study of gene effect at the level of individual cells or as an overall phenomenon in a cell population is doubtless very promising. Only one investigation has so far been carried out by Horikawa *et al.* (1967); the second approach was pursued. Three enzymes were considered (alkaline phosphatase, alcohol dehydrogenase, and xanthin dehydrogenase), using one wild stock (Oregon) and two mutants (rosy² and maroon-like), both lacking xanthin dehydrogenase.

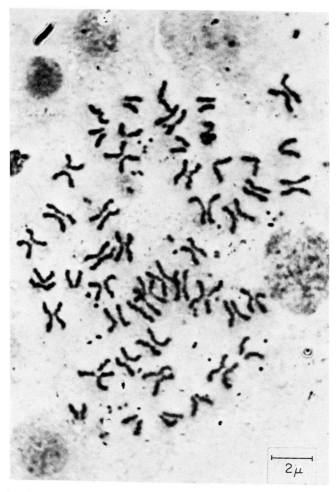

FIG. 8. Anaphase of a 16-ploid embryonic cell of T (Y:3) P80 stock. Short-term culture. (Unpublished. Courtesy of Miss C. Halfer; orcein.)

Each enzyme behaved differently when challenged with an appropriate substrate, but alkaline phosphatase remained constant in the cultured cells, alcohol dehydrogenase increased, and xanthine dehydrogenase decreased.

The study of gene action in relationship to development can take advantage of imaginal discs which have been submitted to development *in vitro* for many years.

The first successful researches were carried out by Fischer and Gottschewski (1939), who cultured isolated eye anlagen together with cephalic ganglia from late pupae. Development was normal including pigmentation. Concerning the latter, cinnabar proved autonomous, while vermilion. and lozenge required an addition of vermilion substance to produce pigmentation. Positive results, as far as development *in vitro* is concerned, have been obtained culturing discs (Demal, 1955), wing discs of both wild and vestigial flies (Kuroda, 1959), and eye–antennal complexes of *D. melanogaster* and *D. virilis*, according to Kuroda and Yamaguchi (1956) and Schneider (1963, 1964).

There is, thus, the possibility of studying the differentiation of the discs of different organs *in vitro*. Especially the use of appropriate experimental devices to interfere with the development of mutant patterns seems to offer fruitful lines of work.

A related line is that of comparing the growth curve of cultured cells derived from different inbred stocks. Differences in curve shapes indicate that the ability to multiply *in vitro* is genotypically controlled. Findings of this type have been obtained by Castiglioni and Rezzonico Raimondi (1963) using nervous and lymph cells; more recently this has been confirmed by Rezzonico Raimondi and Gottardi (1967) using embryonic cells. Interstock hybrids give intermediate results or show a certain degree of dominance of one of the parental stocks (Rezzonico Raimondi *et al.*, 1964). Similar results, showing a genetic control upon growth *in vitro* of interstock hybrids have been found by Ottaviano Gottardi (1969), who was also able to demonstrate a maternal effect in reciprocal crosses.

E. Genetic Problems of Cell–Virus Systems

Although the relationship between arbovirus and cultured cells of insects pose many problems of the greatest interest, the only case known so far where genetic problems have been considered is that of the CO_2 sensitivity factor, known as σ and generally considered as a viral agent. Work on this interesting and still puzzling phenomenon has been carried out, until recently, using the technique of injecting σ *in vitro* (for reference, see L'Héritier, 1962). Only very recently Ohanessian (1971) has taken advantage of the availability of continuous cultures for the study of the multiplication of σ in cells in different genotypes.

V. GENERAL CONCLUSIONS AND PERSPECTIVES

After many years of attempts to grow cells and tissues of invertebrates *in vitro*, some successes can be recorded. Since the aim of the present contribution is confined to the study of genetic problems by means of *in vitro* cultures, we have taken into consideration only those groups which are genetically already well known or those whose cytological background seems to be particularly promising. This is why the whole chapter has been centered mainly on *Drosophila.*

If we try a balance between the problems which have been, or are about to be, resolved by means of tissue and cell cultures of vertebrates and what has been found so far with invertebrates, we find that most of these problems have also been studied in insects (especially in *Drosophila*) although the results are still only preliminary and, in some cases, merely indicative.

The fine analysis of chromosomes can now easily be carried out with insects (especially Diptera with their low chromosome numbers) as well as with vertebrates. A special advantage is given by the large number of well-mapped genes in *Drosophila* and in the possibility of utilizing its numberless chromosome aberrations. The anomalous chromosomal mechanisms which are responsible for the diversification of chromosome sets *in vitro* are likely to become better known in insects than in mammals. Gene action and the genetic background of cell differentiation are also problems which have a better prospect of being solved with these organisms than with vertebrates. The turning point in invertebrate tissue and cell culturing is undoubtedly the discovery of how to obtain continuous cell lines. This achievement, which is one of the most recent in this field, has so far provided only a minimal part of the data that it will certainly provide in the near future.

It is easy to predict a brilliant development of the work which will be done with *Drosophila,* but it does not follow that we should discard other species or groups, which may become important. The high chromosome number of Lepidoptera may be a limitation in some aspects but not all. Therefore, the silk worm has good prospects of becoming a fashionable species. Among Diptera, the genus *Aedes* is likely to become excellent material, and among other orders, Orthoptera and Hemiptera seem to be fertile fields for investigation.

REFERENCES

Barigozzi, C., Dolfini, S., Fraccaro, M., Rezzonico Raimondi, G., and Tiepolo, L. (1966). *Exp. Cell Res.* **43**, 231.
Barigozzi, C., Dolfini, S., Fraccaro, M., Halfer, C., Rezzonico Raimondi, G., and Tiepolo, L. (1967). *Atti Ass. Genet. Ital.* **12**, 291.
Castiglioni, M. C., and Rezzonico Raimondi, G. (1963). *Experientia* **19**, 527.
Cooper, K. W. (1959). *Chromosoma* **10**, 535.
Demal, J. (1955). *Bull. Acad. Roy. Med. Belg. Cl. Sci.* [5] **41**, 1061.
Dolfini, S., and Gottardi, A. (1966). *Experientia* **22**, 144.

Dolfini, S., and Tiepolo, L. (1968). *Proc. Int. Colloq. Invertebr. Tissue Cult., 2nd, 1967* p. 182.

Dolfini, S., Tiepolo, L., and Courgeon, A. (1971). *Experientia* **22**, 234.

Echalier, G., and Ohanessian, A. (1968). *Proc. Int. Colloq. Invertebr. Tissue Cult., 2nd, 1967* p. 174.

Echalier, G., and Ohanessian, A. (1969). *C. R. Acad. Sci.* **268**, 1771.

Echalier, G., and Ohanessian, A. (1970). *In Vitro* **6**, 162.

Echalier, G., Ohanessian, A., and Brun, G. (1965). *C. R. Acad. Sci.* **261**, 3211.

Fischer, I., and Gottschewski, G. (1939). *Naturwissenschaften* **27**, 391.

Green, A. E., and Charney, J. (1971). *Curr. Topics Microbiol. Immunol.* **55**, 51.

Halfer, C., Tiepolo, L., Barigozzi, C., and Fraccaro, M. (1969). *Chromosoma* **27**, 395.

Harris, R. J. C. (1964). "Cytogenetics of Cells in Culture." Academic Press, New York.

Horikawa, M., and Fox, A. S. (1964). *Science* **145**, 1437.

Horikawa, M., Ling, L. N., and Fox, A. S. (1967). *Genetics* **55**, 569.

Kakpakov, V. T., Gvozdev, V. A., Platova, T. P., and Polukarova, L. G. (1969). *Genetika (USSR)* **5**, 67.

Krooth, R. S. (1964). "Somatic Cell Genetics." Univ. of Michigan Press, Ann Arbor, Michigan.

Kuroda, Y. (1959). *Med. J. Osaka Univ.* **10**, 1.

Kuroda, Y., and Yamaguchi, K. (1956). *Jap. J. Genet.* **31**, 98.

L'Héritier, P. (1962). *Ann. Inst. Pasteur, Paris* **102**, 511.

Lima de Faria, A., and Jaworska, H. (1968). *Nature (London)* **217**, 183.

Ohanessian, A. (1971). *Curr. Topics Microbiol. Immunol.* **55**, 230.

Ottaviano Gottardi, A. (1967). *Proc. Int. Colloq. Invertebr. Tissue Cult., 2nd, 1967* p. 189.

Ottaviano Gottardi, A. (1969). *Atti Accad. Naz. Lincei, Cl. Sci. Fis., Mat. Natur., Rend.* **47**, 389.

Rezzonico Raimondi, G., and Gottardi, A. (1967). *J. Insect Physiol.* **13**, 523.

Rezzonico Raimondi, G., Ghini, C., and Dolfini, S. (1964). *Experientia* **20**, 1.

Schneider, I. (1963). *Genetics* **48**, 908.

Schneider, I. (1964). *J. Exp. Zool.* **156**, 91.

Siniscalco, M. (1970). *Atti Ass. Genet. Ital.* **15**, 3.

5

INVERTEBRATE ORGAN CULTURE IN HORMONAL RESEARCH

Josette Berreur-Bonnenfant

I. INTRODUCTION

The object of *in vitro* culture is to withdraw the explant from the complex influences of the fluid body in order to try to elucidate the conditions in which the isolated phenomenon occurs.

181

This technique particularly makes it possible to study the endocrine effects. Thus, it can be checked whether the effect of a hormone upon a known effector is direct or indirect and whether an intermediary organ is necessary or not.

Two kinds of cultures can be used for these researches: (a) cultures of cells and (b) cultures of organs.

Although the culture of cells has been taken as a basis for the study of various biological phenomena, the technique of organotypic cultures made it possible to reproduce the phenomena observed *in vivo* with more accuracy.

Organ culture does not alter the structure of explant. Cohesion between the cells and organ tissues is maintained. A balance is established which avoids some elements to be favored at the expense of others. For these reasons, the culture of organs is propitious to the study of hormonal effects.

In vitro culture of organs of invertebrates shows some delay compared to the studies made in the same field for the group of vertebrates. Most of the studies which will be referred to herein have been made during these latest years. This delay must mainly be due to technical difficulties: The great diversity of the species of invertebrates makes it difficult to find a culture medium adapted to each case, the constitution of the body fluid varying a great deal for the various groups and species. Another most important difficulty is to obtain the sterility of cultures because almost all invertebrates are germ carriers.

Researches made on invertebrates show that organotypic culture may help to solve many problems in which hormones participate as in vertebrates. These problems are analyzed below under two main paragraphs. We shall successively look at the researches which concern the development of organs and those which are related to the realization of the sex.

II. ENDOCRINE FACTORS AND MORPHOGENESIS IN POST-EMBRYONIC DEVELOPMENT AND REGENERATION

The methods of organotypic culture have been utilized to analyze the post-embryonic development in Diptera and the regeneration in Blattopteroidea.

A. Morphogenesis in Post-Embryonic Development

During the metamorphosis of Diptera and Blattopteroidea, organs named imaginal discs, which up to then remain at an anlage stage, grow and differentiate; they reach their full development in the adult or the imago. This slow phase of growth of some organs constitutes a post-embryonic development.

The post-embryonic development of insects is controlled by a complex hormonal mechanism of which the reader will find a concise analysis in Wigglesworth's reviews (1964) and Gilbert's (1964) or a more detailed one in those of Novak (1966).

The molting hormone, secreted by the prothoracic gland, has a lot to do in this control. In Diptera the prothoracic gland constitutes a part of the ring gland (or Weismann ring), which organ forms a ring round the dorsal vessel, above the brain. Other glandular tissues exist in the ring gland, i.e., the corpus allatum (which the juvenile hormone is probably secreted by) and the corpus cardiacum; they are associated with pro-thoracic cells. It results from this that the endocrine interpretation of the experiments on the transplantation of the whole ring gland associated with the brain is not always as accurate as one may wish.

1. Studies Concerning *Calliphora*

The first experiments showing the possible action of hormonal factors during *in vitro* differentiation of imaginal discs were probably carried out by Frew (1928). Frew cultured imaginal discs of larvae legs of *Calliphora* in a drop of lymph filtered on collodion. He showed that an evagination of the anlage of the leg comparable with the one which occurs during nymphosis can be realized *in vitro* with the presence of pupal lymph. Evagination does not occur in the larval lymph. This allows one to think that at the beginning of metamorphosis, the hemolymph of Diptera carries a factor favorable to the unfolding of imaginal anlages.

Demal (1956) and Demal and Leloup (1963) undertook to study some aspects of the metamorphosis of Diptera while using the method of organotypic cultures. Their researches needed the setting up of techniques adapted to the material which was being studied. They realized an aseptic breeding of larvae of *Calliphora* and, so, eliminated the risks of infection of the culture medium. Such risks were heavy when the animals studied had lived on meat. Finally, they imagined and compared several culture media. Part of the researches made by Demal and Demal and Leloup aimed at the *in vitro* study of the differentiation of imaginal discs of legs.

The imaginal discs of *Calliphora* legs are taken when the larva which is still whitish comes to a standstill in order to form its puparium. They are cultured as hanging drops in a culture medium containing chick embryo extract and a pupa extract. In this medium the imaginal anlages go on living for three or four days. As from the first hours of the culture and during the first two days, the shape of the leg anlage is altered: The segments devaginate as they do at the beginning of nymphosis. However, cellular reproductions do not appear *in vitro* and the wall of the leg is thinner than the normal animal's. Beyond six days necrosis spreads on the organs explanted. Similar observations were made when discs put in culture had been taken from animals which had been empupated for four or five hours, and in all the cases, it was possible to start the differentiation *in vitro*. Pupa extracts help this differentiation, but they cannot start mitosis.

2. Studies Concerning *Drosophila*

Essential characteristics of post-embryonic development of *Drosophila* are the same as those of *Calliphora,* and many studies have been devoted to the *in vitro* differentiation of the eye–antennal anlage.

Kuroda and Yamaguchi (1956) had in view to study the genetic determinism of the differentiation of the eye anlage. In organotypic culture they could hope to place the organ studied in conditions easier to be checked than is possible *in vivo*. Their studies consist in comparing the development of the eye–antennal anlages coming from normal larvae with the development of anlages of Bar mutants. (In these latter animals the antenna is normal but the imago's eye is reduced to a vertical narrow band.)

The eye–antennal discs and the cephalic complexes are taken out of third instar larvae 95 hours after hatching, then cultured in hanging drops of a synthetic medium by the cover slip technique.

When the eye–antennal discs of normal larva associate together with the cephalic complex of normal larva in the same preparation, the eye disc increases in size and reaches a normal differentiation. When the eye–antennal disc of normal larva is cultured together with the cephalic complex of Bar, it performs the development of the eye disc of Bar larva.

Inversely, the eye discs of the Bar mutants explanted in presence of a normal cephalic complex take the characteristics of a normal eye.

Kuroda and Yamaguchi realized many associations of this kind. They came to the conclusion that imaginal eye–antennal discs have the same possibilities, either coming from larvae of normal *Drosophila* or coming from Bar mutants. Normal animals are different from mutants because of their cephalic complexes, and the character (Bar or normal) of the eyes might be determined by this cephalic complex. The specific activity of these different cephalic complexes might be implied to differences in hormonal secretions.*

---

* In order to compare, here are the results obtained when similar associations were realized *in vivo* (Steinberg, 1941, 1943). Eye discs of 24-hour larvae of the

Carrying on these researches, Horikawa (1958, 1960) experiments on several mutants characterized by an outstanding aspect of the adult eye. Working in a liquid medium for 72 hours, he puts the eye–antennal anlage in the presence of a great number of cephalic complexes. He so points out that 10 cephalic complexes are necessary to ensure a large growth of the eye discs. Cephalic complexes in greater number bring death after 48 hours.

In the same experimental conditions he associates eye–antennal imaginal discs taken from normal animals or from mutants to 10, 20, 50 cephalic complexes coming from animals of other strains. A great number of associations of this kind have been realized. It is interesting to compare some of them to those of Kuroda and Yamaguchi's described above. According to Horikawa, when eye–antennal discs of normal larvae are cultured with 10 cephalic complexes of Bar mutants, the growth and the differentiation of these discs are normal. On the contrary, when the eye–antennal discs of normal larvae are associated with 20 or 50 cephalic complexes of Bar mutants, no growth or differentiation of discs are to be discovered.

From his experiments, Horikawa deducts that the molting hormone secreted from the cephalic complex of Bar mutants (or other mutants he analyzed) seems to be of a quality different from the one produced by normal animals; moreover, he shows that the molting hormone seems to be produced in various quantities by the different mutants analyzed. Finally, he considers that the genes control the differentiation of the eye anlage by working on the level of the cephalic complex as well as on the eye–antennal discs.

A third series of experiments devoted to the determinism of the *in vitro* differentiation of the eye–antennal anlage was realized by Fujio (1960, 1962). Like Horikawa, Fujio first juxtaposed the eye–antennal anlage with a large number of cephalic complexes in a synthetic medium. Later on, Fujio does not juxtapose the organs anymore; he puts the discs in the medium containing the substances secreted from the cephalic complexes which were previously cultured in rotating tubes.

Thus, he studies the effects of the substances secreted from the cephalic complexes of various strains on their own eye–antennal discs, the effects of these substances on the eye–antennal discs of normal animals, and the effects of the substances secreted from the cephalic complexes of normal larvae on the eye–antennal discs of various strains.

Bar mutant were implanted in larvae of wild type. When developing, the eye anlage kept the Bar type. Inversely, the eye anlages of normal larvae implanted in Bar larvae kept a normal aspect.

Fujio observes that the substances favorable to the growth of eye–antennal discs elaborated in the normal larvae brains can spread in the culture medium and work in the absence of cephalic complexes.

These substances would be of several types, i.e., the cephalic complexes of normal larvae would work only by generation of molting hormone. The cephalic complexes of Bar mutants would elaborate substances which would have specific effects, different from those of molting hormone, on the growth and the differentiation of the imaginal discs of Bar mutants. These substances would be accountable for the characteristic shape of the eye.

Gottschewski (1958, 1960) elaborated a complex synthetic medium in which he cultured imaginal discs of *Drosophila*. His studies are essentially devoted to the morphogenesis of the eye. He studies, *in vitro*, the differentiation of ommatidies, the coming out of pigments, and the constitution of post-retinal nervous fibers when connections between the eye disc and the brain bulb have been broken. In the course of these studies, he thus described the constitution of the eye anlage of Bar mutants, Eyeless, Lobe, Glars, Lozenge, and Rough, in organotypic culture. Gottschewski and Querner (1961) ascertain that eye discs taken out at the end of third larval instar can grow *in vitro*, even though they advance little if they come from younger animals (second larval instar or beginning of third larval instar).

The young discs which have kept their normal association with their respective cephalic complexes behave in the same way. Gottschewski and Querner use fluorescent molecules in order to find whether there is, in young animals, a barrier which can stop the spreading of substances produced by the cephalic complex. These molecules injected into a certain point of the explant can spread in the organ studied and pass from the cephalic complex to the disc. In the young instars no barrier seems to stop the substances from going through the explanted organs.

It is also the differentiation of eye–antennal anlages that Schneider (1964, 1966) analyzed *in vitro*. His studies are devoted to *Drosophila melanogaster* and *Drosophila virilis*. Organs taken from larvae of third larval instar (95 hours for *D. melanogaster*, 146 hours for *D. virilis*) are cultured in a synthetic medium less complex than Gottschewski's. Sometimes, various embryonic extracts, mainly extracts from prepupa of *Drosophila*, are added to this medium. Day after day, organs cultured are fixed, cut, and studied histologically, the analysis taking place within 4–48 hours of culture.

Schneider ascertains that all organs do not reach the same state of development in organotypic culture. The antennal anlages undergo the most elaborate evolution *in vitro*. A tripartite antenna, comparable to the imago's, appears in 70% of the explants. Eyes and brain have a more limited development, and in only 10% of the cases, eye anlages are at the base of an eye normally constituted and pigmented.

Three factors might explain these different reactions *in vitro,* i.e., lesions, more or less serious, occurring during dissection, sensitiveness of organs to adverse conditions in an artificial medium, and finally, inherent capacity for autonomous development. The capacity for development of an organ can be characterized if a sequence of development occurs *in vivo* within the 24 hours following the formation of the pupa. This same sequence appears in culture with no delay noticeable. Later on, three- to nine-day delays occur.

The author explains these observations by assuming that a newly explanted organ is in possession of sufficient hormonal and nutritious reserves to keep on developing normally during the first instars of culture. Later on, delays could be due to a deficiency of the medium in nutritious factors or to a bad secretion of cephalic hormone.

Considering the fact that the Diptera studied in these researches belong to groups which are sufficiently similar, comparisons can be contemplated among the results obtained. In most cases, the same conclusion appears. In some given conditions, cephalic complex helps the culture of imaginal anlages. Naturally, this effect is mostly implied to the molting hormone secreted from the ring gland. However, in many cases, the factors, coming from cephalic complex and which participate in these experiments have not been identified. This doubt is explainable when one thinks that the cephalic complex of insects has several endocrine activities and that in the cephalic complex of superior Diptera this character is highly prevailing. The corpus cardiacum, the corpus allatum (origin of juvenile hormone), and the peritracheal gland (origin of the molting hormone) are associated in the cephalic complex under the shape of a ring gland. When entire cephalic complexes are introduced into a culture medium, they bring many endocrine factors, and in these conditions the method of organotypic cultures loses the main interest which it normally has in endocrinology, i.e., it no longer permits a strict checkup of hormonal conditions.

More recently, researches have undertaken to try to specify the action of the growth and molting hormone (ecdysone). However, from the results obtained from various researches, it is not possible to come to a conclusion as to the action of ecdysone *in vitro.*

Oberlander and Fulco (1967), studying *in vitro* metamorphosis of the alar discs of the butterfly *Galleria mellonella,* show that the ecdysone of synthesis added to culture medium promotes growth and partial metamorphosis of these imaginal discs. Nardon and Plantevin (1969), experimenting on the same animal, are unable to bring forth an action of ecdysone on the metamorphosis of the intestine of the last instar larvae.

B. Regeneration

A tissue, an organ, or a set of organs in the course of regeneration have many common points with an embryonic anlage. We have just seen that the development of imaginal anlages of Diptera is under the influence of secretions coming from the cephalic complex. During the regeneration of organs of Insecta previously cut off, similar actions can be pointed out.

Marks and Reinecke (1965) have transplanted, *in vitro*, the stump of regeneration which develops in the *Leucophaea maderae* nympha when the anlage of a leg has just been cut off. In this growing organ they showed that the formation of vesicles and of buds of regeneration is dependent on endocrine phenomena. Two critical periods, of different meaning, characterize the evolution of the organs explanted. One corresponds to tissues which have been explanted before seven days of regeneration have elapsed *in vivo*. These tissues never are the center of the formation of vesicles or of buds, whatever endocrine complex is associated with them in culture. The other period corresponds to tissues which have been explanted after 10 days of regeneration *in vivo* have elapsed. In most cases, they form buds or vesicles of regeneration even if they are not associated with endocrine organs in the culture medium.

Tissues that were allowed to develop from seven to nine days *in vivo* react to an endocrine action in culture. Associated with the brain and with the prothoracic gland, they form buds in 75% of the cultures.

Marks and Reinecke notice that the first coming out of buds on the cultured tissues coincides with the time when, *in vivo*, the prothoracic gland shows its first signs of activity. They also find out that the prothoracic ganglion has the same effect as the brain on the production of vesicles of regeneration. The prothoracic ganglion could mimic the effect of the brain by stimulating the prothoracic gland. These observations are the object of comparisons with the results obtained from another source following the study of the regeneration, *in vivo*, of appendixes of Blattopteroidea and the analysis of the activity of these animals' endocrine glands. The interactions discovered in culture can appear in the synopsis of the endocrine control of regeneration drawn up subsequently to *in vivo* studies.

Marks' recent studies (1968) give more details about the results obtained by Marks and Reinecke (1965). This author made a series of experiments in order to test, *in vitro*, the action of incubates of endocrine glands on the regeneration tissue of the legs of *Leucophaea maderae* E.

Incubates with an increasing number of prothoracic glands produce an increased effect on the tested tissues. The author compares the action of the incubates of prothoracic glands to the action of the 20 hydroxyecdysone, and although results are similar, he does not agree to identify the purified hormone with the substance contained in the incubates of prothoracic glands. Through *in vivo* experiments backed again on the regeneration of the legs of the nymphae of *Leucophaea maderae*, Marks respectively demonstrates the stimulating action of the corpus allatum and the inhibiting action of the brain on the prothoracic glands. The author expresses the hypothesis of the existence of an inhibiting cerebral hormone which would participate in the feedback mechanism which checks the activity of the prothoracic gland.

III. ENDOCRINE FACTORS AND SEXUALITY

We have just seen that *in vitro* culture might make it possible to study the potentialities of differentiation of embryonic organs and to look for factors that can control this differentiation.

The study of gonads is also very interesting. It is a fact that in invertebrates the differentiation of gonads takes place during the whole life of the animal and, moreover, is not always irreversible. Along with gonochoric species, in which sex seems to be determined definitively, there is a spontaneous and normal sexual inversion in many invertebrates. It is the case of hermaphrodite species, in which individuals are male during a certain period of their life, then female for another period.

After having for a long time denied the existence of hormonal factors susceptible of having effects on sexual differentiation and on the maturation of genitalia of invertebrates, it was possible to show these factors in some groups of invertebrates.

Then organotypic culture appeared as a technique giving the possibility of specifying the action of these factors, or even of making them appear in groups in which they only were supposed to be.

In recent years, this problem has been studied by many researchers. Several groups of invertebrates have so been explored, i.e, the Annelida, Mollusca, Crustacea, Insecta, and even the Ascidiacea.

A. Endocrine Factors and Sexuality in Annelida

1. Chaetopoda

In the Polychaeta Annelida a hormonal control of the sexual differentiation was shown *in vivo* (M. Durchon, 1952). In the Annelida, there

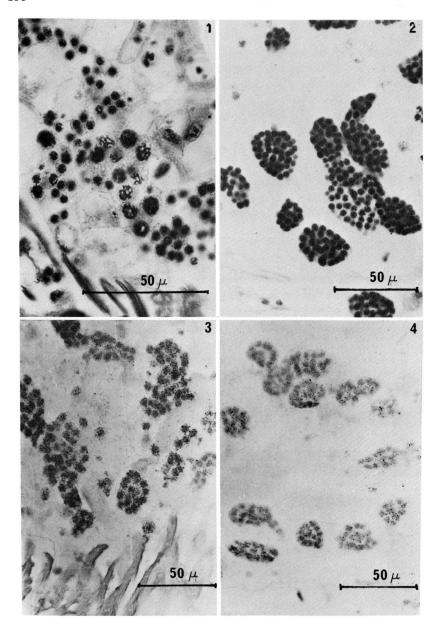

PLATE I

Culture of Male Parapodia of *Nereis diversicolor* (after Durchon)

Fig. 1. Result of culture of male parapod for four days on anhormonal medium. Note maturation divisions.

Fig. 2. Male parapod grown on agar medium with prostomium for 15 days. Spermatogenesis is blocked up.

are no differentiated gonads; sexual tissues formed at the expense of the coelomic wall fall in the common cavity and undergo gametogenesis. In many species of Polychaeta, sexual maturity goes along with a change in the aspect of the animal. Parapodia undergo big morphological transformations: From the atokous form, or immature, the worm goes to the epitokous form, or form of sexual maturity, which is visible by the development of parapodian foliaceous appendixes and a change in the shape and the number of the silks.

After experiments realized *in vivo* M. Durchon (1952) had been induced to think that there was a cerebral hormone secreted at the level of the prostomium; this hormone would inhibit the gametogenesis and the epitoky. In its absence, sexual cells would maturate and the atokous form would be transformed into the epitokous form.

In order to specify the action of the cerebral hormone, Durchon and his assistants realized a culture medium (M. Durchon and Schaller, 1963) which makes it possible to isolate genital tissues from the rest of the animal's body. The part of the body (parapodia) which controls genital tissues is put into culture.

M. Durchon and Schaller (1963) explant parapodia of *Nereis, Perinereis,* and *Platynereis,* either isolated or associated with a prostomium. The study of these parapodia shows that in anhormonal cultures there is an evolution of the parapodia from the atokous form to the epitokous form with differentiation of the typical heteronereidian silks. Besides, in the case of male parapodia, from the fifth day of culture spermatozoa are produced by the nephridian pore. Inside the parapodia many divisions of maturation of spermatocytes are discovered (Pl. I, Fig. 1).

In the presence of prostomium (parapodia and prostomium associated on the same medium rapidly join together) there is no evolution of the parapodia in culture; spermatocytes contained in the parapodia do not advance but keep the same aspect after a month's culture (Pl. I, Fig. 2).

In the female parapodia cultured alone (M. Durchon and Dhainaut, 1964), oocytes increase. Contrarily, when female parapodia are associated with prostomiums, oocytes show only a very slow growth and their structure remains the same as the one they had at the beginning of the culture.

In both sexes, the inhibitory properties of the prostomial hormone are shown in an irrefutable manner.

FIG. 3. Spermatocytes in the grown phase after three days of culture on anhormonal medium with tritiated thymidine (important labeling).

FIG. 4. Spermatocytes cultured for 15 days on a tritiated thymidine medium after two days of culture on an ordinary anhormonal medium (labeling is very slight).

New experiments associating several techniques (*in vitro* culture, historadiography, and electronic microscopy) have made it possible to specify the action of the cerebral hormone at the cellular level.

The historadiographic study (Dhainaut, 1964) of spermatocytes of *Nereis diversicolor* cultured *in vitro* on medium marked with tritiated thymidine, either in the absence or in the presence of cerebral hormone, gave the following results: Spermatocytes examined three days after they were put into culture on tritiated medium in the absence of hormone are in the phase of growth and show heavy labeling (Pl. I, Fig. 3). On the contrary, spermatocytes cultured for two days on anhormonal simple medium and then associated with prostomiums on a medium containing tritiated thymidine, show only little labeling after 15 days of culture. Spermatogenesis is inhibited (Pl. I, Fig. 4).

The cerebral hormone would act as a regulator of the DNA syntheses at the level of the spermatocytes.

In the females of *Nereis diversicolor* this histological examination of the oocytes cultured in the absence of hormone shows that there is increase of stature and storage of ribonucleoprotein fibrillas correlatively with a reject of yolk at the cellular periphery (M. Durchon and Dhainaut, 1964). In the presence of hormone, the structure of oocytes remains unchanged.

M. Durchon and Boilly (1964) ascertain these observations by the study of the ultrastructure of oocytes cultured in the absence of hormone. Correlatively with the formation of RNA, they observe the formation of many ringed lamellae in the perinucleic zone. M. Durchon *et al.* (1965) show that in the presence of cerebral hormone oocytes keep an ultrastructure similar to the one of the oocytes fixed at the beginning of the culture.

In conclusion, in the Annelida, *in vitro* culture has made it possible to analyze the action of the cerebral hormone which had been discovered *in vivo*. At the end of the realized experiments it is possible to assert that it is the cerebral hormone secreted at the prostomium level which is directly responsible for the inhibition of the maturation of genital tissues; the action of this hormone has been investigated at the cellular level, on syntheses of nucleic acids.

2. Hirudinea

In the Hirudinea, thanks to experiments realized *in vivo*, Hagadorn (1966) was able to show that testicular maturation cannot be obtained in the absence of brain.

Malecha (1965, 1967) also studied this problem in *Hirudo medicinalis* by *in vitro* culture. He utilized the medium realized by M. Durchon and Schaller (1963).

Malecha compared two kinds of cultures, i.e., cultures of testes explanted in association with the peripharyngeal nervous mass and cultures of testes explanted alone. The maturation of genital tissue takes place only in the first case. The function of the peripharyngeal nervous mass is confirmed in culture; this nervous mass seems indispensable for the unfolding of spermatogenesis.

Taking into account the results obtained in the Nereidae one may hope that the technique of organotypic cultures will make it possible to specify the action of the hormone secreted by the peripharyngeal nervous mass.

B. Endocrine Factors and Sexuality in Mollusca

Protandrously hermaphroditic (*Calyptraea, Patella*) or simultaneously hermaphroditic (*Helix*) Mollusca have been an object of interest for a long time. After the discovery of hormones in the *invertebrates*, several researchers thought that organotypic culture could make it possible to analyze the mechanisms of the sexual differentiation and of the maturation of the genital products of these animals. Several species are now the object of *in vitro* culture. Streiff studied *Calyptraea sinensis;* Choquet made his researches on *Patella vulgata; Helix aspersa* and *Viviparus viviparus* were studied by Guyard and Gomot; and finally, Durchon and Richard devoted their researches to *Sepia officinalis.*

1. Gastropoda

The Prosobranchiata mollusk *Calyptraea sinensis* is a protandrous hermaphrodite. Young individuals are male. Then, they undergo a phase of sexual inversion to become female during the latter part of their life. When these Mollusca are parasitized, oocytes appear in their gonads even if they are very young individuals which should be males.

Streiff tried to elucidate the mechanism which controls the male and female sexual differentiations of *Calyptraea sinensis*. He elaborated a culture medium (Streiff and Peyre, 1963) which gave him the possibility to study gonads out of their natural medium.

When these gonads are explanted alone, whatever their state may be at the beginning of the culture (nondifferentiated, male, or female gonads) (Pl. II, Figs. 5–8) it is possible to see an invasion of the gonad by female genital tissue.

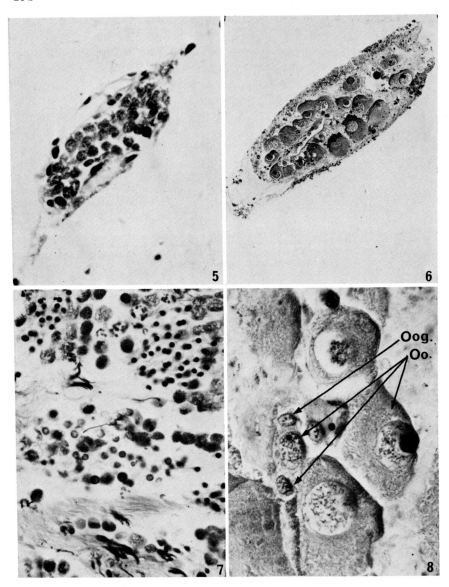

PLATE II

MAINTENANCE AND SEX REVERSAL OF MALE GONADS OF *Calyptraea sinensis*

[Published by W. Streiff in *Ann. Endocrinol.* **28**, 649 (1967)]

FIG. 5. Control gonad (undifferentiated stage).

FIG. 6. Undifferentiated gonad grown on culture medium for 20 days. It is filled with oocytes. There are no spermatogonia or undifferentiated cells.

FIG. 7. Control gonad (male stage). Spermatogenesis is active and oocytes are absent.

Therefore, the female sexual differentiation seems to come about spontaneously without any hormonal influence (Streiff, 1964).

It is not the same with the male sexual differentiation. As we have just seen *in vitro*, gonads of young nondifferentiated individuals become ovaries. Moreover, male sexual tissues degenerate in the male gonads previously well differentiated and explanted alone.

In order that spermatogenesis may be realized normally or that the young gonads may evolve in the male direction, it is necessary to add hemolymph of male to the culture medium.

If the nondifferentiated or male gonads are cultured in association with the central nervous system of a functional male, they realize spermatogenesis (Streiff, 1967b). Spermatogenetic differentiation and maintenance of spermatogenesis seem to be dependent upon a hormonal factor coming from the central nervous system and existing in the hemolymph of functional males.

In *Calyptraea sinensis,* the penis and the seminal groove of males undergo an evolution narrowly linked with the one of the gonad. The genital tract which differentiates in the male direction during the male phase of the individual shows regressive signs at the time of sexual inversion. These signs keep on increasing in the course of the active female phase while female gonoductes differentiate.

Male gonoducts explanted alone on an anhormonal culture medium show no sign of evolution or involution after 212 days of culture. Having no possibility of evolving by themselves toward the male direction or the female one, these gonoductes constitute neutral organs.

To elucidate the determination of differentiation and of retrogression of the male genital tract, some were associated in culture with various secretor organs.

Anlages of immature penes associated with eye tentacles of males differentiate into penes; in the same way, penes already regressed (taken from females) take back their male form if they are cultured in association with eye tentacles of males (Streiff, 1966a).

The evolution of the genital tract in the male direction is, therefore, determined by the eye tentacular.

At the time of sexual inversion the male tract begins its regression. Male genital tracts associated with female nervous systems or cultured on a medium containing hemolymph of female regress (Streiff, 1966a). If one tries to dissociate the action of the various ganglia constituting

FIG. 8. Male gonad grown on anhormonal culture medium for 20 days. Spermatogenesis is stopped. No cells of male lineage remain. Many oocytes appear. Oo, oocytes; Oog, oogonia.

the nervous system (cerebral, pleural, and pedial ganglia), only the pleural ganglia seem to favor the involution of the male genital tract.

A hormonal factor coming from pleural ganglia seems therefore to be responsible for the involution of the male tract during the female phase.

It was also possible to demonstrate that this factor was not characteristic of a species. Cultures realized by indifferently associating pleural ganglia of *Calyptraea sinensis* and pleural ganglia of *Crepidula fornicata* with tracts of males of each of these species show that the male tract degenerates (Lubet and Streiff, 1969), whichever pleural ganglion is associated with it.

As to the female tract, if it is cultured alone on an anhormonal medium, as the male tract, it shows no sign of involution. The presumptive area of this tract does not evoluate on anhormonal medium either. When one adds hemolymph of functional female to the culture medium, no evolution of the anlage towards the female direction is noticed. On the contrary, if one adds hemolymph of an animal in sexual inversion phase to this medium, the anlage of the tract differentiate towards the female direction. Results are the same if, instead of hemolymph, the nervous system of an animal in sexual inversion is added (Streiff, 1967a).

Genital tracts excepted, sexual differentiation in *Calyptraea sinensis* seems to be controlled by a single hormone coming from the central nervous system. In the presence of this hormone, individuals have gonads where spermatogenesis takes place; in its absence, gonads are invaded by ovogenesis. The mechanism which brings the secretion of hormone to a stop at the time of sexual inversion has not been shown.

Patella vulgata, another protandrously hermaphroditic mollusk, seems to evolve in a different manner. Its sexual differentiation does not seem to be dependent upon the same factors as *Calyptraea sinensis.*

It was possible to study gametogenesis of *Patella in vitro* thanks to the elaboration of a culture method on a solid medium (Choquet, 1964). Gonads of *Patella* cultured alone survive on an anhormonal medium; gonia of female gonads develop up to previtellogenesis. In the male gonads, spermatogenesis takes place; it starts again a month after the elimination of gametes. Whereas, in nature, there is always a long rest between two successive gametogenesis.

If gonads and cerebral ganglia of the same male are associated in culture, spermatogenesis is faster than in cultures of isolated gonads. Cerebral ganglia of the male would stimulate the mitotic activity of the cells of the male lineage. On the other hand, if gonad and tentacles of the same male are associated there is no genital activity in the gonad cultured. It is the same if the tentacular complex, cerebral ganglion of a male during a period of genital rest, is associated with a male gonad. Therefore, the period of genital rest would appear to be dependent upon an inhibitory factor of gametogenesis coming from the tentacles (Choquet, 1965).

On the contrary, the stimulation of spermatogenesis seems to come from cerebral ganglia. Choquet arrives at this conclusion by comparing cultures of isolated testes coming from male period of sexual activity, cultures joining together testis and cerebral ganglia, and finally, cultures in which testis and cerebral tentacles–ganglion complex belonging to individuals in sexual action are associated. In these associations spermatogenesis takes place provided there are cerebral ganglia. Tentacles of the male in sexual activity do not stop mitoses if cerebral ganglia are included in the association. However, in the absence of cerebral ganglia, tentacles show an inhibitory activity (Choquet, 1967).

Choquet concludes that the annual genital cycle of *Patella vulgata* during the male phase is under the influence of factors of tentacular and cerebral origin. The stimulating factor coming from cerebral ganglia is preponderant at the time of gametogenesis, whereas, the inhibitory factor coming from the tentacle ensures the period of genital rest.

Helix aspersa is an hermaphroditic mollusk in which male and female gonad gonocytes coexist in every season. However, the snail is a protandrous hermaphrodite since in young individuals germinal cells which differentiate first are spermatogonia; the first oocytes appear later on only.

In order to define the factors which participate in changing the direction of the differentiation of the germinal epithelium, Guyard and Gomot (1964) elaborate a method of *in vitro* culture. They can thus show that, in the ovotestis cultures on nutritious medium, only oocytes keep on growing, whereas the evolution of the male lineage is stopped at the primary spermatocyte instead.

If hemolymph of adult snail is added to the culture medium, one can observe, at the same time, growth of the oocytes and differentiation of spermatogonia in spermatozoa. The factor existing in the hemolymph of adult snails seems to favor the maturation of genital products.

To check the possible action of the various substances existing in the medium more carefully, associations were realized in a medium exclusively containing mineral salts, synthetic organic substances, and yeast extracts. If one explants on such a medium the gonad and the nervous collar coming from the same young animal, the male lineage is favored and a new spermatogenetic rising is observed in the culture. If a cerebral ganglion of adult snail in ovogenesis period is associated with the juvenile gonad, the male lineage degenerates and ovogenesis is

preponderant. In these associations the germinal cells of the indifferen-
tiated gonads taken from very young individuals evolve toward the
stade of oocytes (Guyard, 1967).

The gonads of young snails were also associated with cerebral ganglia
of Paludina, Gastropoda with separate sexes. If the cerebral ganglion
associated with the juvenile gonad of the snail comes from a male
Paludina, no lineage is favored by the association. If the ganglion comes
from a female, the lineage of female germinal cells is favored.

These researchers do not make it possible yet to define whether the
differentiation of male and female germinal cells is the result of the
action of a single hormone or of two different ones. The authors merely
reach the conclusion that the endocrine behavior of the hermaphrodite
Helix aspersa varies and that in the ovogenesis period it can be compared to a female's behavior.

More recently, the phenomenon of ovarian autodifferentiation, which
was shown for the first time in *Calyptraea sinensis* by Streiff (1964),
has just been demonstrated with *Crepidula fornicata* (Lubet and
Streiff, 1969), with *Viviparus viviparus* by Griffond (1969), and with
Helix aspersa by Guyard (1969).

2. Cephalopoda

In their study on the Cephalopoda mollusk, *Sepia officinalis*, M.
Durchon and Richard (1967) analyzed maturation of genital tissues.

The function of the optic gland in the maturation of genital tissues
having been shown in the Octopus (Wells, 1960), M. Durchon and
Richard, thanks to the method of *in vitro* culture, were able to show
a similar action in the cuttlefish.

The optic gland of young cuttlefish (3–4 cm) is first cultured alone
in vitro to test its secretory activity. From the fourth day of culture, a
secretory activity is noticed, whereas these glands remain inactive in
the individuals of the same age but maintained in breeding during the
culture.

The suppression consequent on the absence of brain in culture would
account for the secretory activity observed *in vitro*. Experiments of
ablation made on the octopus lead to the same conclusion (Wells and
Wells, 1959).

Moreover, M. Durchon and Richard compared immature cells of
ovaries of *Sepia* explanted alone to cultures of ovaries explanted by the
side of active optic glands. The endocrine influence of the optic gland
is necessary to the growth of oocytes from the moment they go into
previtellogenesis.

All researches made in the Mollusca class lead to the analysis of several distinct phenomena: (a) phenomenon of orientation of the gonia to the male or female direction, (b) phenomenon of acceleration of gametogenesis, and (c) phenomenon of acquisition of secondary sexual characters. These researches are not numerous enough, compared to the great variety existing in the world of mollusks, to make it possible to propose a synopsis of the hormonal control of these phenomena. At best, one is allowed to ascertain that the hormonal mechanisms in action are probably not identical in all the species studied.

C. Endocrine Factors and Sexuality in Crustacea

In Crustacea a series of experiments realized *in vivo* showed that a hormone secreted by androgenic glands, most often situated along the vas deferens, must be present during the whole life of the Crustacea in order that the male sex may develop normally and that spermatogenesis may occur (Charniaux-Cotton, 1954). In the absence of hormone and although the animal had already a male activity, spermatogenesis does not occur anymore.

From these data, the place of action of the androgenic hormone in the gonad has been specified; the existence of an endocrine factor coming from the protocerebrum and having also a function in the realization of gametogenesis has been shown.

The putting into action of techniques of organotypic culture has proved fruitful for the study of these problems. Utilizing the results obtained by Wolff and Haffen (1952), Berreur-Bonnenfant (1962a,b, 1963a,b) succeeded in realizing a culture medium in which the gonads and the endocrine glands of Crustacea could develop. The endocrine researches which followed this realization concerned the Peracaridea crustaceans *Orchestia gammarella* and *Talitrus saltator* (Amphipoda Talitridae) and *Meinertia oestroidea* (Isopoda Cymothoidea).

In *Orchestia gammarella*, cultures of isolated testes were compared with cultures joining together testes and androgenic glands. Two conclusions appear from this comparison. The spermatogenesis initiated before putting isolated testes into culture reach completion *in vitro*. Therefore, the androgenic hormone is not necessary to the unfolding of spermatogenesis. Moreover, it can be observed that in the germinal zones of cultured testes (areas where primary gonia appear) mesodermic cells and gonia show signs of degeneracy at various degrees. This degeneracy occurs after four days culture when testes are isolated; it occurs only after seven days when testes and androgenic glands are cultured side by

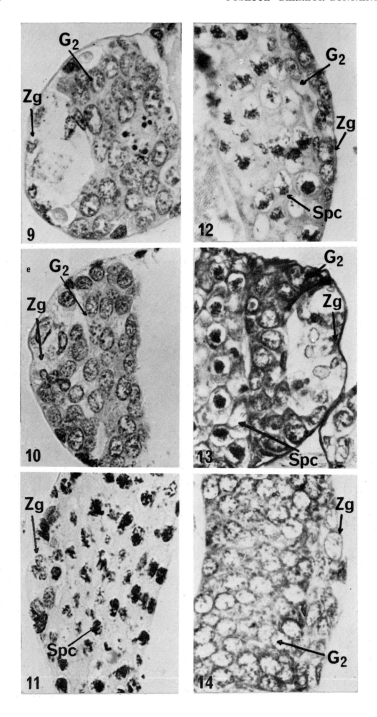

side and in most cases a new rising of spermatogenesis can be initiated *in vitro* (Pl. III, Figs. 9 and 10). If the androgenic hormone acting by itself is sufficient to start a new outlet of male gonia from the germinal zone, it is not, however, sufficient to maintain the cultured testes in a normal state (Berreur-Bonnenfant, 1967).

In *Talitrus saltator* results are not identical since the testis has ovogenesis cyclically; it then has the aspect of an ovary (Fried-Montaufier, 1967).

The evolution of the testes cultured depends on the state in which they were at the time of the explantation.

If testes put into culture are in full spermatogenetic activity and do not show oocytes, some oocytes appear as from the 14th day following the explantation (Berreur-Bonnenfant, 1963b), whereas the spermatogeneses which are started carry on. This observation has since been verified on more than 100 testes cultured.

Montaufier (1963) observed that in *Talitrus saltator* ovogenesis and spermatogenesis do not come from the same side of the germinal zone. It is the same in culture. One side of the germinal zone seems to release ovogonia, whereas the other side releases spermatogonia, with the result that in the gonad oocytes and spermatozoa are separated by the germinal zone.

In the presence of androgenic hormone, testes put into culture do not show ovogenesis; rather, spermatogenesis occurs.

As for *Orchestia gammarella*, the androgenic hormone is sufficient to

---

PLATE III

CULTURE OF DIFFERENTIATED TESTES OF *Orchestia gammarella*

(AFTER BERREUR-BONNENFANT)

FIG. 9. Section through a testis of *Orchestia gammarella* cultured for seven days on anhormonal agar medium. Germinative zone (Zg) degenerates; G_2, secondary spermatogonia.

FIG. 10. Section through a testis of *Orchestia gammarella* cultured for eight days on agar medium with androgenic gland. Beginning of degeneracy of the germinative zone.

FIG. 11. Section through a testis of *Orchestia gammarella* cultured for seven days on agar medium with brains of males. Germinative zone is in good state; Spc, spermatocyte.

FIG. 12. Section through a testis of *Orchestia gammarella* cultured for seven days on agar medium with brains of males and androgenic glands. Germinative zone is in good state; new spermatogenesis is done.

FIG. 13. Section through a testis of *Orchestia gammarella* cultured for four days on agar medium with brains of females. Germinative zone is degenerated.

FIG. 14. Section through a testis of *Orchestia gammarella* cultured for seven days on agar medium with brains of masculinized females and androgenic glands. Germinative zone is in a good state, new spermatogenesis occurs.

start the release of male gonia, but it cannot make possible the maintenance the germinal zone in its state of normal activity during all the
culture (degeneracy as from the 10th day of culture).

In *Meinertia oestroides*, an Isopoda with protandrous hermaphrodism,
young males already show a gonad with hermaphrodite characteristics.
It includes a testicular area formed with three utricules and an ovarian
area. Each of these two areas has its proper germinal zone. The usual
activity of the testicular germinal zone is to release spermatogonia;
exceptionally, it can produce oocytes at the time of sexual inversion.
On the contrary, the female germinal zone produces only oocytes (Berreur-Bonnenfant, 1962a). These animals, during the male phase of their
life, have an androgenic gland which degenerates at the time of the sexual
inversion (Bonnenfant, 1961).

As for the species previously studied, two kinds of cultures are compared: cultures of gonads coming from young functional males and cultures of gonads identical to the former ones and associated to androgenic
glands taken from functional males.

The gonads cultured alone, after 15 days of experiments, take the
aspect of gonads of an animal at the end of sexual inversion, i.e., testicular utricules are degenerating and ovogenesis has taken place in the
ovarian zone.

On the other hand, in the gonads cultured in association with androgenic glands, a new wave of spermatogenesis is realized *in vitro* and the
ovarian germinal zone shows no activity at all. However, the existence
of androgenic hormone is not sufficient to maintain the testicular germinal zone in good condition for more than 10 days.

In these three species of Crustacea, androgenic hormone starts the
liberation of spermatogonia and releases new waves of spermatogeneses.
It inhibits ovogenesis when this one tends to realize but cannot maintain
the germinal zone in good conditions for more than 7–10 days.

Experiments on associations of brain with testes show that a factor
coming from the central nervous system of males acts on the male germinal zone and, so, participates in the control of spermatogenesis (Berreur-Bonnenfant, 1967, 1968).

In *Orchestia gammarella,* it is sufficient to associate, in culture, androgenic glands and brains of males with testes to obtain a working of the
testis identical to the one which it has *in vitro* (Pl. III, Fig. 12).

Comparing the experiments which have just been described with others
in which testes are associated only with brains of males, it is possible
to specify which function is fulfilled by the brain in the control of the
testicular activity. Germinal zone of testes cultured in the presence of

male brains only shows no sign of degeneracy, but no new gametogenesis appears in it (Pl. III, Fig. 11).

The brain coming from a female has not the same properties. Testes explanted in association with a female brain behave as if they were explanted alone (Pl. III, Fig. 13). Experiments on implantation of testes in females show the same results (Berreur-Bonnenfant, 1967, 1968).

On the other hand, female brains which received an implant of androgenic gland two months before the explantation behave as male brains (Pl. III, Fig. 14). In fact, these females show external male characteristics and gonads in which spermatogenesis occurs (Charniaux-Cotton, 1954). The activity of these female brains seems to have changed under the action of the androgenic hormone.

It is the same with male brains which are surgically deprived of their androgenic glands. Two months after the operation the brains of these males behave as female brains.

These experiments lead to the conclusion that two substances seem to act on the male germinal zone. The factor coming from the male brains is necessary to prevent the cells of the germinal zone from degenerating and to maintain a normal aspect. The androgenic hormone ensures the liberation of the gonia and directs them towards the male sex. Moreover, the factor coming from the brain is itself dependent upon the androgenic hormone.

Similar experiments were made on *Meinertia oestroides;* they lead to the same conclusion (Berreur-Bonnenfant, 1968).

Only male gonads associated with male brains and with androgenic glands keep their initial structure. Male gonads associated with female brains or with brains of animals in sexual inversion take the structure of a female gonad after 15 days of culture.

The male brain associated with gonads entering into the phase of sexual inversion has the possibility of stopping the inversion phenomenon. The germinal zone remains in good condition and the spermatogenesis already started carries on. In order that a new spermatogenesis may take place, it is necessary to add active androgenic gland to the culture.

Comparing this with what was obtained in *Orchestia gammarella* and considering the observations realized in culture in *Meinertia oestroides,* the hermaphrodism of this latter species could be explained in the following manner: The androgenic hormone exists during the first phase of the animal's life. It stimulates the brain, which, in its turn, ensures the activity of the male area of the gonad. Afterwards, the androgenic gland disappears; the brain gets the characteristics of a female brain and the testicular area of the gonad degenerates. These modifications start the sexual inversion of the animal, which, little by little, becomes a functional female.

In final analysis with in Crustacea, it is the androgenic hormone which constitutes the essential factor of the realization of the sex. It acts on both the brain, to stimulate the secretion of the factor necessary to the male germinal zone, and the gonia, to induce the sex. The activity of the androgenic glands is discontinuous, but the factors which govern their rhythm of secretion are not known at the present time.

D. Endocrine Factors and Sexuality in Insecta

In Insecta, the influence of hormonal factors on the maturation of genital products has been extensively studied. Growth of oocytes and vitellogenesis would be under the dependence of a gonadotropic hormone secreted by the corpora allata. This hormone would be without any effect on the production of oocytes (ovogenesis) at the level of the germarium. However, some species of Lepidoptera and of Diptera are an exception to this rule. For instance, *Bombyx mori* (Bounhiol, 1942), *Hyalophora cecropia* (Williams, 1959), *Galleria mellonella* (Röller, 1962), and *Calliphora erythrocephala* (Possompès, 1955). Cassier (1967), in an important survey of this problem, concludes that the corpora allata would be necessary to vitellogenesis among Insecta in which the ovarian activity starts only after the imaginal ecdysis. The maturation of male genital products does not seem to be subject to the control of the corpora allata.

Demal (1961) cultures testes of young pupae of *Calliphora* and of *Drosophila*. In these gonads, spermatogenesis goes on actively on a medium without any hormone of insects. Spermatocytes divide and undergo all the phases of differentiation until they form a great number of fascicles of spermatozoa.

Cultures of ovaries of pupae of *Calliphora* have also been realized (Leloup, 1964; Leloup and Demal, 1968). However, vitellogenesis, which is late in this Diptera, cannot be studied. Only the differentiation of the ovary and the division of the gonia are observed. Within eight days of culture on anhormonal medium, Leloup (1964) obtained differentiation of the ovaries of nymphae; the last egg-room of the ovarioles detachs from the others, which usually happens only a few days after the hatching of the imago. If cerebral complexes from pupae are added to the culture medium, the evolution of ovary anlages is better.

In *Galleria mellonella* (Lepidoptera), spermatogenesis and vitellogenesis occur early. Duveau-Hagège (1962) studied its evolution by culturing gonads of last instad larvae.

Testes and ovaries of the last instad caterpillars survive in good condition for seven days on anhormonal medium. Gametogeneses started at the beginning of culture goes on normally. In the cultured ovaries, oocytes enter a phase of large increase and follicles get organized. After differentiation of the nutritious cells and of the follicular cells, previtellogenesis occurs. However, the deposit of vitellus is never realized (Lender and Duveau-Hagège, 1963).

Crushings of adult heads (Duveau-Hagège, 1964) or of male and female nymphae allow better cultures. In these conditions, the follicles formed are in greater number and the increase in the oocytes is heavier. However, no neat vitellogenesis is obtained.

These researches do not make it possible to specify whether vitellogenesis is controlled by an hormonal factor or a trophic factor is necessary for the deposit of vitellus.

In *Tenebrio molitor,* the comparative study of vitellogenesis and of the variations of volume of the corpora allata suggested to Lender and Laverdure (1964, 1965) that the ovarian development of this insect could be dependent upon secretions coming from the brain and from the corpora allata.

Lender and Laverdure (1967) verified this hypothesis by realizing several series of organotypic cultures to which they added various extracts of fat body.

In the ovary explanted on anhormonal medium, only the nutritious cells and the oocytes of the transitory zone lived long. However, they showed no growth or vitellogenesis.

The nonlipidic fraction of the fat body, added to the culture medium, ensures the survival of the young oocytes of the germanium but does not make possible the growth of oocytes or vitellogenesis (Pl. IV, Fig. 18). With the lipidic fraction of the fat body, growth of the oocytes and vitellogenesis occur, but young oocytes degenerate. (Pl, IV, Fig. 17).

To obtain survival of all the elements of the ovary, growth of the intermediate oocytes, and deposit of vitellus *in vitro,* it is necessary to add to the culture medium extracts joining together lipid and nonlipid fractions of the fat body or to explant the ovaries by the side of cephalic complexes (Pl. IV, Fig. 16).

Following these experiments, Laverdure arrives at the hypothesis that the hormones coming from the cephalic complex, of lipid nature, would exist in the lipid fraction of the fat body.

Laverdure (1969) also cultured ovaries of *Tenebrio molitor* nymphae. She showed that growth of the ovaries is stimulated *in vitro* by the

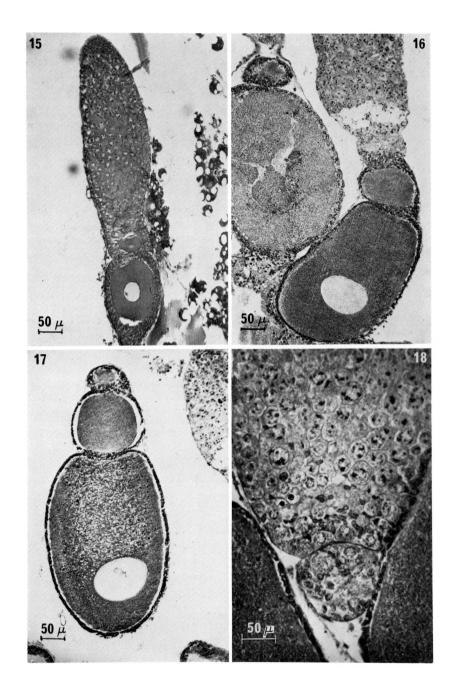

presence of ecdysone. However, ecdysone does not seem sufficient to start the differentiation of the transitory area and of the vitellarium.

E. Endocrine Factors and Sexuality in Ascidiacea

Demonstration of the action of the neural gland on the production of gametes of two Ascidiacea (Ciona and Phallusia) (Carlisle, 1951) was the starting point of researches made *in vitro* by Sengel and Kieny (1962, 1963). These researches show that the neural gland, nervous ganglion, and vibratile organ complex has effects on the maturation of the sexual products of the Ascidiacea.

Experiments were done on gonads of *Molgula manhattensis*. These animals' gonads are hermaphroditic and this fact permits following the evolution of both the male and the female parts.

In this species the sexual cells are very sensitive to the composition of the culture medium and the nutritious contribution seems to have a lot to do in the survival of these cells.

Even with an optimal medium, when gonads are cultured alone, they never progress in the sexual differentiation. On the contrary, if one associates a neural gland, nervous ganglion, and vibratile organ complex to the gonad, the female sexual cells advance: The number of the cells of the testa increases; the diameter of the oocytes grows; the basophily of their cytoplasm decreases; there is differentiation and thickening of the secondary follicle and the coming out of acidophilous refracting particles in the follicular cells in process of vacuolization.

The male lineage seems unsensitive to the action of the neural gland, nervous ganglion, and vibratile organ complex. It does not advance in culture.

Sengel and Kieny noticed that the complexes put into culture have the same activity whatever the age of the animals from which they were taken. The complex coming from a mature Ascidia has the same effects as the one coming from an immature specimen.

PLATE IV

SURVIVAL AND DIFFERENTIATION OF LARVAL GONADS OF *Tenebrio molitor*

(AFTER LAVERDURE)

FIG. 15. Control ovary.

FIG. 16. Ovary grown for five days on glucose standard medium with cephalic complexes.

FIG. 17. Ovary grown for five days on glucose standard medium with the soluble in ether fat body extract.

FIG. 18. Ovary grown for five days on glucose standard medium with insoluble in ether fat body extract.

Other organs associated with the gonad are incapable of ensuring a better survival of the gonocytes and have no effect whatsoever on their differentiation.

The complex formed by the neural gland, nervous ganglion, and vibratile organ really seems to favor the differentiation and the maturation of the oocytes.

IV. SUMMARY

Considering the state of the studies made in this field, it seems useful to recall, hereupon, the object of the various researches accomplished in culture on the subject of sex. Really, they all concern sexual differentiation, but this concept is complex, and distinct phenomena are sometimes designated under this same definition. Three among them were approached by the method of organotypic cultures.

1. The orientation of the gonia toward a male or female evolution: This problem is especially interesting among invertebrates since the cases of hermaphrodism on the evolution of which one may hope to experiment exist in great number. The work realized on the mollusks *Calyptraea sinensis* (Streiff) and *Helix aspersa* (Guyard) and on the Crustacea (Berreur-Bonnenfant) concerns this problem.

2. The maturation of the genital products: Durchon analyzed the endocrine determinism and the mechanism of it in some Polychaeta. Researches made on Hirudinea (Malecha), Cephalopoda (Durchon and Richard), Insecta (Duveau-Hagège, Laverdure), and Ascidiacea (Sengel and Kieny) concern the endocrine control of vitellogenesis.

3. The differentiation of the external sexual characteristics, phenomenon which essentially concerns the somatic cells: On this subject, Streiff studied the differentiation of the genital tract in the mollusk *Calyptraea sinensis*.

Our knowledge of the endocrine mechanisms which control the various phases of the realization of the sex in the invertebrates is yet too fragmentary to make it possible, at the present time, to propose a synopsis summing up the whole of our information in this field. The method of the organotypic cultures certainly will give the possibility of bringing new contributions to the study of these mechanisms.

REFERENCES

Berreur-Bonnenfant, J. (1962a). *Bull. Soc. Zool. Fr.* **87**, 253.
Berreur-Bonnenfant, J. (1962b). *Bull. Soc. Zool. Fr.* **87**, 377.

Berreur-Bonnenfant, J. (1963a). *C. R. Acad. Sci.* **256**, 2244.
Berreur-Bonnenfant, J. (1963b). *Bull. Soc. Zool. Fr.* **88**, 235.
Berreur-Bonnenfant, J. (1967). *C. R. Soc. Biol.* **161**, 9.
Berreur-Bonnenfant, J. (1968). *Arch. Zool. Exp. Gen.* **108**, 521.
Bonnenfant, J. (1961). *C. R. Acad. Sci.* **252**, 1518.
Bounhiol, J. J. (1942). *C. R. Acad. Sci.* **215**, 334.
Carlisle, D. B. (1951). *J. Exp. Biol.* **28**, 463.
Cassier, P. (1967). *Ann. Biol. Anim., Biochim., Biophys.* **6**, 595.
Charniaux-Cotton, H. (1954). *C. R. Acad. Sci.* **239**, 780.
Charniaux-Cotton, H. (1957). *Ann. Sci. Nat. Zool. Biol. Anim.* **19**, 411.
Choquet, M. (1964). *C. R. Acad. Sci.* **258**, 1089.
Choquet, M. (1965). *C. R. Acad. Sci.* **261**, 4521.
Choquet, M. (1967). *C. R. Acad. Sci.* **265**, 333.
Demal, J. (1956). *Ann. Sci. Nat. Zool. Biol. Anim.* **18**, 155.
Demal, J. (1961). *Bull. Soc. Zool. Fr.* **86**, 522.
Demal, J., and Leloup, A. M. (1963). *Ann. Epiphyt.* **14**, 91.
Dhainaut, A. (1964). *C. R. Acad. Sci.* **259**, 461.
Durchon, M. (1952). *Ann. Sci. Nat. Zool. Biol. Anim.* **14**, 119.
Durchon, M., and Boilly, B. (1964). *C. R. Acad. Sci.* **259**, 1245.
Durchon, M., and Dhainaut, A. (1964). *C. R. Acad. Sci.* **259**, 917.
Durchon, M., and Richard, A. (1967). *C. R. Acad. Sci.* **264**, 1497.
Durchon, M., and Schaller, F. (1963). *C. R. Acad. Sci.* **256**, 5615.
Durchon, M., Boilly, B., and Dhainaut, A. (1965). *C. R. Soc. Biol.* **159**, 106.
Duveau-Hagège, A. M. (1962). *Bull. Soc. Zool. Fr.* **87**, 380.
Duveau-Hagège, A. M. (1964). *Bull. Soc. Zool. Fr.* **89**, 66.
Fried-Montaufier, M. C. (1967). *C. R. Soc. Biol.* **161**, 2104.
Frew, J. C. H. (1928). *J. Exp. Biol.* **6**, 1.
Fujio, Y. (1960). *Jap. J. Genet.* **35**, 361.
Fujio, Y. (1962). *Jap. J. Genet.* **37**, 110.
Gilbert, L. I. (1964). *In* "The Physiology of Insecta" (M. Rockstein, ed.), Vol. 1,
    pp. 149–225. Academic Press, New York.
Gottschewski, G. H. M. (1958). *Naturwissenschaften* **45**, 400.
Gottschewski, G. H. M. (1960). *Wilhelm Roux' Arch. Entwicklungsmech. Organismen*
    152, 204.
Gottschewski, G. H. M., and Querner, W. (1961). *Wilhelm Roux' Arch. Entwick-
    lungsmech. Organismen* **153**, 168.
Griffond, B. (1969). *C. R. Acad. Sci.* **268**, 963.
Guyard, A. (1967). *C. R. Acad. Sci.* **265**, 147.
Guyard, A. (1969). *C. R. Acad. Sci.* **268**, 966.
Guyard, A., and Gomot, L. (1964). *Bull. Soc. Zool. Fr.* **89**, 48.
Hagadorn, I. R. (1966). *Amer. Zool.* **6**, 251.
Horikawa, M. (1958). *Cytologia* **23**, 468.
Horikawa, M. (1960). *Jap. J. Genet.* **35**, 76.
Kuroda, Y., and Yamaguchi, K. (1956). *Jap. J. Genet.* **31**, 98.
Laverdure, A. M. (1967a). *Bull. Soc. Zool. Fr.* **92**, 629.
Laverdure, A. M. (1967b). *C. R. Acad. Sci.* **265**, 505.
Laverdure, A. M. (1969). *C. R. Acad. Sci.* **269**, 82.
Leloup, A. M. (1964). *Bull. Soc. Zool. Fr.* **89**, 70.
Leloup, A. M., and Demal, J. (1968). *Proc. Int. Colloq. Invertebr. Tissue Cult., 2nd,
    1967* p. 126.

Lender, T., and Duveau-Hagège, J. (1963). *Develop. Biol.* **6**, 1.

Lender, T., and Laverdure, A. M. (1964). *C. R. Acad. Sci.* **258**, 1086.

Lender, T., and Laverdure, A. M. (1965). *C. R. Acad. Sci.* **261**, 557.

Lender, T., and Laverdure, A. M. (1967). *C. R. Acad. Sci.* **265**, 451.

Lubet, P., and Streiff, W. (1969). *In* "Cultures d'organes d'Invertébrés." p. 135. Gordon & Breach, New York.

Malecha, J. (1965). *C. R. Soc. Biol.* **159**, 1674.

Malecha, J. (1967). *C. R. Acad. Sci.* **265**, 1806.

Marks, E. P. (1968). *Gen. Comp. Endocrinol.* **11**, 31.

Marks, E. P., and Reinecke, J. P. (1965). *Gen. Comp. Endocrinol.* **5**, 241.

Montaufier, M. C. (1963). D. E. S. Paris.

Nardon, P., and Plantevin, G. (1969). *In* "Cultures d'organes d'Invertébrés," p. 193. Gordon & Breach, New York.

Novak, V. J. A. (1966). *In* "Insect Hormones" Methuen, London.

Oberlander, H., and Fulco, L. (1967). *Nature (London)* **216**, 1140.

Possomprès, B. (1955). *C. R. Acad. Sci.* **241**, 2001.

Röller, H. (1962). *Naturwissenschaften* **49**, 524.

Schneider, I. (1964). *J. Exp. Zool.* **156**, 91.

Schneider, I. (1966). *Embryol. Exp. Morphol. G. B.* **15**, 271.

Sengel, P., and Kieny, M. (1962). *C. R. Acad. Sci.* **254**, 1682.

Sengel, P., and Kieny, M. (1963). *Ann. Epiphyt.* **14**, 95.

Steinberg, A. G. (1941). *Genetics* **26**, 325.

Steinberg, A. G. (1943). *Can. J. Res.* **21**, 277.

Streiff, W. (1964). *Bull. Soc. Zool. France* **89**, 56.

Streiff, W. (1966a). *Ann. Endocrinol.* **27**, 385.

Streiff, W. (1966b). *C. R. Acad. Sci.* **263**, 539.

Streiff, W. (1967a). *Ann. Endocrinol.* **28**, 461.

Streiff, W. (1967b). *Ann. Endocrinol.* **28**, 641.

Streiff, W., and Peyre, A. (1963). *C. R. Acad. Sci.* **256**, 292.

Wells, M. J. (1960). *Symp. Zool. Soc. London* **2**, 87.

Wells, M. J., and Wells, J. (1959). *J. Exp. Biol.* **36**, 1.

Wigglesworth, V. B. (1964). *Advan. Insect Physiol.* **2**, 247–336

Williams, C. M. (1959). *Biol. Bull.* **116**, 323.

Wolff, Et., and Haffen, K. (1952). *C. R. Acad. Sci.* **234**, 1396.

6

PHYSIOLOGY OF THE EXPLANTED
DORSAL VESSEL OF INSECTS

J. David and M. Rougier

I. INTRODUCTION

The circulatory system of insects consists of the dorsal vessel prolonged in its anterior part by the aorta. Working both as a suction pump and a force pump, this vessel, often called a heart, ensures, by its contractions, the circulation of the hemolymph.

It has long been known that the dorsal vessel continues to contract when severed from its connections with the central nervous system and explanted in a physiological solution. Several studies dealt with the physiology of the dorsal vessel of insects (Beard, 1953; Jones, 1964; McCann, 1970) but the results obtained with the whole insect were not distinguished from those obtained with the isolated organ. However, research on explanted hearts needs particular techniques, more or less akin to the methods used for organ culture. That is why, although it was possible to consider a variety of physiological problems with explanted dorsal vessels (study of the automatic nature of the contraction, influence of various experimental factors), all these experiments have some features in common. A detailed examination of these experimental studies shows that they are generally heterogeneous and that our knowledge is still very insufficient. It, therefore, appeared of interest to review, systematically, the reported information in order to emphasize what is still lacking and to point out those problems which require further analysis.

After a brief morphological recapitulation, explantation techniques will be considered, as well as the composition of various culture media.

Then the contraction wave will be described, with its concomitant electrical phenomena. Next, the effects of various physical factors and of the mineral ions in the medium will be studied. Finally, the action of various organic substances will be analyzed.

II. MORPHOLOGICAL RECAPITULATION AND EXPERIMENTAL TECHNIQUES

The morphology of the dorsal vessel of insects shows considerable variation (in its details) and was the subject of numerous studies. A detailed exposition can be found in a review by Jones (1964), and only general indications will be alluded to here.

A. Morphology

Located in the upper part of the abdomen, the dorsal vessel usually appears as a tube with swellings constituting a succession of cardiac chambers. Each chamber possesses an opening on either side which constitutes an ostial valve or ostia. The hind end of the dorsal vessel is generally closed, while in front it is continued by the aorta (Fig. 1).

Typically, the ostia appear as oblique or vertical slits, frequently prolonged inwards by lips corresponding with invaginations of the chamber walls. The dorsal vessel is held in place against the abdominal tergites by both dorsal suspensory muscles and lateral aliform muscles.

1. Structure

The walls of the cardiac tube are made up of striated muscles with their fibrillae generally in a circular or an oblique disposition. The histological aspect of the tissues is uniform and shows no special differentiations. The internal layer of the tube (intima) is formed by the muscular surface; it usually comprises a thin layer of hyalin cytoplasm limited by a fine membrane (sarcolemma). The external face is surrounded by a conjunctive tissue which, by analogy with vertebrates, is called adventice.

2. Innervation

The innervation of the dorsal vessel has been studied in a number of species and it shows considerable variability.

In the most typical cases (*Blatta,* Alexandrowicz, 1926; *Carausius,* Opoczynka-Sembratova, 1936; *Periplaneta,* McIndoo, 1945; Miller and

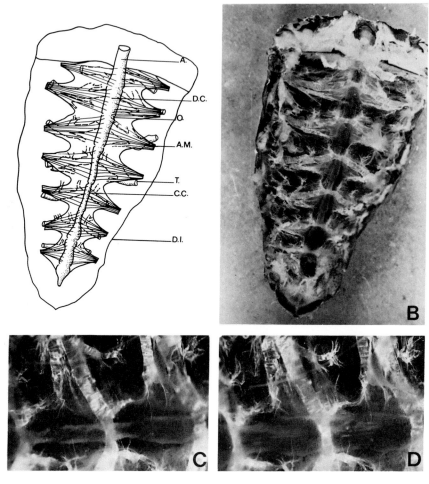

FIG. 1. Explanted dorsal vessel of *Cybister* (Coleoptera, Dytiscidae). (A) General diagram (A., aorta; A.M., alary muscle; C.C., contracted chamber; D.C., dilated chamber; D.I., dorsal integument; O., ostia; T., trachea); (B) the whole dorsal vessel; (C) a contractile chamber in diastole; (D) a contractile chamber in systole.

Thomson, 1968; *Blattella*, Edwards and Challice, 1960) both lateral and segmentary innervations (Fig. 2) are found. Two lateral cardiac nerves containing nerve cells stretch out on either side of the dorsal vessel. At the level of each segment, segmentary nerves branch off the ventral nerve chain and innervate the aliform muscles and the cardiac tube.

This typical disposition may be simplified. In *Aeschna* (Zawarzin,

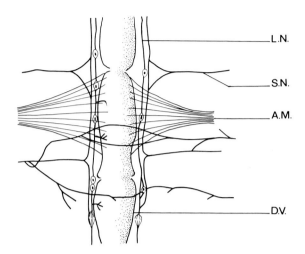

Fig. 2. Diagram of the innervation in the dorsal vessel of a cockroach (from Alexandrowicz, 1926). A.M., alary muscles; D.V., dorsal vessel; L.N., lateral nerve; S.N., segmentary nerve.

1911) and in *Bombyx* (Kuwana, 1932) the lateral nerves are deprived of neurones. In a rather large number of species (*Prodenia* larva, McIndoo, 1945; *Musca*, *Tipula*, *Anopheles*, Jones, 1964; *Belostoma* and *Anax*, Maloeuf, 1935) the lateral nerves are completely lacking. Sometimes the segmentary nerves appear to stop at the level of aliform muscles and the heart appears to be completely uninnervated (*Belostoma* and *Anax*, Maloeuf, 1935; *Tipula* and *Anopheles*, Jones, 1964; *Hyalophora*, Sanger and McCann, 1968).

As we shall subsequently see, the question of innervation is very important for the understanding of the mechanism of heart automatism. But most of the studies on this point are rather dated and should be reinvestigated with more modern methods.

B. Operation Techniques

The explantation of the dorsal vessel is generally carried out on a living insect which has just been decapitated. However, some investigators use preliminary anaesthesia: immersion in water for *Periplaneta* (Yeager and Hager, 1934) [or anaesthesia by carbon dioxide (Ludwig *et al.*, 1957a,b)] and anaesthesia by ether for *Musca* (Brebbia and Ludwig, 1962).

For the dissection, a number of experiments (Levy, 1928a; Yeager and Hager, 1934; Bergerard *et al.*, 1950; Irisawa *et al.*, 1956; McCann,

1963; etc.) open the ventral face and clear the dorsal vessel by removing the viscera. Another technique, more rapid but with the risk of damaging the dorsal vessel, is to cut off the abdominal tergites to which the heart adheres (Dubuisson, 1929; Davenport, 1949; Barsa, 1955; Wilbur and McMahan, 1958; David and Rougier, 1965; etc.). According to the morphology of the species studied, one or the other of these techniques is preferred.

In nearly all cases the dorsal vessel is kept in place attached to the abdominal tergites. This type of preparation is to be called "semi-isolated heart." In fact, as well as the dorsal vessel, the preparation includes the aliform muscles, the pericardial cells, and fragments of the fat body and the tracheae, etc. Many workers fix the preparation with fine needles on a slide of wax or paraffin for better observation of the contractions (Yeager and Hager, 1934; Davenport, 1949; Wilbur and McMahan, 1958; Barsa, 1955; Harvey and Williams, 1958; Brebbia and Ludwig, 1962; McCann, 1963; etc.).

After being isolated, the dorsal vessel is transplanted into a medium of a composition different from that of the hemolymph. The functioning of the organ is always disturbed at first and various authors noted or described this operatory shock (Levy, 1928a; Dubuisson, 1930; Yeager and Hager, 1934; Krijgsman and Krijgsman-Berger, 1951; Harvey and Williams, 1958; Rougier, 1966; etc.).

The dorsal vessel is put into a recipient adapted to its size, and the medium is renewed periodically. Often a certain variation of the rhythm of contractions is observed on renewal (Dubuisson, 1930; Kojantchikov, 1932; Hamilton, 1939; Bergerard et al., 1950; Rougier, 1966; Cazal, 1967). This variation may be attributed to various causes, for example, to a change in the pH or in the oxygen content of the medium.

Following Yeager and Hager (1934), many researchers used a system of perfusion which renews the liquid and allows oxygenation at the same time (Krijgsman and Krijgsman-Berger, 1951; Krijgsman, 1952; Barsa, 1955; Brebbia and Ludwig, 1962; Butz, 1962; etc.).

C. Explantation Media

It is convenient to classify these media as simple media (containing at the most a carbohydrate as organic substance) and complex media (containing numerous organic compounds).

1. Simple Media

The solutions used are essentially made up of chlorides and are derived from Ringer solution. In particular, they contain all the Na, K, and Ca ions. It seemed useful to indicate some of the media used by the authors in Table I. An examination of the table reveals a very remarkable heterogeneity.

Doubtless this variability is due to the use of various species. But even for a given species (*Periplaneta americana*) or neighboring species (butterflies of the *Samia* and *Telea* genera)* considerable differences are observed. This allows two conclusions: first, most of the explantation media must be badly adapted to their object; second, the dorsal vessel is remarkably tolerant to very diverse media.

In relation to this tolerance of the dorsal vessel the workers rarely attempted to develop a medium truly adapted to its needs. In most cases, the choice of the various concentrations appears to be highly empirical. In some recent studies (Barsa, 1955; Butz, 1962; Brebbia and Ludwig, 1962; McCann, 1963), adjusting the concentrations of the cations and the osmotic pressure to the characteristics of the insect's hemolymph has been tried. But this method is not necessarily justified because the hemolymph of insects contains very small quantities of chlorides, it is likely that the cations analyzed are not all under ionized forms, and the hemolymph of insects is always rich in magnesium (Florkin and Jeuniaux, 1964), whereas this cation is often noxious in explantation medium.

Lastly, it must be noted that the pH of the medium is often adjusted at the time of preparation to a value approximately close to that of the hemolymph of the insect studied (Yeager and Hager, 1934; Bergerard *et al.*, 1950; Krijgsman *et al.*, 1950; Krijgsman, 1952; etc). This adjustment is usually done with solution of HCl or NaOH.

2. Complex Media

These media have been used essentially to obtain cultures of cells from the circulatory system. The works of Hirumi and Maramorosch (1964), Mitsuhashi (1965), and Vago and Quiot (1969) on various species may be quoted as examples.

The composition of these media is extremely varied, and data on their constitution are to be found in other sections of this book.

The physiology of the dorsal vessel was rarely studied with such media. However, it may be noted that Mitsuhashi observed the persistence of the contractions in explanted cardiac tissues for about 10 days. Demal and Leloup (1963) indicate that the best medium for the viability of the dorsal vessel of *Calliphora* is a mineral medium with 20% chicken embryo extract added. The addition of insect hemolymph appears to be noxious.

---

* In 1970, McCann replaced *Samia* and *Telea* genera by *Hyalophora*.

TABLE I
Some Media Used in the Study of the Dorsal Vessel Explanted from Insects

Species	Stages	Concentrations in (gm/liter)									Authors
		NaCl	KCl	CaCl$_2$	MgCl$_2$	NaHCO$_3$	Na$_2$HPO$_4$	KH$_2$PO$_4$	Glucose	Others	
Dyctyoptera											
Blatta orientalis	A	9.82	0.77	0.50		0.18	0.01		1		Yeager and Hager (1934)
Periplaneta americana	A	10.93	1.57	0.85	0.17						Yeager (1939a)
Periplaneta americana	N or A	11.0	1.4	1.1							Ludwig et al. (1957b)
Periplaneta americana	N or A	9.00	0.70	0.46							Ludwig et al. (1957a)
Periplaneta americana	A	7.5	0.2	0.2		0.1				a	Unger (1957)
Periplaneta americana	A	9.0	0.2						4		Ralph (1962)
Orthoptera											
Acheta domestica	A	10.8	0.85	0.55		0.2	0.01		1		Bergerard et al. (1950)
Stenopelmatus longispina	A	6.7	0.15	0.12		0.15					Davenport (1949)
Melanoplus differencialis	A	9.0	0.2	0.2		0.2					Crescitelli and Jahn (1938)
Chortophaga viridifasciata	A	13.61	0.54	0.54							
Coleoptera											
Popilius disjunctus	A	6.8	0.2	0.2	0.1	0.12	0.2		7.7		Wilbur and McMahan (1958)
Popilius disjunctus	A	6.8	1.0	0.2	0.1	0.2	0.2		12.87		Collings (1966)
Tenebrio molitor	L	16.0	1.4	1.0							Butz (1957)
Mylabris phalerata	A	9.82	0.77	0.50							Agrawal and Srivastava (1962)
Cybister lateralimarginalis	A	9.82	0.77	0.50		0.2		0.62	1.5	b	Rougier (1966)
Lepidoptera											
Galleria mellonella	L	10.8	0.85	0.55		0.2	0.01		1		Bergerard et al. (1950)

Galleria mellonella	L	8.5	3.4	1.07		0.2	0.01	1	Dreux and Grapin-Poupeau (1963)
Chilo zonellus	L	9.0	0.71	0.46		0.16	0.01		Agrawal and Srivastava (1962)
Dichocrocis ponctiferalis	L	9.0	0.71	0.46		0.16	0.01		Agrawal and Srivastava (1962)
Samia walkeri	P	18.03	0.73	0.61					Barsa (1955)
Samia cecropia	P	7.5	0.35	0.21					Harvey and Williams (1958)[c]
Telea polyphemus	A	9.0	0.2	0.2		0.2			Tenney (1953)
Telea polyphemus	A	0.11	3.36	0.38	1.49				McCann (1963)
Hyalophora cecropia	A	0.11	3.36	0.38	1.49				McCann (1963)
Diptera									
Calliphora erythrocephala	L	9.0	0.71	0.46		0.16	0.01		Levy (1928a)
Phormia regina	L	9.0	0.71	0.46		0.16	0.01		Levy (1928a)
Musca domestica	L	8.51	0.36	0.39					Brebbia and Ludwig (1962)
Musca domestica	P	12.76	0.32	0.12		0.09	0.08	4	Brebbia and Ludwig (1962)
Musca domestica Other organs	A	14.41	0.36	0.29		0.09	0.08	4	Brebbia and Ludwig (1962)
Periplaneta americana	A	14.0	0.2	0.4		0.2			Griffiths and Tauber (1943)

[a] Also for *Blatta orientalis* and *Blattella germanica*

[b] Na_2HPO_4, 0.15 gm; penicillin, 120 mg; steptomycin, 20 mg.

[c] 0.001 M of potassium phosphate and a few milligrams of phenylthiourea and steptomycin.

Caspari (quoted by Jones, 1964) observed contractions for several weeks in fragments of *Bombyx* heart cultured in Trager medium. Larsen (1963) maintained fragments of the *Blaberus* embryo dorsal vessel in activity for several weeks. More recently, Larsen (1967) obtained the persistence of a contractile activity after over two years of explantation.

The use of complex media for a long time requires rigorously aseptic conditions. Thus, the above authors worked with completely isolated fragments and not with whole organs.

Very little research was done on semi-isolated preparations retaining the abdomen's tergites. Cazal (1967) points out that cell culture media are not suitable for observing the contractions of *Locusta* heart. On the other hand, it was possible to keep the semi-isolated dorsal vessel of *Cybister* functioning for two weeks in TC 199 medium whereas the longest periods of viability observed in simple medium seldom exceed five days (Rougier, 1967).

D. Criteria Studied

Most often the activity of the cardiac vessel was characterized by the frequency of the heartbeat. Observations of the amplitude of the contractions and the eventual stopping of the heart in systole or diastole have been reported (Bergerard *et al.*, 1950; Unger, 1957; etc.).

In an attempt to obtain more precise indications, various investigators mechanically recorded the contractions, and different techniques were developed (Uramoto, 1932; Kojantchikov, 1932; Duwez, 1936, 1938; Yeager, 1938, 1939a; Jahn *et al.*, 1937; Hamilton, 1939; Crescitelli and Jahn, 1938; De Wilde, 1947; Krijgsman *et al.*, 1950; Williams *et al.*, 1968).

Several experiments dealt with the electrical activity of the cardiac vessel. Electrograms were studied by Uramoto (1932), Duwez (1936, 1938), Jahn *et al.* (1937), Jahn and Koel (1948), Crescitelli and Jahn (1938), Tenney (1953), Irisawa *et al.* (1956), and Williams *et al.* (1968). More recently, intracellular potentials were recorded and analyzed with microelectrodes (Irisawa *et al.*, 1956; McCann, 1963, 1964a,b, 1965, 1966).

The duration of the contractions, up to now, has been given little attention. Generally, a survival of 24 hours appeared to be satisfactory and the organ was observed for only a few hours. However, the duration of the dorsal vessel's functioning is an interesting criterion, giving particular information on the validity of the explantation medium (David and Rougier, 1965; Rougier, 1966).

III. STUDY OF THE CONTRACTION WAVE

With a few exceptions (Jones, 1964), the cardiac chambers do not contract simultaneously, but successively, thus creating a wave which normally moves forward. During the contraction, the orifice of each ostia is obturated by the blood pressure.

These phenomena are difficult to study on the organ in place unless the teguments are transparent. In contrast, they are easily observed on an explanted vessel, and it is therefore natural that a number of studies were devoted to the physiology of the contraction with an isolated organ.

After describing the aspect of the contraction wave, we shall discuss the part played by the aliform muscles. The electrical phenomena related to the contractile activity will be dealt with next, and finally, we shall examine the problem of the reversal of the heartbeats.

A. Description and Recording of the Contraction Wave

In each chamber a succession of contractions, or systoles, is observed, followed by phases of relaxation or diastoles (Fig. 1C,D). The contractions are coordinated and constitute a peristaltic wave directed forward. The systole is effected by the circular muscle fibrillae of the cardiac vessel. Two mechanisms seem to contribute to the occurrence of the diastole. On one hand, the slackening of the circular fibrillae; on the other, the eventual contraction of the aliform muscles. But, as we shall see, the part played by the latter is controversed.

With succession of electrodes placed along the dorsal vessel of *Telea*, Tenney (1953) was able to record the speed with which the contraction wave moved. He noted that this speed varies considerably from one vessel to another and that it also varies, for the same organ, according to the segment studied.

The mechanical recording of the contractions permitted a more precise knowledge of the phenomena. An example of the results obtained is given in Fig. 3A. For each contraction, the ascending part of the line corresponds to the systole and the descending branch to the diastole. Between two contractions a period of rest or diastasis is noted. In some cases (Yeager, 1939b) the contraction is preceded by a presystolic notch. When the frequency of the beats increases, the length of the resting period decreases. Finally, it should be pointed out that, if all the mecanograms present a quite similar aspect, differences in detail are observed in rela-

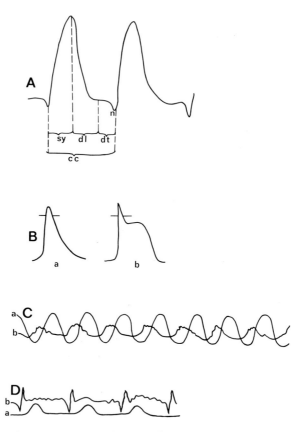

Fig. 3. Examples of recordings of the mechanical and of the electrical activities of the dorsal vessel. (A) Mechano-cardiogram of the *Periplaneta*: cc, cardiac cycle; dl, diastole; dt, diastasis; n, presystolic notch; sy, systole (from Yeager, 1938). (B) Intracellular action potentials in *Telea* (from McCann, 1963). (C) and (D) Simultaneous recording of the mechano-cardiogram and of the electrocardiogram in the *Melanoplus* (from Crescitelli and Jahn, 1938). (C) normal temperature; (D) temperature of 2°. B–D: a, mechano-cardiogram; b, electrocardiogram.

tion to the region of the vessel where the recording is made (Yeager, 1939a; Crescitelli and Jahn, 1938).

B. Role of the Aliform Muscles

The role of the aliform muscles is still debated although it has been the object of numerous studies.

If completely isolated, the dorsal vessel of many insects ceases to contract (Jones, 1964). However, if only the aliform muscles are sectioned, the persistence of the contractions is usually observed although their amplitude and frequency may be modified (*Dytiscus*, Brocher, 1917, and Duwez, 1936; *Periplaneta*, Yeager, 1939a; *Melanoplus*, Hamilton, 1939; *Anopheles*, Jones, 1954; *Cryptotympana*, Irisawa *et al.*, 1956; *Rhodnius*, Jones, 1964). According to Dubuisson (1929), the heart of *Hydrous* ceases to contract if the aliform muscles are sectioned. But this interruption only occurs if the totality of the fibers is cut, i.e., an incomplete section maintains cardiac functioning.

According to the above data, the aliform muscles seem to be useful though rarely indispensable. Various investigators have tried to determine whether these muscles have their own contractile activity. It appears that the answer to this question varies according to the species.

In certain species, i.e., *Dytiscus* (Duwez, 1938) and *Periplaneta* (Yeager, 1939a), no contractile activity was discovered, neither spontaneous nor following electric stimulation. However, a contractile response in *Periplaneta* was observed by Miller and Metcalf (1968).

In other species, i.e., *Chironomus* and *Agrion* larvae (Dubuisson, 1929) adult *Hydrous* (Dubuisson, 1929; De Wilde, 1947), adult *Melanoplus* (Hamilton, 1939), various Lepidoptera larvae (De Wilde, 1947) or adults (McCann, 1970), a contraction of the aliform muscles was observed. In *Hydrous* and *Cossus*, for instance, these contractions are of two types (De Wilde, 1947): contractions synchronized between the left and right muscles, which generally alternate with the cardiac systoles and must facilitate the diastole; and contractions alternating between the left and right muscles, with a slower period, which laterally displace the dorsal vessel. According to Hamilton (1939), these contractions are not observed on the normal dorsal vessel of *Melanoplus*, but they occur after a treatment by acetylcholine or nicotine. In the cicada *Cryptotympana*, Irisawa *et al.* (1956) obtained simplified electrograms after sectioning the aliform muscles. It seems, therefore, that they must be the seat of some contractile activity.

Finally, the part played by the aliform muscles appears to be variable according to the species, and an interesting hypothesis concerning the regulation of their activity is given by McCann (1970). At the least, their natural elasticity should facilitate the diastole and the renewal of the systoles. Often, these muscles present contractions but the contractions rarely seem to be well synchronized with the contractions of the dorsal vessel so that their usefulness remains hypothetical. Lastly, it should be noted that much of the research work on various species is not recent and was done with unsuitable media. Further research is desirable.

C. Spontaneous Electric Activity

1. Electrocardiogram

The electric activity of the heart is manifested by variations of the surface potential, and we indicated that a number of investigators studied the electrocardiograms.

The recordings (Fig. 3C,D) usually show a complex wave with several oscillatory deflexions. The periodicity of this wave coincides with that of the contractions measured by mecanogram (Duwez, 1936, 1938; Crescitelli and Jahn, 1938).

An important characteristic of the recorded waves is their variability, which proceeds from multiple causes. Jahn et al. (1937) emphasized the differences observed from one individual to another. Differences according to the region where the recording is done were noted by Crescitelli and Jahn (1938) and mainly by Tenney (1953) and Irisawa et al. (1956).

The complexity of the electrograms makes them difficult to use. Therefore, various authors tried to obtain simplified curves. According to Duwez (1936), this complexity may be one of the consequences of the explantation since recordings obtained in situ are much more regular. Irisawa et al. (1956) simplified their recording by sectioning the aliform muscles and thereby concluded that the latter had some effect on the electrogram. By lowering the temperature, Crescitteli and Jahn (1938) obtained, in Melanoplus, a more simple wave composed of a large rapid oscillation, just before the systole, followed by a series of slower oscillations (Fig. 3D). However, it is to be noted that in Dytiscus, Duwez (1936) obtained opposite results: a simple curve at 21° and a very complex one at 6°.

2. Intracellular Action Potentials

To date, intracellular action potential has been little studied in insects. In the cicada Cryptotympana, Irisawa et al. (1956) were able to record such action potentials in a localized region of the second segment. In Samia and Telea, McCann (1963, 1964a,b, 1965, 1966, 1967) noted that such potentials may be observed all along the dorsal vessel. The potentials are of two kinds (McCann, 1963) (Fig. 3B): (a) one observed in the forward region of the dorsal vessel and which resembles the recordings made in vertebrate atria; (b) the other observed in the posterior part is close to that observed in vertebrate ventricles. The average characteristics of these two types of potentials are condensed in Table II.

Some of these potentials are of the pacemaker type (slow initial de-

TABLE II

MEAN CHARACTERISTICS OF TWO TYPES OF ACTION POTENTIALS
OBSERVED IN *Telea* BY McCANN (1963)

		a	b
Resting potential	(mV)	−47.0	−47.0
Action potential	(mV)	56.5	59.5
Overshoot	(mV)	9.5	12.5
Rise time	(msec)	69.5	48.6
Maximum rate of rise	(mV/sec)	601	980
Fall time	(msec)	472	665
Maximum rate of fall	(mV/sec)	96.0	71.5
Duration	(msec)	610	750

polarization, then rapid) but they are distributed all along the dorsal vessel and no special structure was noticed (McCann, 1964a). In a more recent work, McCann (1966) indicates that these types of action potentials may not really be pacemakers.

D. Reversal of the Heartbeat

For many species (Gerould, 1933) a temporary reversal of the direction of the contraction wave may be observed in live animals. The physiological significance of this phenomenon was the object of much research work reviewed by Jones (1964).

Under conditions of explantation this phenomenon was observed by various authors among whom we may cite Uramoto (1932) Tenney (1953), Irisawa *et al.* (1956), and McCann (1963). The problem was analyzed thoroughly by Tenney (1953) in adults of *Telea polyphemus,* where these inversions are pratically frequent. By recording electrograms obtained at various points of the dorsal vessel, Tenney was able to distinguish five different types of reversal. Most often, the appearance of an inversed wave is associated with the stopping or slowing down of the main pacemaker, which normally ensures the forward propagation of the wave. It is then supposed that a new anterior pacemaker (ectopic center) goes into action.

The existence of two main pacemakers, one anterior, the other posterior, seems to be confirmed by an experiment by Jones (1964) on adult *Anopheles.* After transversal section of the dorsal vessel, it is often noted that the posterior half keeps the normal direction of beats whereas the anterior half presents inverse contractions, directed backward. But the potential pacemakers are doubtless more numerous. For example,

Gerould (1938) noted that an isolated cardiac chamber of *Bombyx* is not only capable of continuing function but that it presents, like the whole dorsal vessel, periodic reversals of its contractions. In the same line of thought, Dubuisson (1930) noted that mechanical stimulation of the median region of the dorsal vessel of *Hydrous* provokes a persistant contraction of the region stimulated. Often enough, this region becomes the starting point of contractions which move forward in the anterior part and backward in the posterior part.

Thanks to his multiple electrodes, Tenney was able to measure the speed of propagation of the contraction wave. In this way he noted that an inversed wave was always less rapid than a normal one. This polarization of the conduction may explain, at least in part, the fact that the normal propagation of the contractions goes forward. But we must recognize that no special structure of the dorsal vessel explains this polarization. Furthermore, in a neighboring species, *Samia cecropia*, McCann (1964a) found that stimulations produced by electric impulsions are propagated at the same speed in both directions. Finally, it is to be emphasized that, with a mathematical model, McCann (1970) was able to simulate a reversal when certain hypothesis concerning the excitability characteristics of heart cells are made.

E. Conclusion

Observed under conditions of explantation, the automatic functioning of the dorsal vessel makes it possible to understand satisfactorily how blood circulation is ensured in insects. But the physiology of the dorsal vessel itself is still little known. Thus, the part played by the aliform muscles and the importance of their eventual contractions are still debatable. The fact that the dorsal vessel functions when isolated implies the existence of pacemakers. The cells endowed with a pacemaker-type activity are certainly numerous, the reversal of the beats implies the intervention of a secondary pacemaker. Moreover, it is known that in certain species, at least (Duwez, 1936, 1938; Gerould, 1938; Tenney, 1953; Irisawa et al., 1956), each isolated cardiac chamber is capable of spontaneous contractile activity. As the contraction wave is propagated regularly in the whole organ, it must be admitted that only the posterior region presents a pacemaker activity and that the others are inhibited. Along the dorsal vessel, the conduction of the contraction wave is certainly done in an aneural way. The studies of the contractions and the electrical recordings give us no indications as to the muscular or neural nature of the pacemakers. This important problem will be discussed subsequently.

IV. INFLUENCE OF VARIOUS PHYSICAL FACTORS AND OF MINERAL IONS

The effects of various physical factors on the explanted dorsal vessel were studied. But the greater part of the experiments deal with the influence of the mineral ions.

A. Electrical Stimulations

Yeager (1939a) applied electrical stimulations to the explanted dorsal vessel of *Periplaneta*. The organ reacted to isolated shocks by extra-systoles, both when the stimulation was done during the systole and during the diastole. The refractory period is much shorter than in vertebrates, and the contractions achieved in this way are not followed by a compensatory pause. Stimulations closely following each other cause a complete cardiac tetanus. After interrupting them, a short post-tetanic pause is observed, and then normal rhythm is resumed. According to these experiments, the behavior of the dorsal vessel of insects is different from that of the vertebrate heart. In fact, in the latter, save rare exceptions, complete tetanus cannot be obtained and the heart is killed by tetanizing stimulations.

McCann (1964a) studied conduction in the dorsal vessel of *Samia* by means of global electrical impulsions. She noted that increases in the frequency of the cycles of intracellular action potentials are parallel to increases of stimulation. The speed of conduction does not depend on the direction of the wave propagation. More recently, McCann (1966) was able to study the effects of intracellular electrical impulsions on the membrane potentials. A linear relationship exists between the intensity of the electric current and the membrane potential which, therefore, appears to act as a simple ohm resistance.

B. Ionizing Radiations

The effects of ionizing radiations do not appear to have been studied to date in the semi-isolated dorsal vessel of adult insects. However, the work of Larsen (1964, 1965) on *Blaberus* embryo heart fragments should be noted. The persistence of contractions after irradiation was taken as a test of the effects of radiation. With high doses (82,000 rad), a period of initial shock in which half of the fragments cease contracting is observed. After that a period of restoration in which 90% of the fragments resume functioning is observed. Certain radio-protective substances prevent the initial shock and eventually increase the survival time.

C. Temperature

Like all metabolic processes, cardiac contractions are sensitive to temperature and various authors studied the influence of this factor. To set forth the results, it is convenient to separate the effects of temperatures above and below 0°C.

1. Temperature Above 0°C

The variations in the frequency of the contractions as a function of temperature result in curves of highly variable shape: an exponential-like graph in *Melanoplus* (Jahn and Koel, 1948), an "S" curve in *Acheta* (Bergerard and Reinberg, 1947), and a linear function in *Periplaneta* (Richards, 1963). Lastly, in *Dytiscus,* Duwez (1936) obtained a rather curious curve: rapid increase in the frequency of the contractions at low temperatures, slowing down at medium temperatures, new acceleration at high temperature.

The lowest limit of functioning is about 0° for *Dytiscus* and is identical in the intact animal and for the explanted heart (Duwez, 1936). The heart of various locusts stops at about 5° (Walling, 1906). In *Periplaneta,* Richards (1963) showed that the heart stops at 4° in the live insect and at 9° in the isolated organ. According to Richards, this difference is due to stimulations received by the dorsal vessel from the ventral nerve chain in the intact animal, which permit it to function at a lower temperature. In the explanted organ, on the contrary, the mechanisms of automatism may be blocked at a higher temperature.

In *Acheta domestica,* the explanted dorsal vessel ceases to contract at 11°. Bergerard and Reinberg (1947) did not compare this result with the one obtained in the intact animal. But they noted that hypercalcic solutions lower this limit to 6°, whereas hyperpotassic solutions increase it to above 15°.

The highest limit of automatic functioning varies with the species, i.e., 36° for *Acheta,* 50° for *Dytiscus,* 40° for *Periplaneta.* According to Walling (1906), the heart of various locusts still presents contractions at about 60°.

Generally, the effects of a change in temperature are reversible, at least as long as it has not gone beyond the superior limit. However, Jahn and Koel (1948) observed in *Melanoplus* a kind of hysteresis effect, i.e., the exact frequency observed at the beginning of the experiment was not reproduced after increasing the temperature and returning to the initial temperature.

Although the variations of the rhythm of contractions generally do not follow the Van t'Hoff law, some authors tried to adjust the data to the Van t'Hoff–Arrhenius equation. According to Jones (1964), the value of the coefficients in this equation allows one to distinguish the origin (muscle or nerve) of the cardiac automatism. The neurogenic hearts should have a high μ coefficient (12,000 to 14,000) the myogenic hearts a lower coefficient (4500 to 10,000). The experimental results obtained up to now are not clear, i.e., for *Melanoplus*, Jahn and Koel got a μ of 9000; for *Periplaneta* (Richards), μ varies, according to the temperature, from 7000 to 18,000.

Finally, it should be emphasized that temperature seems to modify the electrograms (Duwez, 1936; Crescitelli and Jahn, 1938), but no general rule can be considered and further research is desirable.

2. Temperature Below 0°C

Walling (1906) noted that the isolated dorsal vessel of locusts is capable of withstanding a temperature of $-10°$ for an hour.

Wilbur and McMahan (1958) studied the survival of the dorsal vessel of *Popilius disjunctus* subjected to a temperature below $0°$. Placed in a simple medium (see Table I) for three hours at a temperature between $-6°$ and $-8°$, two-thirds of the organs were incapable of resuming their activity after return to normal temperature. The resistance was much better if glycerol (5–10%) was added to the medium. A return of activity was then observed even after a very long stay in the cold (1–20 days). However, the return of the contractions was usually accompanied by a loss of their coordination in the organs treated in this way.

Asahina *et al.* (1954) studied cold resistance in the hibernating caterpillar of *Cnidocampa*. In the live insect, the hemolymph remains liquid well below its freezing point. Towards $-20°$, the super fusion ceases and the internal milieu crystallizes. However, the insect withstands the freezing of the internal milieu and the dorsal vessel resumes its functions after return to a high temperature. According to Asahina *et al.*, these results prove that the freezing is limited to the extracellular liquids. At still lower temperatures, ice crystals are also produced in the cells, but then the damage is irreversible and the dorsal vessel no longer resumes its activity.

D. Gas

When explanted, the dorsal vessel usually appears to be an aerobic organ. That is why many authors suggest an oxygenation system for the solution. For example, Bergerard et al. (1950) noted rapid failure of functioning if aeration of the medium is stopped. However, artificial aeration seems to be particularly necessary when the dorsal vessel is put into a container with little open surface or if the experiment is done at high temperature. The dorsal vessel of *Cybister,* put in a petri dish in 15 ml of medium, works perfectly at 19° without artificial oxygenation (Rougier, 1966). In contrast, if most hearts are stopped rather rapidly by lack of oxygen, some of them possess considerable anaerobic reserves. Illustrating this, Harvey and Williams (1958) noted that the dorsal vessel of *Cecropia* chrysalids in diapause contract for several hours under nitrogen atmosphere and have a strong resistance against cyanide. On the other hand, the hearts of larvae or adults of this species are very sensitive to the lack of oxygen.

The effects of various gases on locust hearts were analyzed by Walling (1906). CO_2 and CO, like ether, provoke a failure of functioning that is reversible. Hydrogen gives curious results, i.e., initial stimulation, then failure, and then return of functioning for several hours.

The effects of mixed gases have not yet been studied. Jones (1956b) showed that a low CO_2 content accelerates the heart of *Anopheles* in the intact insect. It is possible that similar effects would be observed in explanted organs.

E. pH of the Medium

The study of this factor is difficult since it is hard to distinguish the effects of the pH itself from those of the various chemicals added to the medium to modify the pH. Many researchers adjust the pH of the media with acid or alkaline solutions. But, as the media are not buffered, it is probable that the pH varies in the course of the experiments.

The influence of the pH on the frequency of the contractions was studied in *Blatta* by Kojantchikov (1932) and mainly in *Periplaneta* by Ludwig et al. (1957a), also in *Tenebrio* by Butz (1962). The extreme pH (too acid or too alkaline) turn out to be very noxious and do not allow normal functioning of the organ. When kept within medium values (5.5 to 8.5 for *Tenebrio,* for instance) the frequency of the contractions increases as the pH rises. The maximum is observed at a pH well above that of the hemolymph, which is 6.3 in *Tenebrio.*

If, instead of immediate effects, the functioning of an organ is observed for several days, more interesting results are obtained. In *Cybister* (David and Rougier, 1965) an increase in the frequency of the contractions is observed when the pH passes from 6.45 to 7.70. For pH above 7, the maximum effect is not immediate, i.e., the rhythm of contractions increases during the first two days of explantation as if the organ adapted itself to the conditions of explantation and gradually recovered from the effects of the surgical shock. The survival is maximum at a pH of 7.7 corresponding to that of the insect hemolymph. If a higher pH is used (8, for example) a high frequency of the contractions is indeed observed, but anomalies of functioning appear; the viability decreases so that the medium appears to be unsuitable (Rougier, 1966).

F. Osmotic Pressure

The osmotic pressure of the hemolymph of insects is most variable and usually rather high (Florkin and Jeuniaux, 1964). In most cases, it is equal to that of a NaCl solution of over 9 gm/liter and sometimes goes above the osmotic pressure of a solution of 15 gm/liter. We have already stated that the variability of the explantation media used gives the impression of a great tolerance of the dorsal vessel to osmotic pressure.

Several works analyzed the effects of a variation in the osmotic pressure of the medium, while the same proportions between the various components are maintained. In *Acheta* and *Galleria*, Bergerard (1947b), Dreux (1950b), and Bergerard *et al.* (1950) noted that hypotonic solutions caused an acceleration of the contractions and hypertonic ones, a slowing down.

If the concentration was too high (double of normal), there was failure in diastole; if the concentration was too low, (half normal) there was failure in systole. But all species do not react in the same way. Thus, Barsa (1955) noted that hypertonicity caused a decrease in the frequency of the contractions in *Samia* and, on the contrary, an acceleration in *Chortophaga*. The stage at which the study is made may also influence the results. According to Butz (1957), the effects of variations in osmotic pressure are different when the dorsal vessel of the larva, the nymph, or the adult *Tenebrio* is considered.

If only slight variations are taken into acount, all authors agree on the great tolerance of the dorsal vessel. In *Cybister*, for instance, differences between 80 and 120% from normal do not cause significant variations either in the length of functioning or in the frequency of the contractions (Rougier, 1966).

G. Influence of Cations

Like all contractile organs, the dorsal vessel of insects is very sensitive to the effect of cations and, in particular, to the respective proportions of Na, K, and Ca. Modifications in the quantity of these ions generally have spectacular effects and much work has been done in this field. Most often the investigators dealt with the frequency of the contractions. More recently, a few experiments 'analyzed the electrical activity.

1. Action on the Frequency of the Contractions

a. The Effects of Alkaline Ions. Many researchers showed that the Na/K ratio plays an important part. The following studies are noteworthy: Levy (1928b) on *Calliphora* and *Phormia;* Bergerard (1947a), Bergerard *et al.* (1950), and Dreux (1950a) on *Acheta* and *Galleria;* Barsa (1955) on *Chortophaga* and *Samia;* Jones (1956c) on *Anopheles;* Butz (1957) on *Tenebrio;* Ludwig *et al.* (1957b) on *Periplaneta;* and Brebbia and Ludwig (1962) on *Musca.*

As a rule, a rise in the potassium content increases the frequency of the contractions. If the concentration is too high, it leads to failure in systole. The increase in sodium has an opposite effect, i.e., slowing of the contractions and, finally, failure in diastole. From these experiments it may be concluded that both cations are necessary for the functioning of the dorsal vessel but that the tolerance of the organ to the Na/K ratio may be considerable. Normal functioning is possible if the ratio is between 3 and 34 in *Chortophaga* (Barsa, 1955), between 3 and 30 in *Periplaneta* (Ludwig *et al.*, 1957b), and between 0.07 and 34 in *Samia* (Barsa, 1955). The tolerance of *Samia* to very low sodium contents is confirmed by McCann (1963), who used an explantation medium in which the Na/K ratio was close to that of the hemolymph, i.e., about 0.04.

b. Action of the Alkaline-Earth Ions. Most explantation media contain only calcium. However, some authors studied the effects of adding magnesium and barium.

In *Acheta* and *Galleria* (Bergerard *et al.*, 1950; Fiszer, 1950), the substitution of Mg for Ca slows down the dorsal vessel if the Mg/Ca ratio is above 1. Failure in diastole is observed when the ratio attains 9. Barsa (1955) compared the effects of media with and without Mg, and he noted a better functioning of the dorsal vessel of *Chortophaga* and *Samia* when Mg is absent.

The effects of barium were also studied in *Acheta* and *Galleria* by

Bergerard *et al.* (1950) and by Dreux and Fiszer (1950). The results obtained are comparable to, but more pronounced than, those observed with magnesium.

c. **Proportions between Alkaline and Alkaline-Earth Ions.** Even if only Na, K, and Ca ions are considered, several proportions are possible. In fact, as we shall see, it is mainly the K/Ca ratio which is important.

The Na $+$ K/Ca ratio was studied by Bergerard *et al.* (1950). This ratio is 39 in the basic medium, but the heart functioning is still possible for a value of 3 (*Acheta*) or 5 (*Galleria*).

When this ratio is lowered by an increase in the quantity of Ca, a slowing down of the contractions and an increase in their amplitude is observed, leading finally to a stop in diastole.

In the above ratio, K and Ca are particularly important. Several authors studied the limits of the K/Ca ratio, which are compatible with normal functioning. The results, given in Table III, show that these limits are always rather narrow. In *Musca*, Brebbia and Ludwig (1962) indicate only the optimum ratio, which is 1.3 for the larva, 4.0 for the pupa, and 1.8 for the adult.

Bergerard *et al.* (1950) tried replacing the Ca by Mg and Ba. The antagonistic action of Mg towards K turned out to be less strong than that of Ca. It is when the medium contains only Ca that the heart may sustain the highest quantity of K. Barium has effects similar to those of magnesium.

d. **Relations with Osmotic Pressure.** Bergerard (1947b), Bergerard *et al.* (1950), and Dreux (1950b) noted that in *Acheta* and *Galleria* an increase in the osmotic pressure of the medium tends to slow down the frequency

TABLE III

LIMITS OF THE K/Ca RATIO COMPATIBLE WITH THE FUNCTIONING OF THE EXPLANTED DORSAL VESSEL IN VARIOUS SPECIES

Species	Instar	K/Ca ratio		Authors
		Inferior limit	Superior limit	
Periplaneta americana	N or A	0.9	3.5	Ludwig *et al.* (1957b)
Chortophaga viridifasciata	A	1	3	Barsa (1955)
Tenebrio molitor	L	0.33	3	Butz (1957)
Tenebrio molitor	P	1	3	Butz (1957)
Tenebrio molitor	A	0.5	3	Butz (1957)
Samia walkeri	P	1	3	Barsa (1955)

of the contractions. These authors were thus able to counterbalance the stimulating effects of K by an increase in the osmotic pressure. Conversely, the systolizing effect of hypotonic solutions may be compensated by an excess of Ca. But these results seem difficult to generalize at present since the effects of the osmotic pressure appear to vary, as we have seen, according to the species.

2. Influence of Cations on Electric Activity

Up to now, McCann (1963) has performed the only study in *Telea* and *Samia*. She recorded the variations of the action potentials after complete suppression of Na, Ca, and Mg.

The absence of Na causes no appreciable modification of the recordings. The suppression of K results in a slowing down and then stoppage by hyperpolarization of the membrane and an increase of the resting potential. The absence of Ca also causes slowing down by increase in the period of repolarization; at the same time, the action potential splits into two parts. It is rather curious that the most distinct phenomena are observed in the absence of Mg, i.e., the suppression of the ion causes an acceleration of the rhythm of contraction and a decrease of the value of the resting potential. After three hours the resting potential is very low; there is no longer any action potential recorded and the organ presents feeble and badly coordinated contractions.

According to these results, McCann (1963, 1964a,b, 1965) supposes that, contrary to what happens in vertebrates, the K and Na ions do not play an essential part in the production of the action potentials. On the contrary, the Mg ion appears to be necessary for their production.

3. Conclusion

The study of the frequency of the contractions reveals the action of two main cations, i.e., potassium, which has a stimulating effect, and calcium, which has the opposite effect. The part played by sodium is rather small since the explanted dorsal vessel tolerates concentrations extremes. However, its total absence seems incompatible with long functioning. Finally, magnesium poses a curious problem. Although this ion is always present, often abundant, in the hemolymph, no study has ever showed the use of Mg excepting the work of McCann (1963). It is also to be noted that McCann's observations, made in *Telea* and *Samia*, are in contradiction with those of Barsa (1955) in *Samia*. The latter, noted normal functioning in the absence of Mg. This contradiction may perhaps be explained by the fact that the explantation media used by both these authors were very different (Table I). This suggests the desirability of further research in this field.

H. Influence of Anions

As shown in Table I, the anion almost exclusively used in the explantation media is chlorine. However, small quantities of phosphate and bicarbonates may also be found. The part played by anions in the explantation medium has barely been studied. However, the work of McCann (1964b), who replaced chlorides by bromides, acetates, nitrates, and sulphates, while maintaining the same balance between the cations, may be mentioned.

Bromides have little influence on the dorsal vessel of *Samia*. Acetates cause rapid stoppage by hyperpolarization of the membrane; nitrates have a brief slowing effect followed by an acceleration. Finally, sulfates produce an acceleration, then a depolarization, of the membrane, which leads to a failure in functioning.

From these results, McCann concludes that the permeability of the cellular membrane to anions is of the following decreasing order: acetate > nitrate > bromine > chlorine. The membrane seems impermeable to the sulfate ion.

I. Conclusion

Much work has already been done on the dorsal vessel of insects, dealing with the part played by physical factors or mineral elements of the medium. However, our knowledge is in great part incomplete. Sometimes, the experimental data are scarce (in the case of anions). Sometimes a great variability between species makes it impossible to draw a general conclusion. (The case of osmotic pressure is noted.) Moreover, most researchers studied only hearts functioning for a short period, considering the organ as stabilized. The study of new criteria appears to be absolutely necessary, particularly the duration of functioning and the electrical activity.

In the absence of much wider experimental data, it seems difficult at present to draw precise conclusions about the part played by the various factors and about their interactions.

V. INFLUENCE OF ORGANIC SUBSTANCES

In order to lengthen the functioning of the dorsal vessel, various nutritive organic substances may be added to the medium. But most studies deal either with the influence of pharmacological substances active on the vertebrate heart, or with toxic substances, or else, with hormones.

A. Nutritive Organic Substances and Antibiotics

We have seen that various simple media contain carbohydrate as well as mineral elements. In general, authors indicate a better functioning of the organ. A precise study of the dorsal vessel of *Cybister* was done by Rougier (1966). The addition of glucose to a mineral medium distinctly improves viability which passes from 35 to over 100 hours. The frequency of the contractions does not seem to be affected. The effect of glucose is maximum when its concentration attains 1.5 gm/liter. Higher glucose content does not make any difference. In the same way the addition of antibiotics (penicillin and streptomycin) lengthen the duration of functioning by reducing the risks of infection. The use of antibiotics appears to be a necessity if the viability is to be studied without working in aseptic conditions.

Finally, it is to be recalled that the use of a complex medium (T.C. 199) allows much longer periods of functioning [two weeks for the dorsal vessel of *Cybister* (Rougier, 1967)]. But the effects of the organic elements of this type of medium (amino acids, vitamins, etc.) have not been studied to date.

B. Chemical Mediators of the Nerve Influx

1. Acetylcholine

The cardio-accelerating action of acetylcholine has been observed in many species of insects. In particular, the following may be quoted: *Periplaneta* (Krijgsman and Krijgsman-Berger, 1951; Naidu, 1955, 1958; Unger, 1957; Metcalf *et al.*, 1964; Miller and Metcalf, 1968), *Blatta* (Prosser, 1942; Unger, 1957), *Melanoplus* (Hamilton, 1939), *Stenopelmatus* (Davenport, 1949), and *Tenebrio* (Butz, 1962). However, no effect was observed in Lepidoptera [*Galleria* larva (Millman, 1935, quoted in Jones, 1964), adult *Telea* and *Samia* (McCann, 1965), *Anopheles* (Jones, 1954, 1956d), and the denervated heart of *Periplaneta* (Miller and Metcalf, 1968)].

2. Epinephrine

The effects of epinephrine on intact insects or on explanted hearts appear to be highly variable according to the concentration used and the species considered (Jones, 1964).

In the semi-isolated dorsal vessel it may be noted that a stimulating effect, more or less durable, was observed by Krijgsman and Krijgsman-

Berger (1951), Naidu (1958), and Miller and Metcalf (1968) in *Periplaneta* and by McCann (1965) in *Samia*. In the grasshopper *Stenopelmatus* (Davenport, 1949), epinephrine stops the heart in diastole. In *Anopheles* (Jones, 1954, 1956d) the effects are very slight.

3. Serotonine

A stimulating action of 5,hydroxy-tryptamine was noted by Davey (1961a) on the dorsal vessel of the *Periplaneta* and confirmed by Brown (1965) and by Miller and Metcalf (1968).

C. Alkaloids

Many alkaloids have been used because of their action on the chemical transmission of the nerve influx or on the nerve cells themselves. For instance, ergotamine, which inhibits the adrenergic systems, slows the heart of the *Periplaneta* (Krijgsman and Krijgsman-Berger, 1951; Naidu, 1955). Similarly, atropine, the antagonist of acetylcholine, slows the dorsal vessel of *Periplaneta* (Krijgsman and Krijgsman-Berger, 1951; Naidu, 1955; Metcalf *et al.*, 1964) and of *Stenopelmatus* (Davenport, 1949). Eserine, an anticholinesterasic substance increases the effectiveness of acetylcholine (Davenport, 1949; Krijgsman and Krijgsman-Berger, 1951; Naidu, 1955) but has no effect on the dorsal vessel of *Anopheles* (Jones, 1956d).

Other alkaloids, active upon the vertebrate heart or vegetative system, were tried on the dorsal vessel of insects. For instance, caffeine, morphine, and strychnine were employed. None of these substances had any notable effect on the dorsal vessel of *Periplaneta* (Krijgsman and Krijgsman-Berger, 1951). On the other hand, caffeine and morphine affect the cardiac rhythm of intact *Anopheles* larvae (Jones, 1956b).

Finally, nicotine was often studied because of its insecticidal properties (Jones, 1964). In vertebrates, nicotine stimulates or blocks the functioning of the nerve cells. In insects, many authors observed a stimulation or a cessation of cardiac activity according to the concentration used (Yeager and Gahan, 1937; Hamilton, 1939; Davenport, 1949; Krijgsman and Krijgsman-Berger, 1951; Naidu, 1955; Agrawal and Srivastava, 1962; Miller and Metcalf, 1968).

D. Insecticides

Many insecticides affecting the nervous system were tried on the dorsal vessel of insects. Studies were often undertaken if the death of the insect could be associated with failure of the blood circulation. Actually, many insecticides or toxic substances barely affect the dorsal vessel in whole insects (Jones, 1964).

The effects of nicotine on the explanted heart were described above. Various products, like pyrethrins, organophosphorous compounds, and carbamates, which block cholinesterase, have a cardio-accelerating action by reinforcing the effects of acetylcholine (Davenport, 1949; Krijgsman et al., 1950; Krijgsman and Krijgsman-Berger, 1951; Agrawal and Srivastava, 1962; Metcalf et al., 1964).

According to Naidu and Zaheer (1956), allethrine causes a temporary decrease in the frequency of the contractions. Like nicotine, this insecticide seems to act on the functioning of the nerve centers. Thiocyanates slow or stop the dorsal vessel in *Blatta* (Yeager et al., 1935).

DDT has been dealt with in several studies, Davenport (1949) observed no effect on *Stenopelmatus*. On the contrary, Agrawal and Srivastava (1962) observed an essentially inhibiting effect on the heart of *Mylabris* and in two species of Lepidoptera. In *Blatta* intoxicated by DDT, the hemolymph contains cardio-accelerating substances (Colhoun, 1959; Sternburg, 1960, 1963), but these substances do not seem to be derived from DDT; they are unknown products (perhaps hormones) which are released into the blood by a stress phenomenon.

E. Insect Hormones

Many authors observed cardiotropic effects of substances extracted from insects. Some of these substances may be known products such as acetylcholine (Unger, 1957) or adrenaline (Barton-Browne et al., 1961). Others seem to be specific hormones and are of greater interest. Cameron (1953) showed that the *corpora cardiaca* of *Periplaneta* contains a cardio-accelerating substance which may be an orthodiphenol. This research was resumed by Unger (1956, 1957), who isolated in various parts of the body (and, in particular, in the corpora cardiaca) two cardio-accelerating substances: a substance C, which provokes a constriction of the dorsal vessel and a substance D, which causes dilatation. Later (Gersch et al., 1960), each of these hormones, C and D, turned out to be composed of a mixture of two products respectively, C_1 and C_2 and D_1 and D_2. At least some of these neurohormones are supposed to be peptides. Brown (1965), who isolated two cardio-accelerating factors, also considered them to be peptides.

In *Periplaneta*, Davey (1961a,b, 1962b) also obtained a cardio-accelerating factor from the corpora cardiaca. This factor is supposed to be a peptide, but its way of action is indirect, i.e., the peptide seems to stimulate the pericardial cells which themselves release a true cardio-tropic substance (probably an indolalkylamine). More recently, Davey (1963) contributed precise data on the metabolic mechanisms involved in the stimulation of the pericardial cells.

The cardio-accelerating action of the corpora cardiaca was also observed by other authors (Ralph, 1962; Cazal, 1967). A similar effect was obtained from perisympathetic organs of the ventral nerve chain (Raabe et al., 1966) in various species, e.g., Carausius, Clitumnus, Periplaneta, and Locusta.

Cardio-accelerating substances may be found elsewhere than in the nervous system and the endocrine glands. It is of interest to note that Davey (1960) observed the existence of a substance of this type in the accessory male glands of Periplaneta.

In all the foregoing works an increase in the rhythm of the beats was obtained. Inversely, Ralph (1962) showed that substances which slow down the frequency of the contractions may also be extracted and isolated from the brain and the nerve centers of the ventral nerve chain in Periplaneta. According to Ralph, it is possible that these factors may be five in number.

F. Conclusion

The numerous substances tested on the dorsal vessel of insects provide us with data concerning two main problems: the endocrine regulation of the cardiac activity and the nature of the automatism mechanisms.

The intervention of hormones is interesting for it allows one to suspect the existence of a non-nervous regulating mechanism in the living animal. It is also known that various sensorial stimuli are capable of modifying the cardiac rhythm (Jones, 1956a; Davey, 1962a). Present results let us expect great intricacy since several accelerating and slowing-down substances seem to exist. If the main substances appear to be neurohormones produced by the nervous system it is not to be forgotten that, at least in some cases, a metabolic relay at pericardial cell level seems necessary.

The nature of the cardiac automatism (myogenic or neurogenic) is studied by means of various substances known to have effect on the activity of the nervous system. Taking into account the various results, and particularly the usual cardio-accelerating effects of acetylcholine, Krijgsman and Krijgsman (1950) consider that the heart of insects is a neurogenic heart. Developing this view, Krijgsman (1952) supposes the existence in the dorsal vessel of two types of nerve cells: large pacemaker cells, in relation with the central nervous system, and motor neurons transmitting the nerve influx to the muscle cells. The pacemaker cells are supposed to be cholinergic and the motor neurons, adrenergic.

However, this conception is not accepted by all authors because the results vary according to the species, so McCann (1965) considers that the heart of *Samia* and of *Telea* have a myogenic activity.

VI. GENERAL CONCLUSION

Although much research work was done on the explanted dorsal vessel of insects, our knowledge still remains very fragmentary. In these conditions this review should state the problems and suggest new research rather than draw precise conclusions.

It should be noted that there are several difficulties in interpretation of the bibliographical results, difficulties which are linked to the complexity of the techniques used and the slowness in the realization of the experimentation.

Individual variability may be mentioned first. Various investigators note that organs coming from a homogeneous insect population and placed in the same medium produce considerable difference from one to another. There is little doubt that explantation is at the source of the variability. Under such conditions, any experimental study must be based on the observation of a sufficient number of organs and upon rigorous statistical analysis. Now it is not certain that all the work published was realized with the necessary experimental strictness, and this perhaps explains some contradictory observations. The variability of the results doubtless also comes from the great diversity of species studied. It is probable that the physiology of the explanted dorsal vessel varies not only as a function of the species but also as a function of the stadium. A last difficulty in the comparison of the results resides in the fact that each work was, almost always, limited to the use of a small number of criteria, in particular, the frequency of the contractions. But a number of other characters can be studied in the explanted dorsal vessel.

The results obtained to date seem to revolve around two main centers of interest: On one hand, the knowledge of the nutritional needs of an insect organ; on the other hand, the knowledge of the mechanisms of contraction, of its automatism and its regulations.

The nutritional needs of the dorsal vessel have not been thoroughly studied. Apparently, the organ needs a balanced ionic solution, an energetic source (carbohydrate), oxygen in all cases, an osmotic pressure and a pH adjusted to the values of the hemolymph of the species studied. But even the minimal requirements sufficient to keep the organ functioning for a few days, are not fully known. In fact, the dorsal vessel appears to be very tolerant and functions in a great variety of media which are very different from the hemolymph. It is to be noted that the part played by the anions is virtually unknown, the cations are generally limited to three (Na, K, and Ca), and the utility of Mg, always present in the hemolymph, is a debated problem.

Longer viability of the explanted organ must require more complex media. It is to be regretted that, up to now, no work has dealt with the functioning of the dorsal vessel in this type of medium; researchers have more often tried to obtain cellular migrations outside the dorsal vessel and then cultures of cells. It is probable that, according to the purpose followed (cell culture or, on the contrary, maintaining the integrity and function of the organ), different media will have to be used.

The greater part of research work has dealt with the physiology of the dorsal vessel. About this it seems well to make two reservations. First, the diversity of the media used suggests that most of them were not well adapted. So it may be supposed that this faulty adaptation of some media is at the source of some contradictions. Second, the explants achieved include not only the cardiac vessel and its eventual innervation but also a complex of other tissues (muscles, pericardial cells, fat body, etc). More or less important interactions between these various tissues are always possible.

The results obtained are often discussed in terms of what is already known of the vertebrate heart. Several differences were noted, and it is possible that the dorsal vessel of insects may present many particularities. In this respect, finding that in certain species function is possible with a very small quantity of sodium is to be particularly recalled. According to McCann (1965, 1970), sodium is little involved in the realization of the action potentials of phytophagous insects.

In attempting to compare the totality of the results, it is often very difficult to draw a conclusion applicable to all species. The problem of the regulation of cardiac activity and its automatism illustrates this case.

In most insects, the cardiac contractions seem to be regulated by hormonal and nervous mechanisms. The existence of a nervous regulation seems proven by the frequent presence of a specific innervation, by the surgical shock ensuing after explantation and by the fact that semi-isolated organs generally contract less rapidly than in the intact insect. The dorsal vessel, when explanted, continues its activity, which establishes the existence of pacemakers. But the very nature of these (nervous or muscular) is debated. If the detail of the results is examined, it seems possible, however, to come to at least a temporary conclusion: in primitive insects, such as cockroaches, the automatism seems neurogenic; in more evolved insects, such as certain Lepidoptera and Diptera, a purely myogenic automatism appears. It must also be considered that the existence of an intermediate type (myogenic and innervated dorsal vessel) is possible and further complicates interpretations (Jones, 1954, 1964).

Finally, many problems remain open. To solve them, it is desirable that future research should be limited to a small number of species while the number of criteria studied simultaneously should be increased. A particular effort must be made to improve the composition of the media of explantation and to adapt them to each of the species studied.

ACKNOWLEDGMENTS

We thank Professor G. Le Douarin, who agreed to read the manuscript of this review, and Professor H. A. Bender, who corrected the English form of the text.

REFERENCES

Agrawal, N. S., and Srivastava, S. P. (1962). *Agra. Univ. J. Res., Sci.* **11**, 23.
Alexandrowicz, J. S. (1926). *J. Comp. Neurol.* **41**, 291.
Asahina, E., Aoki, K., and Shimozaki, J. (1954). *Bull. Entomol. Res.* **45**, 329.
Barsa, M. C. (1955). *J. Gen. Physiol.* **38**, 79.
Barton-Browne, L., Dodson, L. F., Hodgson, E. S., and Kiraly, J. K. (1961). *Gen. Comp. Endocrinol.* **1**, 232.
Beard, R. L. (1953). *In* "Insect Physiology" (K. D. Roeder, ed.), pp. 232–272. Wiley, New York.
Bergerard, J. (1947a). *C. R. Soc. Biol.* **141**, 1079.
Bergerard, J. (1947b). *C. R. Soc. Biol.* **141**, 1081.
Bergerard, J., and Reinberg, A. (1947). *C. R. Soc. Biol.* **141**, 1083.
Bergerard, J., Dreux, P., and Fiszer, J. (1950). *Arch. Sci. Physiol.* **4**, 225.
Brebbia, R., and Ludwig, D. (1962). *Ann. Entomol. Soc. Amer.* **55**, 131.
Brocher, F. (1917). *Arch. Zool. Exp. Gen.* **56**, 347.
Brown, B. E. (1965). *Gen. Comp. Endocrinol.* **5**, 387.
Butz, A. (1957). *J. N. Y. Entomol. Soc.* **65**, 22.
Butz, A. (1962). *Ann. Entomol. Soc. Amer.* **55**, 480.
Cameron, M. L. (1953). *Nature (London)* **172**, 349.
Cazal, M. (1967). *C. R. Acad. Sci.* **264**, 842.
Colhoun, E. H. (1959). *Can. J. Biochem. Physiol.* **37**, 259.
Collings, S. B. (1966). *Ann. Entomol. Soc. Amer.* **59**, 972.
Crescitelli, F., and Jahn, T. L. (1938). *J. Cell. Comp. Physiol.* **11**, 359.
Davenport, D. (1949). *Physiol. Zool.* **22**, 35.
Davey, K. G. (1960). *Can. J. Zool.* **38**, 39.

Davey, K. G. (1961a). *Nature (London)* **192**, 284.
Davey, K. G. (1961b). *Gen. Comp. Endocrinol.* **1**, 24.
Davey, K. G. (1962a). *J. Insect Physiol.* **8**, 205.
Davey, K. G. (1962b). *Quart. J. Microsc. Sci.* **103**, 349.
Davey, K. G. (1963). *J. Exp. Biol.* **40**, 343.
David, J., and Rougier, M. (1965). *C. R. Acad. Sci.* **261**, 1394.
Demal, J., and Leloup, A. M. (1963). *Ann. Epiphyt.* **14**, 91.
De Wilde, J. (1947). *Arch. Neer. Physiol.* **28**, 530.
Dreux, P. (1950a). *C. R. Soc. Biol.* **144**, 803.
Dreux, P. (1950b). *C. R. Soc. Biol.* **144**, 804.
Dreux, P., and Fiszer, J. (1950). *C. R. Soc. Biol.* **144**, 818.
Dreux, P., and Grapin-Poupeaux, D. (1963). *C. R. Soc. Biol.* **157**, 1000.
Dubuisson, M. (1929). *Arch. Biol.* **39**, 247.
Dubuisson, M. (1930). *Arch. Biol.* **40**, 83.
Duwez, Y. (1936). *C. R. Soc. Biol.* **122**, 84.
Duwez, Y. (1938). *Arch. Int. Physiol.* **46**, 389.
Edwards, G. A., and Challice, C. E. (1960). *Ann. Entomol. Soc. Amer.* **53**, 369.
Fiszer, J. (1950). *C. R. Soc. Biol.* **144**, 812.
Florkin, M., and Jeuniaux, C. (1964). *In* "The Physiology of Insecta" (M. Rockstein, ed.), Vol. 3, pp. 109–152. Academic Press, New York.
Gerould, J. H. (1933). *Biol. Bull.* **64**, 424.
Gerould, J. H. (1938). *Acta Zool.* **19**, 297.
Gersch, M., Fiszer, F., Unger, H., and Koch, H. (1960). *Z. Naturforsch. B* **15**, 319.
Griffiths, J. T., and Tauber, O. E. (1943). *J. Gen. Physiol.* **26**, 541.
Hamilton, H. L. (1939). *J. Cell. Comp. Physiol.* **13**, 91.
Harvey, W. R., and Williams, C. M. (1958). *Biol. Bull.* **114**, 23.
Hirumi, H., and Maramorosch, K. (1964). *Contrib. Boyce Thompson Inst.* **22**, 259.
Irisawa, H., Irisawa, A. F., and Kadotani, T. (1956). *Jap. J. Physiol.* **6**, 150.
Ishikawa, S. (1959). *J. Sericult. Sci. Jap.* **28**, 295.
Jahn, T. I., and Koel, B. S. (1948). *Ann. Entomol. Soc. Amer.* **41**, 258.
Jahn, T. L., Crescitelli, F., and Taylor, A. B. (1937). *J. Cell. Comp. Physiol.* **10**, 439.
Jones, J. C. (1954). *J. Morphol.* **94**, 71.
Jones, J. C. (1956a). *J. Exp. Zool.* **131**, 223.
Jones, J. C. (1956b). *J. Exp. Zool.* **131**, 257.
Jones, J. C. (1956c). *J. Exp. Zool.* **133**, 124.
Jones, J. C. (1956d). *J. Exp. Zool.* **133**, 573.
Jones, J. C. (1964). *In* "The Physiology of Insecta" (M. Rockstein, ed.), Vol. 3, pp. 1–107. Academic Press, New York.
Kojantchikov, I. V. (1932). *Leningrad Inst. Control Farm Forest Pests Bull.* **2**, 149.
Krijgsman, B. J. (1952). *Biol. Rev.* **27**, 320.
Krijgsman, B. J., and Krijgsman, N. E. (1950). *Nature (London)* **165**, 936.
Krijgsman, B. J., and Krijgsman-Berger, N. E. (1951). *Bull. Entomol. Res.* **42**, 143.
Krijgsman, B. J., Dresden, D., and Berger, N. E. (1950). *Bull. Entomol. Res.* **41**, 141.
Kuwana, Z. (1932). *Bull. Seric. Exp. Sta., Govt.-Gen. Chosen* **8**, 109.
Larsen, W. (1963). *Ann. Entomol. Soc. Amer.* **56**, 720.
Larsen, W. (1964). *Life Sci.* **3**, 539.
Larsen, W. (1965). *Utah Acad. Sci., Arts Lett., Proc.* **42**, 53.
Larsen, W. (1967). *J. Insect Physiol.* **13**, 613.
Levy, R. (1928a). *C. R. Soc. Biol.* **122**, 1482.
Levy, R. (1928b). *C. R. Soc. Biol.* **122**, 1485.

Ludwig, D., Tefft, E. R., and Suchyta, M. D. (1957a). *J. Cell. Comp. Physiol.* **49**, 503.
Ludwig, D., Tracey, K. M., and Burns, M. L. (1957b). *Ann. Entomol. Soc. Amer.* **50**, 244.
McCann, F. V. (1963). *J. Gen. Physiol.* **46**, 803.
McCann, F. V. (1964a). *Comp. Biochem. Physiol.* **12**, 117.
McCann, F. V. (1964b). *Comp. Biochem. Physiol.* **13**, 179.
McCann, F. V. (1965). *Ann. N. Y. Acad. Sci.* **127**, 84.
McCann, F. V. (1966). *Comp. Biochem. Physiol.* **17**, 599.
McCann, F. V. (1967). *Comp. Biochem. Physiol.* **20**, 399.
McCann, F. V. (1970). *Annu. Rev. Entomol.* **15**, 173.
McIndoo, N. E. (1945). *J. Comp. Neurol.* **83**, 141.
Maloeuf, N. S. R. (1935). *Ann. Entomol. Soc. Amer.* **28**, 332.
Metcalf, R. L., Winton, M. Y., and Fukuto, T. R. (1964). *J. Insect Physiol.* **10**, 353.
Miller, T., and Metcalf, R. L. (1968). *J. Insect Physiol.* **14**, 383.
Miller, T., and Thomson, W. W. (1968). *J. Insect Physiol.* **14**, 1099.
Mitsuhashi, J. (1965). *Jap. J. Appl. Entomol. Zool.* **9**, 217.
Naidu, M. B. (1955). *Bull. Entomol. Res.* **46**, 205.
Naidu, M. B. (1958). *Indian J. Entomol.* **20**, 147.
Naidu, M. B., and Zaheer, S. H. (1956). *Indian J. Entomol.* **18**, 57.
Opoczynska-Sembratova, Z. (1936). *C. R. Acad. Cracow,* p. 411.
Prosser, C. L. (1942). *Biol. Bull.* **83**, 145.
Raabe, M., Cazal, M., Chalaye, D., and De Besse, N. (1966). *C. R. Acad. Sci.* **263**, 2002.
Ralph, C. L. (1962). *J. Insect Physiol.* **8**, 431.
Richards, A. G. (1963). *J. Insect Physiol.* **9**, 597.
Rougier, M. (1966). *Diplôme d'Etudes Supérieures, Lyon.*
Rougier, M. (1967). Unpublished results.
Sanger, J. W., and McCann, F. V. (1968). *J. Insect Physiol.* **14**, 1105.
Sternburg, J. (1960). *J. Agr. Food. Chem.* **8**, 257.
Sternburg, J. (1963). *Annu. Rev. Entomol.* **8**, 19.
Tenney, S. M. (1953). *Physiol. Comp. Oecol.* **3**, 286.
Unger, H. (1956). *Naturwissenschaften* **43**, 66.
Unger, H. (1957). *Biol. Zentralbl.* **76**, 204.
Uramoto, S. (1932). *Bull. Imp. Seric. Exp. Sta., Jap.* **8**, 121.
Vago, C., and Quiot, J. M. (1969). *Ann. Zool. Ecol. Anim.* **1**, 281.
Walling, E. V. (1906). *J. Exp. Zool.* **3**, 621.
Wilbur, K. M., and McMahan, E. A. (1958). *Ann. Entomol. Soc. Amer.* **51**, 27.
Williams, G. T., Ballard, R. C., and Hall, M. S. (1968). *Nature (London)* **220**, 1241.
Yeager, J. F. (1938). *J. Agr. Res.* **56**, 267.
Yeager, J. F. (1939a). *J. Agr. Res.* **59**, 121.
Yeager, J. F. (1939b). *Ann. Entomol. Soc. Amer.* **32**, 44.
Yeager, J. F., and Gahan, J. B. (1937). *J. Agr. Res.* **55**, 1.
Yeager, J. F., and Hager, A. (1934). *Iowa State Coll. J. Sci.* **8**, 391.
Yeager, J. F., Hager, A., and Straley, J. M. (1935). *Ann. Entomol. Soc. Amer.* **28**, 256.
Zawarzin, A. (1911). *Z. Wiss. Zool.* **100**, 245.

7

INVERTEBRATE CELL AND ORGAN CULTURE IN INVERTEBRATE PATHOLOGY

C. Vago

I. INTRODUCTION

During the past 10 years, the development and perfecting of invertebrate tissue culture has made it increasingly possible to study infection
processes and physiological changes *in vitro*. Recent technical reviews
(Flandre, 1967; Vago, 1967; Vago and Bergoin, 1968) have shown that,
until now, most publications in this field only concern insects; but new
data allow us to extend the application of tissue cultures to the pathology
of other groups of invertebrates.

It is only natural that viruses and rickettsiae constitute virtually the
total of the microorganisms studied so far. Indeed, these organisms are
necessarily parasitic and intracellular, and their development requires live
cells in good, healthy condition.

The advantage of *in vitro* studies consists in the possibilities offered
for investigations of intracellular pathogeny on the ultramicroscopic
scale and of cellular and tissue affinities. They also enable one to maintain, for a prolonged period of time, a particular strain of a microorga-
nism by subculturing infected cells and to favor, by the same method,
the interspecific adaptation of pathogens.

At present, studies of infection of tissue cultures with invertebrate
pathogens are being made increasingly in specialized laboratories. They
constitute an indispensable stage in the complete study of a disease by
invertebrate pathologists. It is probable that these methods will have
the same interest for them as vertebrate tissue cultures have for micro-
biologists in medical investigations.

II. TISSUE AND ORGAN CULTURES USED

The tissues and organs of invertebrates explanted and maintained in
culture were for a long time, in the majority of cases, the ovarian sheaths

of the female, and naturally enough it is this tissue that has been used most frequently as a substrate for *in vitro* infection with pathogenic microorganisms. However, during recent years, there has been a tendency to make use of a wide range of cultured tissues and cells.

A. Cell Cultures

The fibroblasts of ovarian sheaths are very suitable for infection by different types of viruses, and these cells present the advantage of developing relatively easily in culture (Trager, 1935; Aizawa and Vago, 1959; Gaw *et al.*, 1959; Vago and Bergoin, 1963; Vaughn and Faulkner, 1963; Vago and Luciani, 1965).

The hemocytes, which have a tendency to migrate to form monocellular layers, have been used with success (Medvedeva, 1960; Martignoni and Scallion, 1961; Sen Gupta, 1963a,b; Mitsuhashi, 1966a). Martignoni and Scallion have, in particular, described the possibilities of utilizing these cells for the study of nuclear polyhedrosis viruses. Their investigations concern blood cells of the "plasmatocyte" type in the Lepidoptera *Peridroma saucia* Hübner.

Recently, cardiac cells sampled from a dozen different species of arthropods have been cultured (Vago *et al.*, 1968), and those of the cricket and the cockroach have been infected successfully with rickettsiae and *Tipula* iridescent virus (Meynadier *et al.*, 1968).

Finally, embryonic cells from *Bombyx mori* L. or *Drosophila* have been used on rare occasions (Takami *et al.*, 1966; Ohanessian and Echalier, 1967).

Most pathological studies have been carried out in cell cultures derived from tissue explants in liquid medium. This is largely due to the prevalence of this type of culture.

Infection was also performed in dissociated cell cultures obtained by submitting the excised tissue or organ to an enzymatic (Aizawa and Vago, 1959; Gaw *et al.*, 1959; Grace, 1962; Vago and Bergoin, 1963; Bellett and Mercer, 1964) or mechanical action (Meynadier *et al.*, 1968).

Cell culture in plasma coagulum is interesting in the sense that the cells are more easily maintained in position than in a liquid medium (Flandre and Vago, 1963). Hence, it would seem to be a particularly suitable technique for tracing the different stages of the cytoplasmic action of the virus inside these same cells. Vago and Bergoin (1963) used it with success for infection of the gonads and ovaries of several Lepidoptera by different viruses.

Various types of invertebrate cell cultures have been used for *in vitro* infection tests, i.e., hanging drop, placed drop, roll tube, tubes with slides, reversible flasks, and plastic flasks.

B. Organ Cultures

Most of pathological studies have been carried out in cultures of individual cells. The first case of the utilization of organ cultures in invertebrate virology involved lobes of fat body of Lepidoptera (*B. mori*), maintained in a liquid medium (Vago and Chastang, 1960). There was little migration of cells from this organ, which remained receptive to nuclear polyhedrosis viruses.

Lastly, suspensions of a large quantity of tissues and fragments of organs in a liquid medium (BM 22), maintained with antibiotics for only three days, allowed Vago et al. (1970) to produce large quantities of polyhedra and nuclear polyhedrosis viruses.

Isolated cases of the use of whole lepidopterous embryos explanted *in vitro* have been reported (Takami et al., 1966).

Whole organ cultures on solid media [Wolff and Haffen (1952) system] were used with success in invertebrate pathology by Vago and Quiot (1970), Devauchelle et al. (1969), and Quiot et al. (1970) to obtain multiplication of polyhedrosis, densonucleosis, and iridescent viruses and of mycoplasma.

III. INFECTION

There are several ways of preparing infective material and of administering it to tissue or organ cultures, from living material or using a purified pathogen. The choice of one or the other of these methods depended on the type of virus used, its abundance, and localization in the tissues.

A. With Nonpurified Material

In the first experiments of contamination of cultures, the infective material consisted of hemolymph from diseased insects and that had been treated in different ways.

For viruses, the development of which takes place in the tissues of visceral cavity, the hemolymph is sampled aseptically, centrifuged, and the supernatant taken as inoculum (Trager, 1935; Aizawa and Vago, 1959; Vago and Chastang, 1960; Grace, 1962; Vaughn and Faulkner, 1963; Vago and Bergoin, 1963).

The polyhedral or granular inclusion bodies being noncontagious at the cellular level, it would seem useful to determine the phase of the disease when the free viruses, as yet unenclosed in the inclusion bodies, are most abundant in the tissues and hemolymph. In the case of the nuclear polyhedrosis virus of *Galleria mellonella* at 30°C this stage is found about the 4th to 6th day following infection *per os* of the larva (Heitor, 1963; Quiot and Luciani, 1968).

In the case of infection by viruses whose cycle does not include the formation of protein inclusion bodies, the process is more direct. Hemolymph of *G. mellonella* larvae infected by densonucleosis virus was introduced directly into cell cultures of Lepidoptera (Vago and Luciani, 1965). Cultures of ovarian cells of *Antheraea eucalypti* were infected in the same way with hemolymph from larvae of *G. mellonella* infected either by the *Sericeshis* iridescent virus or by the *Tipula* iridescent virus (Hukuhara and Hashimoto, 1967).

As far as rickettsiae are concerned, infection of cardiac cell cultures of Orthoptera was obtained with a drop of fat body from this insect infected by *Rickettsiella grylli* (Meynadier *et al.*, 1968).

Infections with protozoa were carried out in some cases before the explantation of organs. Thus, the alimentary canal and fat body tissue of *Pieris brassicae* (Lepidoptera) were already infected by the microsporidian *Nosema mesnili* Paillot at the time they were cultured by Sen Gupta (1964). On the other hand spores of *Nosema bombycis* Naegeli were suspended in 0.1 *M* solution of KOH for 40 minutes before their introduction into four- to five-day cultures of ovarian sheath of *B. mori* (Ishihara and Sohi, 1966).

B. With Purified Germs

The first experiments using, as inoculum, purified viruses from polyhedroses produced doubtful or negative results. This failure can certainly be attributed to loss in infectivity of virions which have suddenly been liberated from their inclusion bodies by chemical means (Vago and Bergoin, 1963; Vaughn and Faulkner, 1963).

Later on, noninclusion virus infections were chosen in preference, and experiments on the infection of invertebrate cell cultures with several of these purified viruses gave positive results. In order to obtain viral particles it is necessary to use differential centrifugation of homogenates of infected larvae, or pupae, followed by membrane filtration. In certain cases centrifugation in density gradients produce pure viral material. The suspension of purified viruses thus obtained was administered directly or preserved at low temperature (−30°C). Hence, it was possible to obtain the infection of ovarian cell cultures of *A. eucalypti* by the *Sericethis* iridescent virus, SIV (Bellett and Mercer, 1964) ; of fibroblasts of the ovarian sheath of *B. mori* and *G. mellonella* by the densonucleosis virus (Vago *et al.*, 1966) ; of the cells of several tissues of the lepidopterous larvae *Chilo suppressalis* Walker by the iridescent virus of this species (Mitsuhashi, 1966) ; of embryonal cells of *Drosophila* by the sigma virus (Ohanessian and Echalier, 1967) ; and of cardiac cells of a cockroach by the *Tipula* iridescent virus (Meynadier *et al.*, 1968).

C. Infection in Series

A virus obtained for the first time in culture can be perpetuated by transferring a portion of infected tissue or medium into healthy cultures. Trager (1935) maintained a strain of the virus of nuclear polyhedrosis of *B. mori* through nine successive passages by depositing a fragment of contaminated tissue in each new culture.

Centrifugation of the culture medium following the development of the virus and introduction of the supernatant into a new culture is a method which has often been used for infection in series. A strain of nuclear polyhedrosis virus has thus been maintained (Vago and Bergoin, 1963) through numerous subcultures. Grace (1962), in the case of a polyhedral cytoplasmic virus, performed three passages in series, and Bellett and Mercer (1964), with *Sericesthis* iridescent virus, performed seven passages in series in ovarian cell cultures of *A. eucalypti*.

IV. STUDY OF THE DEVELOPMENT OF PATHOGENS IN CULTURES

For observation of the development of different microorganisms in invertebrate tissue cultures, the principles applied in vertebrate pathology are used with slight modification of the techniques.

Cellular alterations are demonstrated by means of current histological and histochemical strains (Giemsa, hemalum, Mallory, Feulgen, etc.) and viral material by means of specially adapted techniques, e.g., the nuclear areas of densonucleosis viruses are stained a greyish-blue color by hemalum-erythrosine (Vago and Luciani, 1965) and a special stain (Vago and Amargier, 1963) enables one to distinguish the inclusion bodies of nuclear or cytoplasmic polyhedroses or of spheroidoses which are stained bright red (Vago, 1964; Quiot and Luciani, 1968).

In order to detect rickettsiae in cultures of cardiac cells of Orthoptera, the Pappenheim method was used (Meynadier *et al.*, 1968). The use of the phase contrast microscope is also recommended.

Microcinematography by compressing the time scale, enables one to trace more easily the successive phases of pathogenesis (Vago, 1964, 1965). Nuclear changes and formation of polyhedral inclusion bodies are particularly well demonstrated.

The electron microscope makes the study of viral particles possible both in cells and medium. Such studies on sections of infected cultures date back about 10 years (Vago, 1959). Osmic acid is generally used for fixation (or the double method of glutaraldehyde–osmic acid), and cells are subsequently embedded in Araldite, methyl metacrylate, Epon, or other resins. Ultrathin sections are stained with uranyl acetate or lead citrate (see Chapter 3 of Vol. I).

Different viruses, with or without inclusion bodies, have been studied using ultrathin sections, i.e., cytoplasmic polyhedrosis virus (Grace, 1962), nuclear polyhedrosis virus (Vago and Croissant, 1963), *Tipula* iridescent virus (Meynadier *et al.*, 1968; Hukuhara and Hashimoto, 1967), *Sericesthis* iridescent virus (Bellett and Mercer, 1964), and *Chilo* iridescent virus (Mitsuhashi, 1966, 1967a,b; Hukuhara and Hashimoto, 1967).

The titration of viruses (Bellett, 1965a) and the determination of the infective dose, i.e., $TCID_{50}$ and LD_{50} (Sen Gupta, 1963a; Vaughn and Faulkner, 1963), one carried out according to the methods used in general virology. However, as far as infection by the sigma virus of tissue cultures of *Drosophila* is concerned (Ohanessian and Echalier, 1967), the virions are detected and determined in the supernatant medium by a method that has been specially developed for this virus by Plus (1954); it is based on the relationship between the time of incubation and the number of infective units injected.

The method of fluorescent antibodies has made it possible to detect the viral antigen, e.g., of *Sericesthis* iridescent virus in cell cultures of *A. eucalypti* (Bellett and Mercer, 1964) and of nuclear polyhedrosis virus of *B. mori* (Krywienczyk and Sohi, 1967). Autoradiography is used to investigate the site of viral nucleic acid synthesis (Bellett, 1965b).

In order to detect the development of protozoa, the usual techniques of fixation and staining have been employed, i.e., Heidenhain's haematoxylin (Sen Gupta, 1964) and Giemsa (Ishihara and Sohi, 1966).

V. DEVELOPMENT OF INVERTEBRATE VIRUSES

Attempts at infection of invertebrate tissue cultures by invertebrate viruses have been made for a long time, the first being contemporary with the first experiments on cell cultures of the Lepidoptera.

A. Nuclear Polyhedrosis Viruses

R. W. Glaser (1917) did not succeed in infecting *in vitro* the hemocytes of *Malacosama americanum* Fabricius larvae with the nuclear polyhedrosis virus of the same species. The formation of inclusion bodies only occurred in transplanted cells *in vitro* 10–12 days following *in vivo* infection of the insect.

Later, Trager (1935) noted the formation of nuclear polyhedral inclusion bodies in explanted tissues of the gonads of the silkworm *B. mori* larvae which were infected with the grasserie virus. A strain of this virus has been maintained by successive infection of nine cultures throughout a period of three months. Alterations in the cells and the nuclear polyhedral bodies appeared within two days following inoculation. The ingestion of infected cultures at the third, fourth or fifth passage by larvae of *B. mori* resulted in disease. The number of cells showing polyhedral bodies was variable, i.e., it depended on the condition of the tissue culture at the time of infection and during the days that followed and not on the source of the virus. In certain cultures the first polyhedra appeared after 24 hours and the cells began to die approximately one week following infection of the culture. The author notes great differences from one cell to another, with regards to the number and size of the polyhedra.

Having observed the intranuclear development of polyhedra in a cell culture of the ovarian sheath of *Hemerocampa leucostigma* SM and ABB (Lepidoptera) larvae, after addition to the culture medium of pupal hemolymph of *Callosamia promethea* DRURY, Grace (1958) considered that the larvae that had provided the tissue were the site of a latent infection by a polyhedrosis virus. The latter would manifest itself as a result of the physiological stress created by a sudden change in the culture medium.

The development of nuclear polyhedrosis viruses in modern cell cultures has been traced by Aizawa and Vago (1959). Cultures of fibroblasts of the ovarian sheath of *B. mori* maintained at 30° are infected with the virus of the same species by introducing hemolymph from a diseased larva. Continuous observation under the phase contract microscope shows that after 48 hours of incubation the morphology of the fibroblasts is modified. The pseudopodia retract, the cells become rounded and nearly immobile, while the nuclei increase in size. The first corpuscles are seen in the nucleus on about the fourth day following infection, but their polyhedral outline only appears later on. After six days the polyhedra have invaded the nuclei of the cells and the cytoplasm becomes lysed.

In size and morphology, these inclusion bodies are identical with those formed in live larvae, but generally, the number per cell is reduced and sometimes there is only one giant body. The virulence of the viruses is maintained even after repeated passage in cell cultures (Fig. 1).

Other observations (Gaw et al., 1959) made on cell cultures of the ovarian sheath and trachea of the silkworm infected by grasseria virus are in agreement with those previously concerning the cellular alterations and formation of polyhedra.

According to Medvedeva (1960), polyhedra obtained in cultures of cells and of hemocytes of the Lepidoptera B. mori and Antheraea pernyi infected with nuclear polyhedrosis viruses of the same species are polymorphous. More recently, other Soviet authors, working with the same material, have studied the possibilities of activating latent viruses in cell cultures by means of ultraviolet irradiation (Miloserdova et al., 1966).

Infection by the nuclear polyhedrosis virus of Galleria mellonella and Mamestra brassicae L. seems more intense in the cells in vivo than in vitro according to Sen Gupta (1963a), who explains this fact by the necessity for viral development of having cells with an active metabolism. The same author (Sen Gupta, 1963b) noted that the degree of multiplication of nuclear polyhedrosis virus in fat body and in blood cells of G. mellonella was variable, depending on the nutritional and metabolic

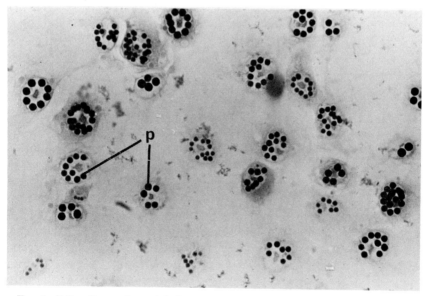

Fig. 1. Cell culture of ovarial sheath of Bombyx mori (Lepidoptera) three days after infection with nuclear polyhedrosis virus. p, polyhedral inclusion bodies. (Vago–Amargier; ×260.)

conditions of the culture, the more active ones being the more liable to viral attack.

Apart from cell cultures, other methods of maintaining tissues *in vitro* have been used for studying nuclear polyhedral viruses. Thus, Vago and Chastang (1960) obtained pathogenesis of *Borrelinavirus* and formation of nuclear polyhedral inclusion bodies in female larval gonads and in whole ovarian chains of Lepidoptera kept in culture medium for only three days. These results show that the virus can accomplish its development and replication in cells whose metabolism is on the decline. Furthermore, they open up a special field of experimental investigation since pathogenesis and the production of polyhedral bodies may also be obtained under conditions that are not strictly aseptic, in large quantities, in a relatively short space of time as shown recently by Vago *et al.* (1970) (see, also, Section I).

The development of nuclear polyhedrosis viruses in organ cultures on solid media was achieved recently by Vago *et al.* (1969) and Quiot *et al.* (1971) by means of the Wolff and Haffen technique (1952). The ovaries and dorsal vessels of *B. mori* and *G. mellonella* larvae and chrysalids were infected at the moment of explantation, or just afterwards, with the corresponding viruses. Histological examination made after 8–20 days of culture showed the development of polyhedral inclusion bodies in most of the nuclei of the sensitive tissues (Fig. 2).

Not only tissues but whole embryos have made it possible to obtain *in vitro* infection with nuclear polyhedrosis viruses. Takami *et al.* (1966) infected, with *Borrelinavirus bombycis*, embryos of *B. mori* which developed in eggs freed from their chorion and maintained in culture medium. Polyhedrosis viruses multiply in the nuclei of a great variety of tissues (serosa, yolk cell, amion, epidermis, muscle, tracheal epithelium, ganglion, and gut gonad), including those in which no mention has ever been made of the formation of inclusion bodies either in the larvae or the pupae. These results also indicate a diminution of the specificity of the virus since polyhedral formation has only rarely been observed in the intact eggs of *B. mori* and did not occur even if they were injected with the virus.

Culture of cells explanted from insects at different periods in the development of diseases has made it possible to study, *in vitro*, the evolution of pathogenesis. Female gonads were sampled from fifth instar larvae of *B. mori*, previously infected orally with a suspension of nuclear polyhedrosis virus (Aizawa and Vago, 1959). If the gonads are cultured shortly after infection of a larva, migration of the fibroblasts is normal and the culture proceeds without alteration. Hence, the viruses do not affect the gonads during a period that rarely exceeds three days at 22°.

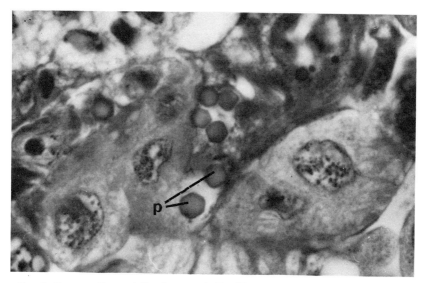

FIG. 2. Organ culture of *Bombyx mori* (Lepidoptera) dorsal vessel five days after infection with nuclear polyhedrosis virus. p, polyhedral inclusion bodies. (Vago–Amargier, ×425.)

After this time, the cells often show no lesions at the time they are placed in culture. Their migration is normal and slow mitoses are observed. However, these fibroblasts subsequently show signs of degeneration; after 24–48 hours in culture they round off and their nucleus is progressively invaded by polyhedra. When inclusion bodies are noted in the explanted tissues, migration of the fibroblasts is slight and irregular. Typical alterations appear rapidly and destruction of the cell occurs within two to three days.

In cell and tissue cultures of insects, several fundamental problems concerning invertebrate viroses have been studied.

Details of viral pathogenesis have been traced by electron microscopy of ultrathin sections of fibroblast cultures of the ovarian sheath of the Lepidoptera *Bombyx mori*, *Antheraea pernyi*, and *Lymantria dispar* fixed at 12-hour intervals (Vago and Croissant, 1963). About 24 hours after infection with nuclear polyhedrosis virus, the mitochondria and perinuclear vacuoles assume unusual shapes. In the hypertrophied nucleus, next to the nucleolus, dense areas appear in which one can gradually distinguish short rod-shaped elements approximately 25 to 30 mμ wide and 50 to 200 mμ long. After 72 hours, in transverse sections, the rod-shaped formations are surrounded by a clear substance and, progressively, by a membrane. These structures coexist within the nucleus with irregular protein masses, the beginnings of the future polyhedral bodies. From this moment onwards, all the stages of incorporation of the viruses into the crystalline structure of the polyhedra can be observed and, particularly, the details of the association of the viruses with the developing mass of protein of the polyhedral formations can be noted. The cytoplasm is gradually invaded by voluminous vacuoles, and mitochondria are no longer visible.

The pathogenesis of this virus traced *in vitro* confirms the observations that have been made in live insects and provides additional details particularly with regard to the initial phase (Figs. 3 and 4).

The formation of polyhedral inclusion bodies being a continuous, gradual process, the method of microcinematography is likely to provide interesting results in its study. Hence, Vago's films (1964, 1965) enables one to trace the progression of pathogenesis *in vitro* of a nuclear virus infection in the fibroblasts of the ovarian sheath of *L. dispar* and *B. mori*. Forty-eight hours after the introduction of a small quantity of hemolymph containing the virus into the culture medium, a gradual rounding off of the cells may be observed, followed by a change in the structure of the nucleus and the progressive growth of refractive bodies, the polyhedral outline of which gradually becomes more and more distinct. Changes in the structure of the cytoplasm become visible, and the dead cell is phagocytosed by other fibroblasts.

One preoccupying problem with regard to insect virology is the origin and determinism of the shape of the polyhedral inclusion bodies. In the case of live insects, morphological variations in the polyhedra are often noted (Gershenzon, 1960; Aizawa, 1961; Bergold, 1963). Examination of the shape of the polyhedra in terms of variations in the genetic material of the virus or the host species of the Lepidoptera are likely to provide information on the formation of the inclusion bodies. Tissue culture seems to provide conditions that are particularly convenient for such experimental observations. Hence, Vago and Bergoin (1963) noted in cultures of fibroblasts of the ovarian sheath of *L. dispar* infected with the nuclear polyhedrosis virus of *A. pernyi* that most of the polyhedra have a triangular cross section characteristic of most of the polyhedra of *A. pernyi*, whereas those of *L. dispar* are typically polygonal.

The question was also studied by Quiot and Luciani (1968) with two other species of Lepidoptera. The nuclear polyhedrosis of *G. mellonella* is characterized by the formation of inclusion bodies that are almost all cubic. This virus has been inoculated into tissue cultures of *B. mori*, a species in which the virosis can be distinguished from that of *G. mellonella* by its polygonal polyhedra. The inoculum must contain only free viruses without inclusion bodies. It consists of the supernatent of larval hemo-

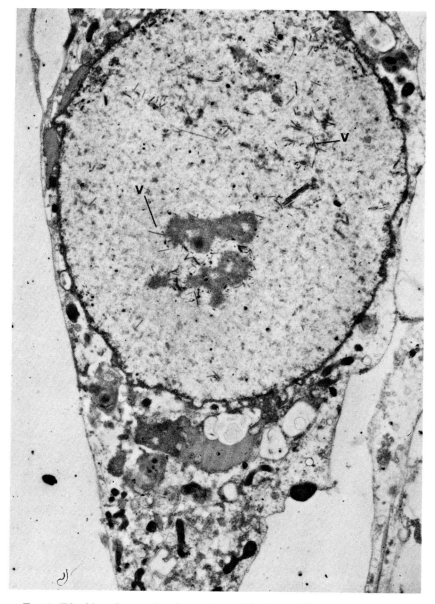

FIG. 3. Fibroblast from cell culture of ovarial sheat of *Bombyx mori* (Lepidoptera). Initial stage of polyhedrosis in the hypertrophied nucleus. v, free virus rods. (Electron microscope, ×10,000.)

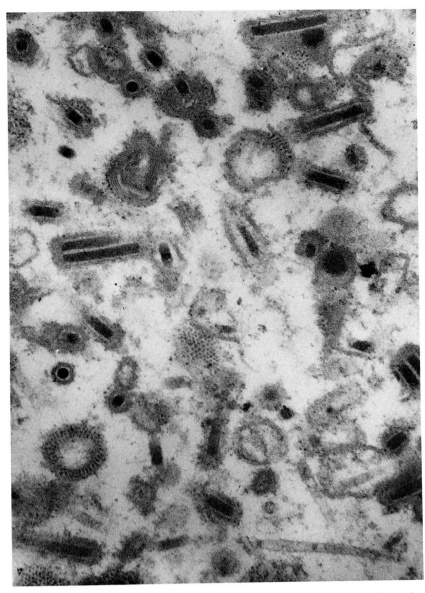

Fig. 4. Different stages of polyhedrosis virus development in the nucleus of a cultured fibroblast of *Bombyx mori* (Lepidoptera). (Electron microscope, ×80,000.)

lymph centrifuged at 4000 rpm for five minutes. After 24 hours the volume of the nucleus increases and inclusion bodies of cubic outline appear progressively in the nucleus. Preparations stained by the method of Vago–Amargier (1963) confirm the square cross section of these bodies. This is particularly evident in certain nuclei that contain only one or two very large polyhedra.

These results show experimentally *in vitro* the influence of the viral genetic material on the formation of polyhedral inclusion bodies.

Another general problem has been raised by experiments on viral infection of cell cultures. Although the hemolymph of lepidopterous larvae, suffering from a nuclear polyhedrosis and containing free viruses, infects live insects just as easily as cells cultured *in vitro*, the virions obtained by disintegration of the polyhedral inclusion bodies are pathogenic for the insects but only slightly so for cells in culture. The reasons for this have not been completely elucidated; it seems that viruses freed by rather violent chemical means no longer possess the ability to penetrate the cells. However, it is possible that the occluded virions are subsequently dissociated into subunits either by the hemolymph of the live organism or under the influence of the gastric juice (Vago and Bergoin, 1963). Infection also seems to require the presence of polyhedral proteins (Vaughn and Faulkner, 1963; Faulkner and Vaughn, 1965).

B. Cytoplasmic Polyhedrosis Viruses

The RNA cytoplasmic polyhedrosis viruses, of cubic symmetry, multiply in the epithelial cells of the midgut of several species of Lepidoptera. Cultivation of intestinal cells of insects having been performed only fairly recently, *in vitro* study of viruses of this group has not been undertaken for very long. However, cytoplasmic polyhedra were observed in some cultures of fibroblasts of *A. eucalypti* following infection with the nuclear polyhedrosis virus of *B. mori* (Grace, 1962). The appearance of polyhedra 9–22 days after infection was accompanied by cell hypertrophy and development of intracytoplasmic vacuoles. These polyhedra are approximately the same size as those formed in the intestinal cells of larvae of *Antheraea* but, in culture, cubic forms predominate. Sections of these polyhedra show occluded viruses similar to those of other cytoplasmic polyhedroses (Fig. 5). Cases of spontaneous occurrence of cytoplasmic polyhedral formations were noted in a cell culture of *Antheraea* established for 10 months, as well as in subcultures derived from it.

Infection of cell cultures with the virus of cytoplasmic polyhedrosis from a live insect was performed by Vago and Bergoin (1963), who

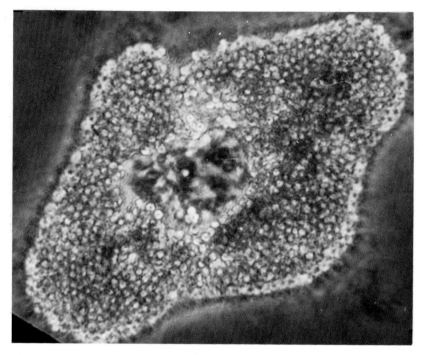

FIG. 5. Cell from a culture of *Antheraea eucalypti* (Lepidoptera) ovarian tissue. The cytoplasm is filled with polyhedra. (Phase contrast, ×200; from Grace, 1962.)

inoculated a fibroblast culture of the ovarian chain of *L. dispar* with the virus of the same species. Virus pathogenesis occurred but in smaller amounts than in infections with nuclear polyhedrosis virus. The cytopathological process is identical to that observable *in vivo*. Polyhedra are formed in the cytoplasm of the cells, and the nuclei only show slight hypertrophy. Hence, the fibroblasts of ovarian origin that are insensitive towards the cytoplasmic polyhedrosis virus in the live insect can serve to support the multiplication of a virus that develops *in vivo* only in the endothelial cells of the midgut.

In surviving *Bombyx mori* embryos (Takami *et al.*, 1966), cytoplasmic polyhedra developed in gut cells after infection of the cultured embryo with this virus.

C. Granulosis Viruses

The only publication concerning the infection of tissue cultures by a granulosis virus is that of Vago and Bergoin (1963). The virus chosen was that described by Vago *et al.* (1955) from a disease of *Pieris brassicae* observed by Paillot in 1926. The tissue cultured consisted of fibroblasts of *L. dispar*, a species in which a virosis of the granulosis type had never been reported, and the source of the inoculum was larval fat body of infected cabbage worms.

Experimental infection was not obtained regularly; however, in a certain number of cultures, characteristic cellular lesions were observed on examination by phase contrast and in stained cultures, i.e., increase in size of the nucleus, increase in the density of the cytoplasm, and presence of granules similar in size and shape to those noted in the larvae of *Pieris brassicae*. Electron microscopy of these cells prepared with phosphotungstic acid confirmed the characteristic structure of granules and of viruses.

These results are interesting from the point of view of the modification of viral specificity in cell cultures. The virus used is of a type which has never yet been found in *Lymantria;* moreover, the insect that furnished the cultured cells and that providing the virus do not belong to the same genus.

D. Densonucleosis Virus

This new type of virosis discovered a few years ago (Vago *et al.*, 1964) in *G. mellonella* (Lepidoptera) is characterized by the formation in the cell nuclei of dense masses consisting of virus particles of cubic symmetry and small in size (20 mμ). This DNA virus appears to be similar to the parvoviruses which included, until now, only vertebrate viruses. The densonucleosis virus seems to have a high specificity since no species other than *G. mellonella*, even among Lepidoptera, could be infected.

The first investigations of the propagation of the densonucleosis virus in insect tissue cultures were carried out shortly after the discovery of the virus (Vago and Luciani, 1965). Cell cultures of ovarian sheaths of *G. mellonella* and *B. mori* pupae are maintained at 26°, taking in account the fact that the virus develops in the wax moth at a temperature of approximately 28°. Three or four days after inoculation with a drop of infectious larval hemolymph of *Galleria*, a nuclear hypertrophy appears and the nucleus is invaded by basophilic Feulgen positive areas. The pseudopodia retract and after six days there is only slight growth of the cells, whereas the control cultures continue to proliferate. Vago *et al.* (1966) also succeeded in infecting cultures of the same types by using purified virus as the source of infection. The cytopathic action is comparable to that described in the insect (Amargier *et al.*, 1964).

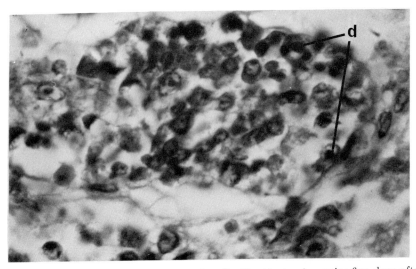

Fig. 6. Organ culture of *Galleria mellonella* (Lepidoptera) ovaries five days after infection with densonucleosis virus. d, nuclei typically altered by the virus. (Hemalum-eosine-orange G., ×1100.)

The nuclear lesions due to densonucleosis virus do not resemble any of the cellular alterations observed during infection of tissue cultures by other types of invertebrate viruses, but they present some similarity with the lesions provoked by certain adenoviruses. The fact that this virus is able to infect *in vitro* the cells of species (*B. mori*) other than that of the obligatory host contrasts with the strict specificity observed in live animals. This fact is also emphasized by the recent success of Giauffret *et al.* (1968b) in obtaining the development, with characteristic lesions, of this virus in cell cultures of the bee, an insect systematically very different from the natural host.

Recently, the development of densonucleosis virus was also achieved in organ culture (Vago *et al.*, 1969; Quiot *et al.*, 1971). The ovaries and dorsal vessels of *G. mellonella* (Lepidoptera) maintained on a solid medium (see Chapter 2, Vol. I) showed eight days after infection with DNV nuclear lesions characteristic of the densonucleosis (Fig. 6).

E. Iridescent Viruses

Following the observation in *Tipula paludosa* Meigen (Diptera) of "iridescent" viroses (Xeros, 1954), diseases of this type have been described in Coleoptera, e.g., *Sericesthis pruinosa* Dalman (Steinhaus and Leutenegger, 1963); Diptera, e.g., *Aedes taeniorhynchus* Wiedemann

(Clark *et al.*, 1965); and Lepidoptera, e.g., *Chilo suppressalis* Walker (Fukaya and Nasu, 1966). The DNA viruses, of cubic symmetry and 100–180 mμ, accumulate in closely packed aggregations in different tissues, particularly fat body, and they confer an iridescent color to the organs affected. Cultures of ovarian cells of *A. eucalypti* have been infected with the *Sericesthis* iridescent virus (SIV) (Bellett and Mercer, 1964) purified by differential centrifuging and filtration on millipore membranes of tissue homogenate from infected *G. mellonella* larvae. The cells lose their mobility, their cytoplasm becomes granular, and the nuclei become hypertrophied. Accumulation in the cytoplasm of the SIV antigen and of its DNA has been demonstrated by immunofluorescence and acridine orange staining.

Under the electron microscope, virus particles are visible scattered throughout the cytoplasm at different stages in development, ranging from empty capsids to mature nucleocapsids. They are located in a narrow evagination of the cytoplasm and they seem to acquire membranes derived from the cell membrane.

Seven successive passages of SIV in cell cultures of *A. eucalypti* were obtained at 21°, but at 25° infectivity disappears on the third passage. The optimum temperature for the production of antigen of SIV by *Antheraea* tissue culture was 20° (Bellett, 1965a).

By using purified virus, the ratio of particle infectivity was approximately 80/1, whereas for the same virus Day and Mercer (1964), using a method of titration *in vivo*, found ratios of less than 10/1.

Combining fluorescent antibody techniques and autoradiography by incorporation of tritium labeled thymidine, Bellett (1965b) found that viral DNA was synthesized in the cytoplasmic foci which also contained antigen. The average period of latency of SIV was four to five days. The infective virions are then produced continually and subsequently freed by the cells up to the seventh or eighth day.

Experiments on multiplication of the iridescent virus of *Chilo suppressalis* (CIV) in cell cultures were carried out (Mitsuhashi, 1966a,b) in hemocytes of larvae in diapause of the same species. The purified virus was inoculated either in subcultures, or in primary 10-day-old monocellular layers of plasmatocytes. In both cases, after 24 hours, one observes an increase in size of the cells that join together. Ten days later, an iridescence appears in the aggregated cells, the cytoplasm of which is highly vacuolated. Electron microscope study using ultrathin sections of infected cells reveals the presence in the cytoplasm, but not in the nucleus, of numerous virus particles arranged in regular clusters. The infectivity of the affected cells and of the culture medium was demonstrated by injection into the hemocele of *Chilo* larvae.

In an infected culture, the cells continued to multiply and subcultures were made (Mitsuhashi, 1967a). Multiplication of CIV in this cell line was observed by examination under the electron microscope. This revealed numerous virus particles in the cytoplasm of hypertrophied cells. Ten months after inoculation, a homogenate of the cells or the cell-free culture medium infected healthy larvae of C. *suppressalis* by inoculation into the hemocele.

Various tissues of *Chilo* larvae were found to be receptive to CIV after inoculation *in vitro* (Mitsuhashi, 1966a). As already mentioned, the site of multiplication of the virus is detectable by the iridescence which its accumulation causes in the infected tissues. This iridescence appears after an average of 10 days in the dorsal vessels, the fat body, the testes, the ovaries, and most remarkable of all, the brain, an organ which is never affected in the whole larva even when it is highly infected (Mitsuhashi, 1966b).

Some experiments on the transmission of CIV to insects, which are not its natural host, were performed by Japanese authors. Mitsuhashi (1967b) orally infected pupae of the leafhopper *Nephotettix cincticeps* with purified CIV in a 5% sucrose solution. A 50-day-old culture of epithelial cells of the same insect was infected with an inoculum obtained from iridescent larvae of C. *suppressalis*. Twenty-four hours after inoculation, granules of various sizes and vacuoles appeared in the cytoplasm of the epithelial cells. Iridescence occurred after seven days. Some cells degenerated about the 10th day. In cultures infected for 14 days, a few cells were still alive, whereas most of the others had disintegrated. Ultrathin sections of infected cells revealed the presence in the cytoplasm of numerous particles of CIV and some empty membranes. Viral particles were lacking both in the nucleus and in symbiotic bacteria located in the cytoplasm.

Cultures of ovarian cells of A. *eucalypti* (Grace, 1962) were susceptible to infection with CIV at a temperature of 20°–21° (Hukuhara and Hashimoto, 1967). After staining with Harris' haematoxylin, spherical inclusions are observed in the cytoplasm of most of the cells, which strongly take up this stain and are never found in healthy cells. Examination under the electron microscope confirms the presence of viral particles in the cytoplasm and protrusions of the cytoplasmic membrane containing virions. Again, one can see on the surface of the infected cells small extracellular bodies consisting of from one to three viruses enclosed in a common membrane. Another characteristic feature of cultures infected by CIV is the presence of spherical bodies, isolated in the medium, measuring 2–5 μ, and entirely occupied by numerous viral particles.

Among the group of iridescent viruses, that of the dipteran *T. paludosa*

(TIV) was the first to be described, but it is only recently that it has been chosen to infect insect tissues cultured *in vitro*. It is known that the specificity of the TIV in the whole insect is low with regard to both the tissue localization of the virus and the systematic position of the host. Indeed, besides the fat body, which is the initial site of development of the virus, various other tissues were found to be experimentally sensitive to TIV in several species of Diptera, Lepidoptera, and Coleoptera (Smith *et al.*, 1961; Gershenzon, 1964). This limited specificity is confirmed under *in vitro* conditions since the first experiments mentioned concern the multiplication of the TIV in cultures of ovarian cells of Lepidoptera and the cardiac cells of Dictyoptera.

Hukuhara and Hashimoto (1967) inoculated ovarian cell cultures of *A. eucalypti* with a fraction of hemolymph from larvae of *G. mellonella* infected with TIV. In the cytoplasm they observed inclusions that were intensely stained with haematoxylin and surrounded by a lighter halo. An electron microscope study revealed the presence of virus particles of TIV grouped in masses in the cytoplasm.

Cultured cardiac cells of an insect belonging to Dictyoptera, *Blabera fusca* Brunner, was found to be receptive to TIV (Meynadier *et al.*, 1968). Aseptically sampled fat tissue of infected *Tipula* was suspended in the culture medium. The supernatent served as inoculum. The effect of the virus on cardiac cells was very rapid. Less than one day after infection, the cells lost their pseudopodia, rounded off, and became vacuolated. The culture was entirely destroyed three days after infection. A normal explant pursued its rhythmic pulsations for 15 days in the same medium. Two days after inoculation, TIV virions of 120 mμ appeared in the cytoplasm of the cultured cells. The multiplication of the viruses in the cells and their liberation following lysis of the culture were proved by infection of fresh cell cultures with the supernatent of the contaminated culture diluted 100 times. The results of these experiments provide a new example of the reduction of tissue specificity under the conditions of culture *in vitro*. They also prove that cultures of cardiac cells of insects constitute an excellent substrate for virological studies and suggest that it would be well worthwhile to extend their use.

F. Other Nonoccluded Viruses

Although several viruses without inclusion bodies have been isolated in various invertebrate diseases, their *in vitro* study in organ and cell cultures has rarely been considered. A short time after obtaining the first cultures of bee cells (Giauffret *et al.*, 1968a) cultures of bee embryo cells were infected (Giauffret *et al.*, 1968b) with several invertebrate viruses, notably with the "black disease" virus and with an RNA virus isolated from healthy bees (Bailey *et al.*, 1963) and experimentally causing acute paralysis (ABPV). In both cases cytopathic action, often accompanied by the formation of cytoplasmic inclusions, was observed and the development of the cultures was seriously impaired.

G. Sigma Virus of *Drosophila*

The virus of *Drosophila melanogaster* Meigen is a nonpathogenic hereditary virus which confers to its host a particular sensitivity to carbon dioxide. Numerous studies have been devoted to the relationship between the virus and its host (reviewed by L'Héritier in 1962) and to genetics of this virus (Brun, 1963; Ohanessian-Guillemain, 1963; Vigier, 1966; Gay and Ozolins, 1968). From the point of view of its structure, the sigma virus resembles very closely rabies virus and vesicular stomatitis virus (Berkaloff *et al.*, 1965; Printz, 1967).

With a view to obtaining a better understanding of the relationship between the σ virus and its host, Ohanessian and Echalier (1967) infected primary cultures of embryonal cells of *D. melanogaster* with a suspension of purified σ virus. After a period of absorption of one hour, the infective suspension was replaced by culture medium. Virions were detected and determined in the supernatant medium by a method (Plus, 1954; Brun, 1963) based on the relationship between the time of incubation and the number of infective units injected. No cytopathic effect was observed in the cultures. The cells appeared to be quite normal under the phase contrast microscope and showed regular mitoses. However, multiplication of the σ virus was recognized. Cultures that received 4×10^3 infective units furnished a total of about 6×10^5 units two months after infection. Viral production took place regularly for several weeks, and if the culture had not been purposely stopped after 69 days, it would probably have continued.

H. Interference of Virus Infections

This problem of topical interest in general virology was recently approached by using invertebrate tissue cultures. Vago and Quiot (1971) infected cell cultures of *B. mori* ovarian sheaths eight days after formation of a continuous layer, with purified densonucleosis virus. After 18 hours of contact, the culture was rinsed with virus-free medium and then covered with medium containing free *B. mori* nuclear polyhedrosis virus.

The virus strain used was particularly adapted to cells maintained in culture and starts the formation of nuclear polyhedra in 58 hours. In the culture previously infected with densonucleosis virus, the first polyhedra appeared only 8–10 days after infection, and 15 days later their number was still rather low. Moreover, the polyhedra were generally badly formed and often replaced by an amorphous mass reacting positively to the Vago–Amargier (1963) staining procedure. The various stages of interference by these viruses are studied in the same cells because both viruses are nuclear, contain DNA, and have affinity for the same tissues.

I. Virus Production

For certain biochemical or serological studies and for the experimental or practical use of viruses in biological control against harmful insects, the production of relatively large quantities of viruses and of inclusion bodies is indispensable. Such production is generally achieved by extracting infective matter from diseased insects collected in nature or infected in the laboratory. However, to make production more homogeneous and to avoid certain drawbacks of rearing, it has lately been thought possible to produce viruses in nymphs immobilized during the development of the virosis (Vago, 1957) and even in cell cultures.

The relatively intricate techniques of tissue culture and of organ removal, the small quantity of cells produced *in vitro* compared to those infected in the living animal, and the frequent loss of sensitivity of the viruses in the repeatedly subcultured cells seemed to make the efficiency of such cultures improbable. However, the rapid production of relatively large quantities of insect viruses was recently achieved *in vitro* by Vago *et al.* (1970). The technique is based on the fact that a strain of *Bombyx mori* nuclear polyhedrosis virus adapted to *in vitro* pathogenesis produces polyhedra in cultured *Bombyx* cells within 48 hours and on the observation (Vago and Chastang, 1960) of the ability of cells of Lepidoptera kept alive for only a short time to insure a metabolism sufficient for complete viral pathogenesis.

A suspension of larval or nymphal tissues is prepared by a simple and rapid sampling technique, without special precations. All tissues (excepting the alimentary canal) are mixed with approximately 10 times their volume of an adequate medium (here, BM 22, Vago and Chastang, 1958). Antibiotics, eventually ascorbic acid, and a drop of virus suspension (for approximately 100 ml of medium) are added, and the suspension is kept at 25° in gently shaken Erlenmeyer flasks. After three days, the nuclei of the majority of the sensitive cells are full of inclusion bodies and it is possible to extract polyhedra and free purified viruses by centrifugation.

Recently, nuclear polyhedrosis virus of *Lymantria dispar* (Lepidoptera), harmful to agriculture, was produced by a similar technique. This virus is important for microbiological control projects.

VI. DEVELOPMENT OF INVERTEBRATE RICKETTSIAE

Few publications exist concerning the infection of invertebrate tissue cultures by invertebrate pathogens other than viruses. This is true of rickettsiae although, at present, rickettsioses are known to exist in Coleoptera, Diptera, Orthoptera, and Dictyoptera (Dutky and Gooden, 1952; Wille and Martignoni, 1952; Müller-Kögler, 1958; Vago and Martoja, 1963; Huger, 1964; Vago and Meynadier, 1965). The experiments of Pourquier *et al.* (1963) and Suitor (1964) only dealt with infections of vertebrate cell cultures, by *Rickettsiella melolonthae* Krieg and *Rickettsiella popilliae* Dutky and Gooden respectively; hence they do not enter the field of our discussion.

Infection of an invertebrate tissue culture by an invertebrate rickettsia

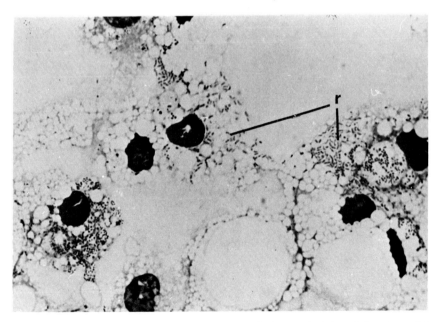

Fig. 7. Heart cells of *Gryllus bimaculatus* (Orthoptera) infected *in vitro* with *Rickettsiella grylli*. r, rickettsia in cytoplasma. (Giemsa; ×1100.)

was performed for the first time by Meynadier *et al.* (1968). The tissues of the dorsal vessel of *Gryllus bimaculatus* De Geer (Orthoptera) were cultured for several months either in the form of explants or following mechanical dissociation. The rickettsia used was *Rickettsiella grylli* (Vago and Martoja, 1963), a naturally pathogen for *Gryllidae*, and by inoculation for other Orthoptera, certain Dictyoptera, and Coleoptera. Infection *in vitro* was performed in two-day-old cultures by introducing into the culture medium a suspension of rickettsiae obtained from the fat body of *Gryllus* suffering from rickettsiosis.

After 48 hours of infection at 28°, one observes, by phase contrast, a tendency for the pseudopodia to retract and a hypertrophy of the cytoplasm. After four days, compact vacuoles appear in the cytoplasm and contain numerous small size rickettsiae (approximately 800 mμ). Their form subsequently changes; One notes a majority of rod-like structures. Eight days after infection, the cytoplasm of the infected cells loses its cohesion: The rickettsiae are set free into the medium, they may remain loosely grouped together in masses or dispersed in the medium, or they often remain in small chains consisting of a maximum of about ten units. The different phases of pathogenesis obtained in cell cultures thus correspond to those observed in the diseased insect (Fig. 7).

VII. DEVELOPMENT OF PROTOZOA

Although several authors have studied the development of vertebrate pathogenic protozoa in surviving tissues of the vectors, relatively few publications concern the study *in vitro* of these pathogens.

In ovarian tissue cultures of the silkworm *B. mori*, Trager (1937) observed the partial development of amoebuli of the microsporidian *Nosema bombycis* Naegeli up to the planonte stage.

An experiment on the same material has been reported by Ishihara and Sohi (1966), but this time the infective agent consisted of spores of *N. bombycis* suspended in a 0.1 *M* solution of KOH for 40 minutes, then introduced into 4- or 5-day-old cultures either by inoculation or by replacing the old medium by a new one containing the spores. At the time of contamination, many of the cells had already migrated from the explant. The vegetative stages of *N. bombycis* were first observed and several young schizonts showed the presence of a flagellum. The sporoblast appeared 72 hours after inoculation. Twenty-one days after contamination, spores which showed an increase in size and nuclear hypertrophy filled the cells. Many of the sporoplasms which emerged from the

270   C. VAGO

spores did not succeed in invading other cells. Usually, they decreased in volume and degenerated. However, a few sporoplasms, the carotype of which was undergoing a change, could be observed. The cytoplasm had increased in volume, but the significance of this phenomenon has not been established.

Sen Gupta (1964) studied the infection of tissue cultures of *Pieris brassicae* by the microsporidian *Nosema mesnili*. Fourth instar larvae of this insect were fed with cabbage leaves covered with a suspension of spores of the protozoan. After 24 hours of feeding, gut and fat body were placed in roll tubes containing growth medium and the culture was incubated at a temperature of 26°. At the beginning of the culture, numerous sporozoites were seen. After seven days of incubation, the tissues were full of mature spores and showed extensive damage. These spores were identical in size and shape to those observed *in vivo*.

The same author (Sen Gupta, 1965) described the development *in vitro* of two schizogregarina, i.e., *Farinocystis tribolii* Weiser in the tissue of *Tribolium castaneum* Hbst. and *Mattesia dispora* Naville in the tissues of *G. mellonella* (Lepidoptera). The larvae of *Tribolium* were infected by ingestion of flour containing oocytes of *F. tribolii*. One day later, their gut and fat body were cultured in roller tubes. At the beginning of the culture, many spores were visible. Between the second and fourth days, sporozoites could be observed, and from the eighth day, the schizont stage was achieved. The protozoa also developed in multiplying cells.

Cultures of hemocytes of *G. mellonella* were contaminated with the hemolymph of larvae of the same species infected with *M. dispora*. The infective material contained, mainly, gregarinoid schizonts. Four days after infection, these schizonts multiplied in the cultures. On the sixth day both schizonts and gamonts were to be seen; on the 10th day, schizonts, gametes, and spores; on the 12th day the spores and schizonts degenerated; and on the 16th day mainly spores were present. The parasite at the different stages of its development was similar in size and shape to that observed in the live insect.

VIII. DEVELOPMENT OF MYCOPLASMATA

The existence of mycoplasmata in invertebrates being known only for a few years, *in vitro* experiments on them are both very recent and few in number. There are mainly two types of relations between invertebrates and mycoplasmata.

The first is concerned with the mycoplasma agents of phyllody and of yellows in plants (Doi *et al.*, 1967; Maramorosch *et al.*, 1968; Giannotti *et al.*, 1968; Maillet *et al.*, 1968; Ploaie *et al.*, 1968). These microorganisms were observed also in the tissues of their leafhopper vectors (Maramorosch *et al.*, 1968; Giannotti *et al.*, 1968).

The first successful attempts to obtain multiplication of mycoplasmata *in vitro* cultures of vector tissues were made by Giannotti *et al.* (1970). These authors used embryo organ cultures of the leafhopper *Euscelis plebejus* obtained by an agar technique similar to the Wolff and Haffen vertebrate organ culture method (1952). The embryo cultures were infected immediately after explantation of the organs, or 24 hours later, with suspensions of mycoplasmata isolated either from phyllody diseased clover or from infectious *Euscelis*. Inoculation of leafhoppers with 15-day-old cultures made them vectors of phyllody, whereas uncontaminated culture material had no effect. Moreover, the inoculation with culture medium contaminated with mycoplasma, but without organ culture and maintained 15 days, did not make the leafhopper infectious either. These results show an active multiplication of mycoplasmata in the organ culture of embryos of the vector.

The second type of relationship between invertebrates and mycoplasma observed in tissue culture is concerned with pathological cases. Devauchelle *et al.* (1969) were the first to note mycoplasma in invertebrates showing pathological symptoms. They observed in *Melolontha melolontha* (Coleoptera) larvae mycoplasma infected dorsal vessel and fat body. Organ cultures of *M. melolontha* could be infected with mycoplasma of this insect. These microorganisms were observed in the cytoplasm of cardiac tissue of *Melolontha* maintained *in vitro* for 15 days (Devauchelle *et al.*, 1969).

IX. IMMUNOLOGICAL STUDIES

With regard to the pathology of invertebrates, serological studies are aimed at demonstrating an antigenic relationship between different viruses, localizing the site of multiplication of virus material, and the titration of virus suspensions. Invertebrate tissue culture has only recently provided a contribution to this subject.

Bellett and Mercer (1964) inoculated tissue cultures of *A. eucalypti* with purified iridescent virus of *Sericesthis* (see Section V, E). By means of the technique of immunofluorescence, they traced the multiplication of the SIV antigen in the infected ovarian cells. Antibodies produced in rabbits inoculated with SIV were labeled using fluorescein isthiocynanate.

After three days incubation at 25°, a few cells showed a pale green fluorescence in the region of scattered foci in the cytoplasm when they were stained with fluorescent antibody. The multiplication of the antigen was reflected by an increase in size and intensity of the fluorescence of the cytoplasmic foci, which joined together and subsequently invaded the cytoplasm of numerous cells (after five days). At 21° the same course of events was observed, but development was slower, the cytoplasm of the infected cells becoming filled with viral material seven to eight days after infection. Electron microscopy showed the presence of masses of viral particles (either complete or in the process of formation) in the region of the cytoplasmic foci. By combining the techniques of fluorescent antibodies and autoradiography, Bellett (1965a) showed that viral DNA was synthesized in the region of the cytoplasmic foci containing the antigen.

The fluorescent antibody method also enabled the titration of SIV particles in cell cultures of *A. eucalypti* by counting stained cells (Bellett, 1965a,b). The virus disappeared for four days following inoculation of the cultures. The infective virus was then regularly produced and liberated from the cells up to the eighth day. The final yield was 500 IU/cell, of which 360 IU/cell were liberated.

Immunofluorescence has also been applied to the study of a virus disease with inclusion bodies, i.e., the nuclear polyhedrosis of the silkworm in ovarian tissue culture of that insect (Krywienczyk and Sohi, 1967).

Polyhedral inclusion body protein antiserum was labeled with fluorescein isothiocyanate and used to stain a culture previously inoculated with a viral suspension (Krywienczy, 1963). Several foci of infection were vsible in monocellular layers 24 hours following inoculation. The infected cells first showed a slight fluorescence in the perinuclear region of the cytoplasm, while a bright yellowish-green fluorescence appeared in the nucleus. The fluorescence in the cytoplasm decreased gradually as that of the nucleus increased; then it disappeared when the polyhedral structures began to form. Results obtained with cells cultured *in vitro* were thus altogether comparable with those reported in 1963 by Krywienczyk for whole larvae.

X. CELLULAR REACTIONS

As in vertebrates, cellular immunity reactions have been known for a long time in invertebrates. They are initiated by the introduction of foreign bodies, particularly pathogenic germs, into tissues and hemolymph of the animal. Phagocytosis is fairly widespread, but is not the only type of reaction. *In vitro* study of cellular reactions constitutes a further facet of work with cell cultures.

Elimination of dead cells by phagocytosis in cell cultures of Lepidoptera can be seen in a film by Grace and Day (1963), and the different stages of phagocytosis of bacteria by cultured insect phagocytes can be followed in the film by Vago (1964) on invertebrate tissue culture.

Observations on phagocytosis in cell culture have been made mostly in mollusks, particularly in oysters. The ingestion of bacteria (*Escherichia coli*) appeared to be just as quick and efficient *in vitro* as in the living animal (Tripp *et al.*, 1966). In certain cases the phagocytosed bacteria seem to remain viable for a long time in the cytoplasm of the phagocytes. This fact may partly explain difficulties encountered in oyster cell culture because of the presence of contaminating bacteria.

The type of reaction that is most characteristic, especially among anthropods, consists of the aggregation of cells around bacteria, mycelia, or protein bodies, forming masses that sometimes attain about 1 mm in size. These formations have received various names, i.e., cysts, nodules, and Knötchen (Speare, 1920; Toumanoff, 1925; R. Glaser, 1926; Metalnikov, 1927; Hollande and Gely, 1929; Paillot, 1933; Sussman, 1951). Recently, they have been defined as "granuloma" by analogy with the formations of the same name known concerning vertebrates (Flandre *et al.*, 1968). The study of this process in live animals, which is essentially histological, provides only descriptive results. For this reason, recent attempts have been made to reproduce the phenomenon *in vitro* in order to trace the development of the reaction under controlled conditions.

Conidia of *Aspergillus* (*A. niger* and *A. flavus*), used as a stimulating agent, were placed together with hemocytes of *G. mellonella* in a medium devised for lepidopteran cell culture. (Vey *et al.*, 1968; Vey, 1969). Very rapid execution of the operation enables one to limit the time of contact of the hemolymph with air and thus avoid melanization. Observations and cinematographic records were obtained under phase contrast, cultures being performed on special perfusion slides.

Hemocytes aggregated around the conidia, sometimes after 5–10 minutes. It seems that one can attribute this to attraction, at least for short distances, of the blood cells by the elements of the cryptogamic agent since movement of the hemocytes occurs in a certain direction (Fig. 8a). In the hours that followed, the hemocytes grouped around the spores gradually drew closer together. Cells became modified as a result of the more compact structure of the mass of cells, and the total volume of the granuloma decreased (Fig. 8). The external cells became extremely stretched, and their cytoplasm spread out into a thin layer. The granuloma thus acquired a neat, regular outline already noted in the histological studies on animals (Fig. 8c). Only blood cells of the plasmatocytic type (micro- and macroplasmatocytes) seem to participate in this cellular reaction).

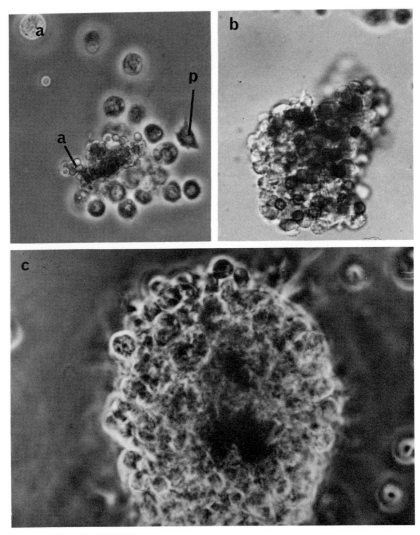

Fig. 8. Formation *in vitro* of granuloma by *Galleria mellonella* (Lepidoptera) hemocytes against conidia, *a*, of *Aspergillus niger*. (a) Beginning of the accumulation of plasmatocytes. (p). (b) Grouping and packing of plasmatocytes around conidia in 20 minutes. (c) Granuloma formed in four hours. (Phase contrast, ×525; from Vey *et al.,* 1968.)

Melanization of the foreign body, characteristic of the hemocytic reaction in the insect, has also been observed *in vitro* with several variations. One sometimes observes the conidia of *Aspergillus* surrounded by a brownish halo, which only really becomes visible after several hours.

Another type of *in vitro* test consisted in sampling aseptically some of the granulomas and transferring them to medium BM 22 (Vey, 1969). Sampling was performed by means of two fine-pointed needles that were sufficiently rigid so as not to damage the granuloma. The culture was performed on calibrated slides. On the edge of the granuloma, 20–30 minutes after placing in culture, certain hemocytes formed excrescences mixed with nonstructured spheres consisting probably of lipid material. Some hemocytes were stretched out into a narrow strip and tended to spread particularly in the distal region; as soon as they encountered the cover slip, they took on a fibroblast form. After five hours numerous cells were found around the explant remaining in contact with the mass of the granuloma. The spreading of the detached cells enabled them to be observed by phase contrast under good conditions. The hemocytes were characterized by a fairly limited number of fine granulations of irregular diameter, mostly gathered around the nucleus.

The duration of the experiment was limited to one day since fungi developed extremely rapidly and the cells become altered. When the experiments were carried out on older granulomas, sampled 4–14 days after infection, the detachment of cells occurred only infrequently.

REFERENCES

Aizawa, K. (1961). *Entomophaga* **6**, 197.

Aizawa, K., and Vago, C. (1959). *Ann. Inst. Pasteur, Paris* **96**, 455.

Amargier, A., Vago, C., and Meynadier, G. (1964). *Mikroskopie* **19**, 309.

Bailey, L., Gibbs, A. J., and Woods, R. D. (1963). *Virology* **21**, 390.

Bellet, A. J. D. (1965a). *Virology* **26**, 127.

Bellet, A. J. D. (1965b). *Virology* **26**, 132.

Bellet, A. J. D., and Mercer, E. H. (1964). *Virology* **24**, 645.

Bergold, G. H. (1963). *J. Ultrastruct. Res.* **8**, 360.

Berkaloff, A., Bregliano, J. C., and Ohanessian, A. (1965). *C. R. Acad. Sci.* **260**, 5956.

Brun, G. (1963). Thesis, Biol. Exptl., Fac. Sci., Orsay.

Clark, T. B., Kellen, W. R., and Lum, P. T. M. (1965). *J. Invertebr. Pathol.* **7**, 519.

Day, M. F., and Mercer, E. H. (1964). *Aust. J. Biol. Sci.* **17**, 892.

Devauchelle, G., Vago, C., Giannotti, J., and Quiot, J. M. (1969). *Entomophaga* **14**, 457.

Doi, Y., Teranaka, K., Yora, K., and Assuyana, H. (1967). *Ann. Phytopathol. Soc. Jap.* **33**, 259.

Dutky, G. R., and Gooden, E. L. (1952). *J. Bacteriol.* **63**, 743.

Faulkner, P., and Vaughn, J. L. (1965). *Proc. Int. Congr. Entomol., 12th, 1964* p. 718.
Flandre, O. (1967). *Proc. Int. Colloq. Insect Pathol. Microbiol. Control, 1966* p. 6.
Flandre, O., and Vago, C. (1963). *Ann. Epiphyt.* **14**, 161.
Flandre, O., Vago, C., Secci, J., and Vey, A. (1968). *Rev. Pathol. Comp. Med. Exp.* **5**, 101.
Fukaya, M., and Nasu, S. (1966). *Appl. Entomol. Zool.* **1**, 69.
Gaw, Z.-Y., Liu, N. T., and Zia, T. U. (1959). *Acta Virol. (Prague), Engl. Ed.* **3**, 55.
Gay, P., and Ozolins, C. (1968). *Ann. Inst. Pasteur, Paris* **114**, 29.
Gershenzon, S. M. (1960). *Probl. Virol. (USSR)* **6**, 720.
Gershenzon, S. M. (1964). *C. R. Colloq. Int. Pathol. Ins., 1962.* Entomophaga Mem. No. 2, p. 361.
Giannotti, J., Devauchelle, G., and Vago, C. (1968). *C. R. Acad. Sci.* **226**, 2168.
Giannotti, J., Quiot, J. M., Vago, C., and Schwemmler, W. (1970). *C. R. Acad. Agr.* p. 59.
Giauffret, A., Quiot, J. M., Vago, C., and Poutiers, F. (1968a). *Proc. Int. Colloq. Invertebr. Tissue Cult., 2nd, 1967* p. 74.
Giauffret, A., Poutiers, F., Vago, C., and Rousseau, M. (1968b). *Bull. Apic.* **11**, 13.
Glaser, R. W. (1926). *Ann. Entomol. Soc. Amer.* **19**, 180.
Glaser, R. W. (1917). *Psyche* **24**, 1.
Grace, T. D. C. (1958). *Science* **128**, 249.
Grace, T. D. C. (1962). *Virology* **18**, 33.
Grace, T. D. C., and Day, M. F. (1963). *Film C.S.I.R.O.*
Heitor, F. (1963). *Ann. Epiphyt.* **14**, 213.
Hollande, A. C., and Gely, M. (1929). *C. R. Soc. Biol.* **102**, 384.
Huger, A. (1964). *Naturwissenschaften* **51**, 22.
Hukuhara, T., and Hashimoto, Y. (1967). *J. Invertebr. Pathol.* **9**, 278.
Ishihara, R., and Sohi, S. S. (1966). *J. Invertebr. Pathol.* **8**, 538.
Krywienczyk, J. (1963). *J. Insect Pathol.* **5**, 309.
Krywienczyk, J., and Sohi, S. S. (1967). *J. Invertebr. Pathol.* **9**, 568.
L'Héritier, P. (1962). *Ann. Inst. Pasteur, Paris* **102**, 511.
Maillet, P. L., Gourret, J. P., and Hamon, C. J. (1968). *C. R. Acad. Sci.* **266**, 2309.
Maramorosch, K., Shikata, E., Granados, R. R. (1968). *Trans. N. Y. Acad. Sci.* **30**, 841.
Martignoni, M. E., and Scallion, R. J. (1961). *Biol. Bull.* **121**, 507.
Medvedeva, N. B. (1960). *Entomol. Obozr.* **39**, 77.
Metalnikoff, S. (1927). "Monographies de l'Institut Pasteur." Masson, Paris.
Meynadier, M., Quiot, J. M., and Vago, C. (1968). *Proc. Int. Colloq. Invertebr. Tissue Cult., 2nd, 1967* p. 218.
Miloserdova, V. D., Karpov, A. E., and Landau, S. M. (1966). *Mikrobiol. Zentralbl.* **28**, 67.
Mitsuhashi, J. (1966a). *Appl. Entomol. Zool.* **1**, 3.
Mitsuhashi, J. (1966b). *Appl. Entomol. Zool.* **1**, 199.
Mitsuhashi, J. (1967a). *Nature (London)* **215**, 863.
Mitsuhashi, J. (1967b). *J. Invertebr. Pathol.* **9**, 432.
Müller-Kögler, E. (1958). *Naturwissenschaften* **10**, 1.
Ohanessian, A., and Echalier, G. (1967). *Nature (London)* **213**, 1049.
Ohanessian-Guillemain, A. (1963). *Ann. Genet.* **5**, 1.
Paillot, A. (1926). *C. R. Acad. Sci.* **182**, 18.
Paillot, A. (1933). *In* "L'infection chez les insectes," p. 84. G. Patissier, Trevaux.

Ploaie, P. G., Granados, R. R., and Maramorosch, K. (1968). *Phytopathology* **58**, 1063.

Plus, N. (1954). *Bull. Biol. Fr. Belg.* **88**, 248.

Pourquier, M., Mandin, J., and Vago, C. (1963). *Ann. Epiphyt.* **14**, 193.

Printz, P. (1967). *C. R. Acad. Sci.* **265**, 169.

Quiot, J. M., and Luciani, J. (1968). *Proc. Int. Colloq. Invertebr. Tissue Culture, 2nd, 1967* p. 233.

Quiot, J. M., Vago, C., Luciani, J., and Amargier, A. (1970). *Bull. Soc. Zool. Fr.* **95**, 341.

Sen Gupta, K. (1963a). *Ann. Epiphyt.* **14**, 187.

Sen Gupta, K. (1963b). *Indian J. Exp. Biol.* **1**, 222.

Sen Gupta, K. (1964). *Curr. Sci.* **33**, 407.

Sen Gupta, K. (1965). *Curr. Sci.* **34**, 125.

Smith, K. M., Hills, G. J., and Rivers, C. F. (1961). *Virology* **13**, 233.

Speare, A. T. (1920). *J. Agr. Res.* **18**, 399.

Steinhaus, E. A., and Leutenegger, R. (1963). *J. Insect Pathol.* **5**, 266.

Suitor, E. C. (1964). *J. Insect Pathol.* **6**, 31.

Sussman, A. S. (1951). *Mycologia* **43**, 338.

Takami, T., Sugiyama, H., Kitazawa, T., and Kanda, T. (1966). *Jap. J. Appl. Entomol. Zool.* **10**, 197.

Toumanoff, C. (1925). *C. R. Soc. Biol.* **92**, 14.

Trager, W. (1935). *J. Exp. Med.* **61**, 501.

Trager, W. (1937). *J. Parasitol.* **23**, 226.

Tripp, M. R., Bisignani, L. A., and Kennedy, M. T. (1966). *J. Invert. Pathol.* **8**, 2, 137.

Vago, C. (1957). *C. R. Acad. Sci.* **245**, 2115.

Vago, C. (1959). *Entomophaga* **4**, 23.

Vago, C. (1964). "Culture de tissus d'invertébrés." Service du film de Recherche Scientifique, Paris.

Vago, C. (1965). "Viroses à polyèdres des insectes." Service du film de Recherche Scientifique, Paris.

Vago, C. (1967). *Methods Virol.* **1**, 567.

Vago, C., and Amargier, A. (1963). *Ann. Epiphyt.* **14**, 269.

Vago, C., and Bergoin, M. (1963). *Entomophaga* **8**, 253.

Vago, C., and Bergoin, M. (1968). *Advan. Virus Res.* **13**, 247.

Vago, C., and Chastang, S. (1958). *Experientia* **14**, 110.

Vago, C., and Chastang, S. (1960). *C. R. Acad. Sci.* **251**, 903.

Vago, C., and Croissant, O. (1963). *Ann. Epiphyt.* **14**, 43.

Vago, C., and Luciani, J. (1965). *Experientia* **21**, 393.

Vago, C., and Martoja, R. (1963). *C. R. Acad. Sci.* **256**, 1045.

Vago, C., and Meynadier, G. (1965). *Entomophaga* **10**, 307.

Vago, C., and Quiot, J. M. (1970). *Ann. Biol.* **9**, 573.

Vago, C., Quiot, J. M., and Paradis, S. (1970). *Entomophaga* **15**, 437.

Vago, C., Lépine, P., and Croissant, O. (1955). *Ann. Inst. Pasteur, Paris* **89**, 458.

Vago, C., Meynadier, G., and Duthoit, J. L. (1964). *Ann. Epiphyt.* **15**, 475.

Vago, C., Quiot, J. M., and Luciani, J. (1966). *C. R. Acad. Sci.* **263**, 799.

Vago, C., Quiot, J. M., and Luciani, J. (1968). *Proc. Int. Colloq. Invertebr. Tissue Cult., 2nd, 1967* p. 110.

Vago, C., Quiot, J. M., and Amargier, A. (1969). *C. R. Acad. Sci.* **269**, 978.

Vaughn, J. L., and Faulkner, P. (1963). *Entomophaga* **8**, 253.

Vey, A. (1969). *Ann. Zool. Ecol. Anim.* **1**, 93.

Vey, A., Quiot, J. M., and Vago, C. (1968). *Proc. Int. Colloq. Invertebr. Tissue Cult.,* *2nd, 1967* p. 254.

Vigier, P. (1966). *Ann. Genet.* **9,** 4.

Wille, H., and Martignoni, M. E. (1952). *Schweiz. Z. Allg. Pathol. Bakteriol.* **15,** 470

Wolff, E., and Haffen, K. (1952). *Tex. Rep. Biol. Med.* **10,** 463.

Xeros, N. (1954). *Nature (London)* **174,** 562.

8

USE OF INVERTEBRATE CELL CULTURE FOR STUDY OF ANIMAL VIRUSES AND RICKETTSIAE

Josef Řeháček

I. INTRODUCTION

Tissue cultures have become an important tool for study of many problems in different fields of microbiology where experimental animals have been replaced by organs, tissues, and cells maintained *in vitro*. Also, arthropod tissue cultures open the way for such studies and are used in microbiology for the multiplication of viruses and rickettsiae. These investigations could be prospective from various angles. Let us to consider some of them.

1. Findings on the sensitivity of arthropod cells *in vitro* to small doses of pathogens as compared with various vertebrate cell systems could serve as a basis for introducing arthropod cells into experiments on the isolation of viruses and rickettsiae from different materials obtained in nature. In such a case it would be possible to detect agents growing not only on arthropod tissue cultures but also on vertebrate cells.

2. Arthropod cells could be used for everyday laboratory work with viruses and rickettsiae, e.g., titration; studies of different properties, problems of virulence, changes of phase, and pathogenesis; preparation of convenient and sufficient amounts of agents for their illustration by means of electron microscope or fluorescent antibody technique; for investigations on a double infection either with different viruses or rickettsiae or with the members of either of these groups; etc.

3. The different behaviors of viruses and rickettsiae in arthropod cells *in vitro* could be used as a criterion for their diagnosis.

4. The multiplication in arthropod cells of agents for obtaining strains with attenuated virulence for animals and men could be used to prepare strains of viruses and rickettsiae for vaccination purposes.

We have mentioned only a few problems which seem to have prospects in virus and rickettsia research.

The surprisingly little attention which has been paid hitherto to the use of arthropod cells in this field was probably caused by the limited number of suitable cells or cell lines available.

From the point of view of the use of arthropod tissue culture for a routine work with viruses and rickettsiae, it seems easier to work with stable lines of cells than with primary cultures which are neither standardizable nor practical. The difficulties in working with them consist in the mixed population of cells, which vary in type and physiology and in probable difference in sensitivity to agents. The preparation of primary cell cultures from different organs is too laborious and threatens microbial contamination. Another problem is the possible natural contamination of these cells with different microbiotes. These cells *in vitro*, provoked with another infection, are also liable to develop latent infection, as shown, for example, by Filshie *et al.* (1967). For all these reasons, as Jones (1962) points out, care must be taken to avoid confusion when these tissue cultures are employed in experiments designed to test the effects of an introduced virus or rickettsia.

In spite of all these seeming difficulties, we assume that the organ, as well as primary cultures, can be successfully employed in virus and rickettsia research. It is possible to use arthropods from laboratory rearings free from living, undesirable microorganisms for the preparation of cells and to eliminate entirely the eventual microbial contamination by use of antibiotics or specific antisera. The susceptibility of primary cell cultures and of stable cell lines to different pathogens has not yet been established and may be higher in primary cells.

II. USE OF SURVIVING ORGANS AND TISSUES *IN VITRO* FOR THE MULTIPLICATION OF VIRUSES AND RICKETTSIAE

A. Multiplication of Viruses in Mosquito Tissues *in Vitro*

Trager (1938) first succeeded in propagating an arbovirus in explants of mosquito tissue *in vitro*. In these investigations he used the tissue from different organs of *Aedes aegypti* for the propagation of WEE virus.

The method of hanging drops was chosen in these experiments. To each hanging drop containing a small piece of tissue, 0.005 ml of the supernatant of brain suspension prepared from a sick mouse was added together with 0.005 ml of the diluted plasma. Mostly, after seven days of incubation at 28°, the fragment of tissue in the culture was triturated and suspended in 0.5 ml of the nutrient solution. This suspension was then used in amounts of 0.005 ml with 0.005 ml of diluted plasma for the preparation of similar fresh cultures.

The following results were obtained. In controls containing no mosquito tissue, a small amount of virus was still present after four days but not after seven days. In the presence of the midgut tissue from full grown larvae, the virus usually survived for seven days and, in one case, for 14 days with only a slight drop in titer. Subcultures made after seven days and tested seven days later were negative, with the exception of one which gave a virus titer no greater than in the original virus suspension. In the presence of adult ovaries the virus survived seven days in only three cultures out of eight. Of the subcultures made from these three seven days later, virus was present in one and titrated one hundred times higher than in the original suspension. In cultures containing larval thoracic body wall tissue, the virus survived for 7 and 14 days with no loss in titer. A subculture made from a 14-day culture and tested seven days later showed again, neither increase nor decrease in the amount of virus, but a similar subculture tested after 14 days had no virus. A subculture to pupal brain and adjoining head tissue, made from a seven-day culture in larval thoracic body wall, showed no loss in virus when tested 14 days later.

The best results were obtained with a series started in larval thoracic tissue subcultured (a) after seven days to pupal head tissue, (b) seven days later to larval thoracic tissue, and (c) seven days later still to pupal head tissue and tested after a further period of seven days (Table I). At the end of these 28 days *in vitro*, the virus titer was 10^5 times higher than in the original suspension. The results of this experiment have shown the different susceptibility of certain mosquito tissues for the multiplication of virus. The virus, after its 20 days in insect tissue culture, was passed 13 times through mice and was then tested in guinea pigs, producing in all of them typical symptoms of the disease. The virulence and serological properties of the virus as shown in neutralization tests had thus remained unchanged.

Haines (1958) studied the problems of survival of EEE virus in maintained fourth instar larval midgut tissue of *Aedes aegypti*. Success in maintaining larval midgut tissue in synthetically prepared media composed principally of buffered salt solution, amino acids, and sugars

TABLE I
MULTIPLICATION OF WEE VIRUS IN *Aedes aegypti* TISSUES AT 28°[a]

Culture	Tissues	Concentration of 5% brain suspension	Total days in culture when tested	Highest dilution (based on original brain suspension) in which virus was demonstrated
			0	10^{-5} (3 dead in 4 days)
Original culture	Larval thoracic	$\frac{1}{2}$	7	10^{-6} (3 dead in 4 days)
First subculture	Pupal head	$\frac{1}{2} \times 10^{-2}$	14	10^{-6} (2 dead in 3 days, 1 dead in 4 days) 10^{-7} (1 dead in 5 days)
Second subculture	Larval thoracic	$\frac{1}{2} \times 10^{-4}$	21	10^{-8} (3 dead in 3 days)
Third subculture	Pupal head	$\frac{1}{2} \times 10^{-6}$	28	10^{-10} (3 dead in 4 days)

[a] From Trager (1938).

prompted additional tests to determine the survival of this virus in the midgut tissue maintained *in vitro*. Only short-term cultures were attempted. Each hanging drop culture (this method was used in experiments) received 0.005 ml of concentrated suspension from an infectious mouse brain with an equal volume of the culture medium. A previously prepared fourth instar larval midgut was placed in the culture medium on a sterile cover slip. The cover slip was inverted, placed on a depression slide and sealed. All cultures were incubated at 27°–28°. After various intervals of incubation, the midgut tissue was ground in a mortar, together with cover slip, in 1 ml of beef heart infusion, for virus titration by intracerebral inoculation in 10–12 gm of mice.

In controls without mosquito tissue a 10^3 decrease in titer occurred during a 24-hour period. The stability of the virus in the presence of the synthetic culture medium alone appeared greater during the first two hours and slightly less thereafter as compared with the titers obtained with the midgut-virus exposed samples. An average of the four titers at the various test intervals indicated good initial survival of the virus in the presence of midgut tissue. A more rapid decline in titers was noted between the first and fifth day with better survival in the control cultures than in the test series. Since the sensitivity of the titration test is approximately $10^{0.6}$, the effect of larval midgut tissue on the survival of EEE virus appeared negligible.

The second problem concerned the comparison of the development of midgut tissue cells *in vivo* with the development of normal or EEE infected cells *in vitro*.

A series of observations was made to determine the type and extent of growth of midgut cells as judged by measurement of their nuclei during the intervening period between the fourth instar larval stage and early adult stage *in vivo* and *in vitro*. A culture of fourth instar larvae was allowed to develop at insectary temperature (24°), and a total of five specimens were examined daily for nine days to determine normal midgut cell activity. Midguts were excised, halved, and immediately stained with acetocarmine. Nuclear measurements were made by means of an ocular micrometer in a compound microscope at 930 magnifications. At the same time (0 days), 45 larvae of equal size were prepared for *in vitro* studies. Midguts were excised and halved, and equal numbers of hanging drop preparations were made with and without the addition of virus. Nuclear measurements were taken daily from five separate microscopic fields of five explants. The mean diameter of the nuclei, expressed in microns, was then determined.

Different sizes of nuclei were found during the time when the larvae approached the adult stage. Regardless of the method of culturing, there was an initial decrease in percentage of small nuclei (10 μ or less in diameter). A greater decrease in the size of nuclei was observed during the first day of culture in both virus-exposed and nonexposed midgut samples *in vitro* than in the tissue from normally cultured specimens *in vivo*. A minimum size in the diameter of the nuclei was reached on the third day in both normal developing as well as in cultured midgut cells. A decrease in size of the nuclei from samples of midgut tissue exposed to EEE virus was noted during the first two days of culture and appeared somewhat less than in the *in vitro* controls (Haines, 1958).

Price (1956) has reported that Japanese encephalitis virus losts its infectivity for a time when grown in mosquito tissue culture. It is a pity that no more details about this very interesting and important finding were published. From Price's short communication it is also unclear what type of culture was used in the experiment.

Johnson (1967, 1969) developed a Maitland-type tissue culture system from minced, germ-free larvae and pupae of the mosquitoes *Aedes aegypti* and *A. triseriatus* as medium for the multiplication of VEE and EEE viruses. He found that growth of VEE virus in mosquito cultures was influenced by the length of incubation time, the temperature and the virus concentration used in the adsorption process, and the temperature, pH, and agitation of cultures during growth. Ten serial passages of Trinidad strain or five serial passages of ninth strain of VEE viruses in *Aedes aegypti* tissue cultures produced no detectable changes in either the virulence for mice or the distribution of plaque size of these virus strains.

B. Multiplication of Rickettsiae in Louse Tissues *in Vitro*

Nauck and Weyer (1941) and, later, Weyer (1952) used the explants of different tissues from lice (*Pediculus humanus* L.) for the cultivation of various rickettsiae. Of the different organs, i.e., stomach, ovary, salivary glands, and nephrocytes and embryonal cells, only the cells from louse stomach were found to be convenient for the propagation of rickettsiae.

The explants from louse tissues were cultured in hanging drops. As the nutrient medium, human and rabbit plasma plus either extract of lice or of rabbit spleen and testes was used. The explants were cultured at 31°–32°. For the infection of explants different strains of *Rickettsia prowazeki*, *R. mooseri*, and *R. quintana* were used. Nauck and Weyer cultured the stomachs of either infected or healthy lice, the latter having been in contact, for 30–60 minutes, with a suspension prepared from infected louse stomachs or to which the pieces of infected rabbit spleen or testicles were added.

The positive results were obtained only with the explants from stomachs of lice infected more than five days before the explantation and which excreted the rickettsiae in their feces. In that case the propagation of *R. quintana, R. prowazeki,* and *R. mooseri* was noted. In an effort to cause infection of healthy pieces of louse stomachs, the louse stomachs were either placed into the suspension of rickettsiae before culture or they were cultured together with infected louse stomachs or explants of rabbit testicles or spleens. In these experiments the transmission of rickettsiae to the healthy organs did not occur.

The form and the ability for staining of rickettsiae multiplying in the culture remained unchanged for several months. Nevertheless, the rickettsiae were propagated reliably only on a rabbit tissue. *Rickettsia quintana* retained the ability to reproduce up to 34 days following explantation; *R. prowazeki* and *R. mooseri* preserved their primitive pathogenic properties up to 4 or 16 days. In a great proportion of experiments, the introduced rickettsiae, though still virulent, propagated only extracellularly in the louse stomachs so that they did not differ from *R. quintana,* but they lost their pathogenicity for guinea pigs and mice. Such extracellular strains were isolated from *R. mooseri* cultures after as many as 23 and 40 days; from *R. prowazeki* cultures, after as many as 36 days and mostly from combined cultures with rabbit spleen and testicles.

In one experiment a healthy tissue from different organs of adult ticks *Rhipicephalus bursa* was kept in contact with the infected explants of louse stomach with *Rickettsia prowazeki* and *R. mooseri.* Although these rickettsiae were surviving in louse explants up to the eight day after the explantation, the infection of tick organs did not succeed.

C. Multiplication of Rickettsiae and Viruses in Tick Tissues *in Vitro*

1. Development of *Coxiella burneti* from Its Filterable Particles

To find a suitable medium for the investigations on the development of filterable particles of *Coxiella burneti,* the surviving tick tissues *in vitro* were enclosed into these experiments [Kordová and Řeháček (1959, 1965)]. For these purposes the organs of 40 half-engorged females of *Ixodes ricinus* L. and 20 females of *Dermacentor pictus* Herm. were taken and eight tissue cultures were prepared from each tick; of these, four were inoculated with filtrates and four were left as controls. Three kinds of empirical medium were used: Parker 199, Parker 199 (3 parts) and heated horse serum (1 part), and hemolymph of cockroaches plus the serum as in the previous medium.

The inoculum was prepared by mixing 2 ml of medium and 0.5 ml of filtrate with 500 units of penicillin/ml of nutrient medium. From the second to the 18th day after seeding and inoculation of tissue cultures, smears were made from the individual organs. On microscopic examination of native cultures, longitudinal and transversal contractions were observed for a period of 10 days in the malpighian tubuli and ovaries. The fragments of tissues in the cultures only survived; slight proliferation of rather undifferentiated cells, morphologically resembling fibroblasts, was observed only in a few cultures from ovaries.

In the cells of tissue cultures inoculated with filterable particles of *C. burneti* from the third to the fifth day, a massive incidence of inclusions was observed. These inclusions were similar to those found in the organs of ticks infected with filtrates *in vivo*. At this time no inclusions were observed in the control tissue cultures. But in the following days of observation similar formations regularly appeared also in the control cultures independently of the type of medium used. More detailed observation was rendered difficult by deep degenerative changes in the cells (granulation of the cell cytoplasm, etc.). As distinct from preparations from organs of ticks infected *in vivo*, morphologically typical *C. burneti* were found in the tissue cultures only after 14–16 days and in quantities considerably smaller than in ticks *in vivo*.

2. Propagation of Eastern Equine Encephalomyelitis (EEE) Virus

Considering our knowledge on the biology of EEE virus in ticks (Řeháček, 1958a), it seems that the cells of tick tissues *in vivo* do not offer suitable conditions for the reproduction of this virus. On the other hand, we tried to answer the question whether or not tick cells represent a substrate suitable for the multiplication of this virus *in vitro*.

For these experiments the method of the culture of fragments from tick organs was used. The explants were cultured in test tubes at 25° in a synthetic medium (Parker 199) prepared from the dried preparation (Difco). Virus assay was carried out in chick embryo fibroblast cultures by means of color tests on plastic panels (Pešek, unpublished).

In the first two experiments, in which the EEE virus was introduced separately in the ovaries, malpighian tubes, salivary glands, guts, and connective tissue with hypodermis, it was found that virus multiplication occurred only in the connective tissue with hypodermis. The other tissues only manifested a protective effect in that the virus had been inactivated in their presence more slowly than in the medium alone. This protective effect was practically identical with all the tissues examined (Table II).

In further experiments, separate cultures of the connective tissue with hypodermis and of a mixture of tick organs (ovaries, malpighian tubes, and salivary glands) were used.

TABLE II

CULTIVATION OF EEE VIRUS IN SURVIVING TICK *Dermacentor pictus* TISSUES

	Hours			Days					
	0	1/2	3	1	2	3	4	5	6
(a) Medium without tissues	2.50[a]	—	1.66	1.50	0.66	0.00	0.00	0.00	0.00
(b) Connective tissue, hypodermis	—	2.66	2.66	2.66	2.00	2.00	2.50	2.66	4.50
(c) Other tissues	—	2.60	2.58	2.20	1.83	1.37	1.14	1.08	0.84
consisting of									
salivary glands	—	2.66	2.33	2.33	2.50	1.33	1.23	1.50	1.50
guts	—	2.77	3.00	2.33	1.50	1.50	1.33	0.50	0.50
ovaries	—	2.33	2.50	2.50	2.00	1.66	1.00	1.00	0.66
malpighian tubuli	—	2.66	2.50	1.66	1.33	1.00	1.00	1.33	0.66

[a] The figures represent log $TCID_{50}$/ml.

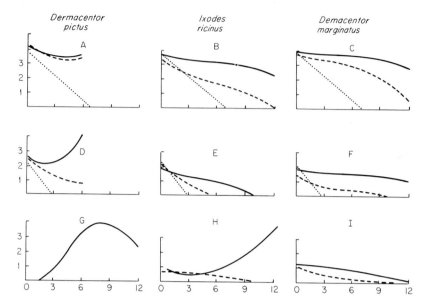

Fig. 1. Cultivation of the EEE virus in surviving tick tissues. A,B,C, large inocula; D,E,F, medium inocula; G,H,I, small inocula. —, connective tissue and hypodermis; - - -, ovary, malpighian, and salivary glands: ·····, medium without tissues. Abscissa: days of culture. Ordinate: log TCID$_{50}$.

The results obtained are illustrated in Fig. 1. The curves in Figs. 1F and 1I were computed from the results of three experiments; that in Fig. 1C, of one experiment; and the remaining curves from the results of two experiments each.

No multiplication of the EEE virus was observed in tissue cultures of *D. pictus* inoculated with 10^4 TCID$_{50}$ of virus. There was only a protective effect of the tissues, which enabled the virus to survive in an almost undiminished titer for six days, whereas the inoculum was inactivated in the controls during this period (Fig. 1A). When the tissues were inoculated with $10^{2.5}$ TCID$_{50}$ of virus, marked multiplication of the virus occurred in the connective tissue and hypodermis, whereas the other tissues only slowed down its inactivation (Fig. 1D). In cultures inoculated with insignificant doses of virus (less than $10^{0.6}$ TCID$_{50}$), the virus multiplied significantly in the connective tissue plus hypodermis, reaching a maximal titer of $10^{4.3}$ TCID$_{50}$ on the ninth day of culture. No virus was found in the controls without tissues or in cultures of the other tissues (Fig. 1G). It was impossible to detect the dose of virus used in this case for infection of the tick tissue cultures by an intracerebral test in mice or in chick embryo fibroblast cultures.

In tissue cultures of *Ixodes ricinus* L. inoculated with $10^{4.0}$ or $10^{2.3}$ $TCID_{50}$ of virus, only a substantially slower inactivation of the virus was observed (Fig. 1B and E). Virus multiplication took place up to the titer $10^{3.3}$ and on the 11th and 12th days but only in the cultures infected with $10^{0.6}$ $TCID_{50}$ of virus (Fig. 1H).

In *Dermacentor marginatus* Sulz. tick tissues inoculated with $10^{4.0}$, $10^{2.3}$, and $10^{1.2}$ $TCID_{50}$, the virus only survived for longer periods than in the controls (Figs. 1C, F, I).

In parallel control experiments we tried to determine (a) whether the EEE virus is able to multiply also in tissues of other arthropods and (b) whether the tissues of *I. ricinus* and *D. marginatus* are suitable for the multiplication of Newcastle Disease Virus (NDV). This virus was chosen because it can be titrated easily in chick fibroblasts by the same method as the EEE virus. In the first case we used as a model the fat body with hypodermis of the *Periplaneta americana* cockroach. The results showed that the EEE virus inoculated in the amount of $10^{4.0}$ and 10^{0} $TCID_{50}$ was inactivated to the same extent as in the control without tissue. In the second case, we found that, in cultures inoculated with $10^{3.0}$, $10^{2.3}$, and $10^{1.6}$ $TCID_{50}$, the NDV did not multiply; its rate of inactivation was slower than in controls without tissues. These experiments were repeated twice with similar results.

The results obtained show that the surviving connective tissues and hypodermis of *D. pictus*, *I. ricinus*, and possibly, *D. marginatus* ticks enable EEE virus to multiply *in vitro*, the virus increment being of the order of 10^4 (Řeháček and Pešek, 1960; Blaškovič and Řeháček, 1962).

III. USE OF PRIMARY CELL CULTURES FOR THE MULTIPLICATION OF VIRUSES AND RICKETTSIAE

A. Multiplication of Viruses in Mosquito Cells

Peleg-Fendrich and Trager (1963a,b) studied the possibility of the multiplication of West Nile (WN) virus in primary cultures of mosquitoes derived from *Aedes aegypti* larvae. In these experiments the growing tissues and cells from the explants of the imaginal discs of fourth instar mosquito larvae were cultured in hanging drops and infected with the Egypt 101 strain of WN virus.

In four experiments representing over 80 cultures, the increase of virus titer of about 10^2 was found in the mosquito tissue cultures, in comparison

TABLE III

AVERAGE AMOUNT OF WN VIRUS PER CULTURE AFTER
INFECTION WITH APPROXIMATELY 5×10^4 LD$_{50}$[a]

	Virus titer log LD$_{50}$[b]					
	Experiment No.				Controls	
Days after infection	1	2	3	4	1	2
2		3.4			2.5	2.0
4		4.2			2.0	1.1
5	4.8	5.5	4.5	5.1	<1.0	<1.0
10	4.5	6.5	4.5	6.5		
15	4.8	6.2	5.5	6.0		
20	4.6	4.7	3.4	7.3		
25		5.8		4.6		

[a] From Peleg-Fendrich and Trager (1963a).

[b] The tissues of three cultures, each of them infected with 0.03 ml of virus suspension containing 5×10^4 LD$_{50}$, were pooled and ground in 1 ml of SSP and titrated. The titer thus obtained was divided by 3 to give the average amount of virus per culture.

with quick decrease of the virus amount in controls without tissues (Table III).

In the second series of experiments the serial passages of WN virus in tissue culture were performed. The authors used the following method. The diluted infected mouse brain suspension giving 8×10^4 mouse LD$_{50}$ virus/0.03 ml was used for the infection. Each of a group of cultures consisting from mosquito larvae imaginal discs was cultured in 0.03 ml of the above-mentioned suspension. After an incubation period of 10 days the virus from three of the cultures was harvested and pooled in 1 ml of culture medium. This procedure therefore gave an 11-fold dilution of the virus present in the first cultures. Part of this suspension was titrated in mice and the remainder served as the medium for growing further larval imaginal disc cultures. This was considered the first of serial passages. This procedure was repeated at 10-day intervals so that the pooled, titrated, and diluted cultures served as the medium for the tissue culture of the next virus passage. Each passage was thus initiated with an 11-fold dilution of the virus yield of the preceding passage. Altogether, four serial passages of the virus were carried out. Since the virus used to infect the first culture was diluted 1.6×10^5-fold during its four serial passages, the amount of virus found in the last passage must be attributed to an active multiplication of the virus in the tissues (Table IV). No indications of cytopathic changes were noted in the infected tissues. Briefly summarizing these results, then: the imaginal disc tissues of Aedes in

TABLE IV
SERIAL PASSAGE OF WN VIRUS IN THE IMAGINAL DISCS OF
Aedes aegypti CULTURED in vitro[a]

| Passage no. | Log LD$_{50}$ of virus | |
	Placed in culture	Recovered per culture (average)
1	5.0	6.2
2	4.7	6.6
3	5.1	5.6
4	4.1	5.0

[a] From Peleg-Fendrich and Trager (1963a).

vitro supported marked proliferation of WN virus for a period of 25 days and in serial passages.

Four experiments with ovarian tissues of Cynthia gonads totaling over 80 cultures, a series of two experiments totaling over 40 cultures, which were incubated at 30°, and one experiment containing about 20 cultures infected seven days after explantation were performed. The titrations were negative except for two sporadic cases. Results were also negative in the one experiment attempted with testes tissue. These investigations were done by Peleg-Fendrich and Trager (1963a,b) as a control to the experiments in maintaining WN virus in mosquito cells.

Peleg (1968a) introduced EEE, SF, and WN viruses in primary tissue cultures of Aedes aegypti. All these viruses multiplied in cultures for as long as 60 days. No cytopathic effect which could be attributed to virus propagation was found in cells. West Nile virus was found to be the slowest growing, EEE was intermediate, and the Semliki Forest (SF) virus grew most rapidly. The experiments on the detection of EEE and SF viruses by fluorescent antibody technique supplied only negative results. The presence of EEE virus in the cultures was verified by the hemagglutination test.

Fujita et al. (1968) multiplied JE virus in primary cultures prepared from Culex molestus ovaries. The virus titers reached a maximum $10^{4.5}$ mouse LD$_{50}$/0.02 ml on the seventh day and was found to persist in cultures with some variation in titer up to 65 days. The virus was successfully passaged in mosquito tissue culture. The original virus inoculum was diluted 10^{27} times during 164 days, and it retained its mouse infectivity throughout these cultures. Virus propagation in cells was shown by immunofluorescent staining. The fluorescence was noted first in the juxtanuclear area and then spread out over the entire cytoplasm.

B. Multiplication of Viruses and Rickettsiae in Tick Cells

1. Different Viruses

When the sufficient amount of *in vitro* growing tick cells is available (Řeháček, 1958b, 1962, 1965a; Řeháček and Hána, 1961a,b; Martin and Vidler, 1962), then it is possible to use them for the investigations with viruses and rickettsiae as mentioned at the introductory part of this paper.

In our first experiments it was found that tick-borne encephalitis (TE) virus multiplied very well in such cells without causing a cytopathic effect (Řeháček, 1962). This stimulated further investigations with this and, also, the other viruses. It was attempted to determine whether other viruses also can be propagated in tick tissue cultures. For these investigations tissue cultures from *Hyalomma dromedarii* Kock ticks were used (Řeháček, 1965b). The cultures were set up in test tubes, and about 24 hours after seeding, when the cells started to grow, the tube cultures were inoculated into the nutrient medium with, usually, three varying doses of different viruses. Samples of medium for extracellular virus assay were taken after two, four, six, and eight days, the withdrawn volumes not being replaced by fresh medium. The viruses were titrated by intracerebral inoculation in mice or by inoculating various tissue cultures, depending on their sensitivity to the given virus. Survey of viruses tested ·in tick tissue cultures is given in Table V.

The kinetics of multiplication of the virus are illustrated in Fig. 2. Each curve was computed from the results obtained in five tube cultures. The samples were taken and titrated from each tube separately.

All the viruses multiplying in tick cultures propagated there without any cytopathic effect detectable by direct observation. Of group A arboviruses, WEE, EEE, Sindbis, and Semliki viruses were chosen and all multiplied significantly. In controls (medium without tissues), there was naturally no multiplication. These viruses multiplied in the cultures approximately at the same speed, by about $10^{0.5}$–10^{1}/day. Using small amounts of virus as inoculum (about 1–10 mouse LD_{50} or CPD_{50}), by the eighth day they reached amounts 10^{4}–10^{5} times greater than those inoculated. Multiplication of group B arboviruses yielded different results. Langat (TP 21), Japanese encephalitis (JE), St. Louis encephalitis, and yellow fever viruses multiplied in tick tissue cultures poorly. Kyasanur Forest Disease (KFD), Powassan, Omsk haemorrhagic fever (OHF), and WN viruses multiplied relatively well. Tickborne encephalitis (western type, TBE), Russian spring–summer encephalitis (RSSE), and louping-ill viruses multiplied excellently. The highest titers of the

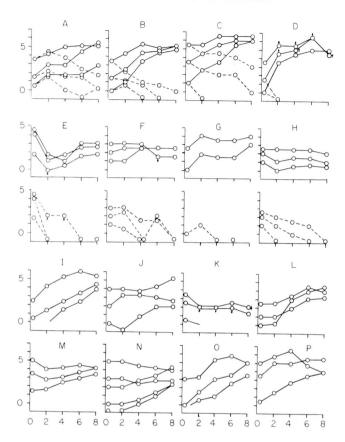

FIG. 2. Multiplication of different viruses in tick tissue cultures. A, WEE; B, EEE; C, Sindbis; D, Semliki forest; E, JE; F, St. Louis encephalitis; G, West Nile; H, yellow fever; I, TBE; J, KFD; K, Langat; L, louping ill; M, OHF; N, Powassan; O, RSSE; P, LCM viruses. —, tick tissue; - - -, cell free medium.

viruses were again obtained with small inocula (1–10 mouse $LD_{50}/0.03$ ml), the virus increment being approximately of the order of $10^{0.5}$–10^1 per day.

Viruses not belonging to arboviruses (Fig. 3), i.e., encephalomyocarditis (EMC), polio, vaccinia, vesicular stomatitis (VS), NDV, and pseudorabies, did not multiply in tick tissue cultures and their titers obtained in samples taken at two-day intervals for eight days were practically the same as in controls. An exception was lymphocytic choriomeningitis virus (LCM), which multiplied in tick cultures very well, similarly to the viruses transmitted in nature as shown by Řeháček (1965b,c).

TABLE V

SURVEY OF VIRUSES TESTED IN TICK TISSUE CULTURES

Virus	Strain[a]	Characteristics[b]	Titrated in[b]	No. tested over No. of tubes	Maximum titer[c]
Arbo A					
EEE	28 N8	M35, adapted to CEC	M,CEC	4/25	5
WEE	0	M21, adapted to CEC	M,CEC	7/41	4
Semliki	0	M31, Ms1	CEC	3/30	5
Sindbis	AR339	Ms7, CE1	CEC	3/40	5
Arbo B					
JE	P3	M65, isolated in 1950	M	3/19	2
St. Louis	0	At IV Brat.—M3	M	3/35	2
West Nile	Egypt 101	M9, isolated in 1957	M	3/18	3
Yellow fever	Asibi	At IV Brat.—M2	M	3/19	1
KFD	W377	M5, isolated in 1957	M	2/29	3
TBE	Hypr	M43	M	10/70	5
Langat	TP21	M8	M	3/37	5
Louping ill	Li59	M3	M	3/18	5
OHF	0	At IV Brat.—M5	M	2/25	2.5
Powassan	0	At IV Brat.—M3	M	2/36	2.5
RSSE	Sofin	At IV Brat.—M9 isolated in 1937	CEC	2/16	5

No arboviruses					
LCM	3b		M	2/21	4
EMC	0	Adapted to MEC	MEC	5/31	—
Polio	MEF 1, type 2	Cultivated in HeLa cells	HeLa	3/37	—
VS	0, type Indiana		MEC	4/18	—
Vaccinia	0	CE33	HeLa	4/29	—
NDV	Herfordshire	Three times purified by plaque isolation, CEC1	CEC	2/16	—
Pseudorabies	BUK	Vaccine strain, CEC625	CEC	2/22	—
	CVHD	CEC9	CEC	2/22	—

[a] Strains obtained without any proper specification are designed 0. IV Brat.: Institute of Virology, Bratislava.
[b] The viruses were passaged and titrated in the following systems; M, adult white mice, by intracerebral inoculation; Ms, baby white mice, by intracerebral inoculation; CEC, chick embryonal cell culture; CE, chick embryo; MEC, mouse embryo cells.
[c] Maximum titer reached in tick tissue culture (log $TCID_{50}$ or log mouse LD_{50}). —: no multiplication.

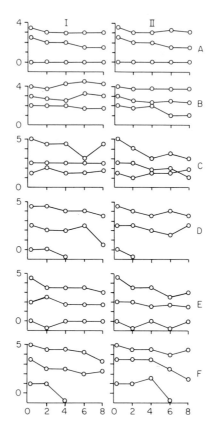

Fig. 3. Survival of viruses in tick tissue cultures. A, polio type 1; B, EMC; C. NDV; D, vesicular stomatitis; E, vaccinia; F, pseudorabies viruses. I, tick tissues; II, cell free medium. Abscissa: days after inoculation. Ordinate: virus titers in log $TCID_{50}/0.1$ ml.

In comparison with vertebrate tissue cultures, the viruses multiplied slowly in tick tissue cultures; perhaps the factor of low temperature (27°) played a role and influenced the multiplication of both cells and virus. In ticks infected with TBE virus, the increase of the virus level was similar (Benda, 1958).

The highest titers of several viruses were reached at the eighth day of their introduction into tick cultures when the experiment was stopped. It may be that in a few cases the virus titers would still increase after this period. Only a single passage of a single strain of the given virus species, except of two different strains of Pseudorabies virus, were examined. It is possible, therefore, that different results could be obtained with other strains, variants, or passages of the same virus species. Regarding the propagation of group A arboviruses in tick tissue cultures, their ability to multiply in various substrates could play a role; for instance, the multiplication of EEE virus in fish embryos *in vitro* (Soret and Sanders, 1954) can be mentioned.

Group B arboviruses behaved differently in tick tissue cultures. Viruses transmitted in nature only by mosquitoes, e.g., yellow fever, St. Louis encephalitis, and JE viruses, multiplied poorly in the cultures, whereas viruses transmitted by ticks, except Langat virus, multiplied readily. Viruses not belonging to arboviruses did not multiply in tick tissue cultures with the exception of LCM virus.

The results obtained indicate that tick tissue cultures represent a suitable substrate for the multiplication of several arboviruses. We assume that these results could be used as a preliminary criterion for a rough classification of viruses. The findings that the propagation of several viruses was demonstrated with very small inocula could apparently be of value when isolating viruses from test materials.

Since tick-borne encephalitis virus (western type) is spread in several parts of Europe, sometimes causing serious disease, our attention in further experiments was paid to this virus. We tried to answer the question of whether or not tick cells *in vitro* are a suitable medium for the multiplication of different strains of this virus (Řeháček, 1963).

Two groups of experiments were carried out. In the first group 40 carrel flasks with cell cultures of *H. dromedarii* ticks including all cell types were inoculated 24 hours after the seeding of the cells with varying doses of TBE virus, strain Hypr, isolated by Pospíšil *et al.* (1954) from man. This virus had undergone, till now, 48 mouse passages. Samples of medium were taken at 48-hour intervals. Titrations of virus were carried out by intracerebral inoculation of mice, weighing 5 gm. Multiplication of virus was observed in all inoculated cultures; the titer of virus increased by 10^1–10^5. It was shown that the multiplication of TBE virus occurs also in cultures composed of hemocytes only, but the titers of virus obtained were not as high as in cultures containing all types of cells.

In the second group of experiments, 20 carrel flasks with cultures of *Dermacentor marginatus* ticks were inoculated 24–72 hours after the seeding of the cells with varying doses of TBE virus, strain Hypr, adapted to Hela cells (Libíková, 1961). Samples of medium were taken at daily intervals, and virus titrations were carried out in Hela cells. Significant multiplication of virus was observed in all the cultures. In eight experiments cells of *H. dromedarii* were used for cultures. It was shown that the cells of this species were susceptible to TBE virus adapted to Hela cells in the same degree as the cells of *D. marginatus* ticks (Řeháček, 1963).

In few experiments the cells from an unknown reason did not adhere to the glass but survived flowing in the medium. TBE virus multiplied here, also, but its quantity increased by about 10^1 only, in comparison with cell cultures in the same experiment in which the amount of virus in settled and growing cells increased by 10^5. This result also indicates that probably only growing cells can support good multiplication of TBE virus. The multiplication of TBE virus both adapted or unadapted to Hela cells in cells of both tick species used occurred without any cytopathic effect. No differences were noted between the rate of multiplication of TBE virus in the cells of both tick species, from which only *D. marginatus* tick is known as vector of TBE virus.

In 22 experiments the doses of virus used for the inoculation of tick tissue cultures were so small that they were not detectable in intracerebrally inoculated mice or on Hela cells. This fact could perhaps be of advantage for virus isolation experiments.

Further problem appeared to establish a degree of the susceptibility of tick cells *in vitro* to TBE virus. To compare the susceptibility of various cell systems to small doses of viruses of the TBE complex, the chick embryo cell cultures appeared the most susceptible (Libíková *et al.*, 1962). In further experiments we compared this method with the detection of smallest amounts of TBE virus in primary tick cultures (Řeháček and Kožuch, 1964).

Tube cultures of chick embryo cells were grown for 24 hours at 37° in medium 199 supplemented with 10% unheated horse serum. Tube cultures of tick cells from *Hyalomma dromedarii* species were prepared as described previously and were grown also for 24 hours. Small amounts of the Hypr strain of TBE virus in its 54th mouse passage were inoculated into 4–10 tubes with chick cells and tick cells separately. After five days, subcultures were made to chick embryo cells in which TBE virus was demonstrated by the method based on interference with WEE virus.

The experiments were repeated six times with almost similar results (Table VI). With inocula of 30 and 3 IFD_{50}, the percentage of infected chick embryo cells and tick cell cultures was about equal. With inocula containing 0.3 and 0.03 IFD_{50} of TBE virus, the respective percentage of infected tick cell cultures was 76 and 18, whereas that of chick embryo cell cultures 34 and 3 only. These results indicate that tick cell cultures provide more sensitive system than chick embryo cell cultures and are actually one of the most susceptible systems for detecting small amounts of TBE virus. Although the results are as yet of only theoretical value, their use in practice on a large scale could become possible provided that more convenient methods of tick cell preparation or a stable tick cell line become available.

The presence of TBE virus in tick cells *in vitro* has been studied by

TABLE VI

COMPARISON OF THE SUSCEPTIBILITY OF TICK (TT) AND CHICK EMBRYO CELL (CEC) CULTURES IN DETECTING SMALL AMOUNTS OF TICK-BORNE ENCEPHALITIS (TE) VIRUS

Inoculum (IFD$_{50}$/0.1 ml)[a]	Type of culture	Experiment						Per cent cultures infected
		1	2	3	4	5	6	
30	TT	5/5[b]	4/4	4/4	5/5	10/10	10/10	100
	CEC	5/5	4/4	4/4	5/5	10/10	10/10	100
3	TT	5/5	4/4	4/4	5/5	9/10	9/10	95
	CEC	5/5	2/4	4/4	5/5	9/10	10/10	92
0.3	TT	5/5	3/4	4/4	5/5	3/10	9/10	76
	CEC	2/5	1/4	2/4	5/5	1/10	2/10	34
0.03	TT	1/5	1/4	1/4	1/5	1/10	2/10	18
	CEC	0/5	0/4	0/4	1/5	0/10	0/10	3

[a] IFD$_{50}$: a dose of TE virus interfering with WEE virus in half of the tubes inoculated.

[b] Numerator: number of tube cultures infected. Denominator: number of tube cultures inoculated.

means of fluorescent antibody staining. The preliminary results have shown that virus propagation occurs only in the cytoplasm and was concentrated around the nucleus (Řeháček, 1965c).

Řeháček and Kožuch (1969) used the *Hyalomma dromedarii* cell cultures for the tick-borne encephalitis virus isolation from different materials in nature. From 187 samples tested, composed of ticks and brains from various small animals, five virus strains were isolated. These strains were isolated also in chick embryo cell cultures as well as in suckling mice. These results indicate that the method of isolation of tick-borne encephalitis virus in tick tissue cultures is equivalent to those of embryo cell cultures and suckling mice.

The growth of Colorado tick fever (CTF) virus in primary tissue cultures of its vector, *Dermacentor andersoni* Stiles, was studied by Yunker and Cory (1967).

These authors have used in two experiments the Florio-2 strain of this virus. This strain had undergone about 30 hamster and 65 mouse passages before use in experiments. In Experiment 3, a low passage strain of CTF virus was used. This strain was isolated from blood of a ground squirrel and passed six times in suckling mice. Tick tissue cultures were prepared from viscera of metamorphosing, engorged nymphal *D. andersoni*. The culture of cells was performed within a 30-ml polystyrene culture flask. Each flask received six viscera and was then allowed to stand undisturbed for one hour, after which 4 ml of culture medium was gently added. The medium consisted of lactalbumin hydrolyzate in Hank's balanced salt solution plus vertebrate serum, chicken egg ultrafiltrate, and bovine plasma albumin. After five days and when the outgrowth of cells occurred, the cultures were infected with the virus. In experiment 1, the flasks in each group received 15.8, 158, and 1580; in Experiment 2, 5020 and 50.20 suckling mouse LD_{50}, respectively. In experiment 3, the explants were prepared from nymphal ticks that had been held at 26–28° for seven days after engorgement on viremic hamsters. Virus assay was performed from tissue culture media two hours after inoculation, at 24 hours, at three- to four-day intervals until the 21st day, and thence at weekly intervals in two- to four-day-old suckling mice by i.c. inoculations.

In experiments 1 and 2, after a postinoculation latent period of 6–10 days was followed by a $10^{4.5}$ increase of extracellular virus at four to five weeks. Virus was recovered in diminishing quantities for as long as 159 days in medium and 166 days in triturated tissues. Tick culture is a very sensitive system for the detection of this virus. It was shown that less than 0.1 of a suckling mouse i.c. LD_{50} can be taken up and propagated to higher levels by tick tissues *in vitro*. In Experiment 3, a more natural route of infecting tick tissues was used to compare virus activity in live nymphal ticks with that of their tissues cultured *in vitro*. Titers in whole nymphs fed on viremic hamsters remained relatively constant from drop-off to molting, but titers in cultures prepared from their tissues increased by approximately 10^5 two weeks after explantation.

The appearance of tick cultures was not affected by viral infection, as shown by comparison with uninfected cultures. Some tests were made to determine whether the long-term propagation of CTF virus in tick tissue cultures altered strain characteristics. The Florio-2 strain differs from other strains of CTF virus in its ability to kill adult mice after intraperitoneal injection. After 98 and 124 days in tick tissue culture, this capability remained unchanged. In addition, average survival time (5 days) of i.c.-inoculated suckling mice was not altered, and titer of mouse brain stock prepared from virus that had been in tick tissue culture for 124 days remained approximately equal.

Řeháček *et al.* (1969) noted the multiplication of Tribec virus in the cultures of *Hyalomma dromedarii* ticks reaching virus titer up to $10^{4.5}$ mouse intracerebral $LD_{50}/0.01$ ml from 4–14 days after the infection. The fluorescent antibody technique revealed the viral antigen in the form of confluent and brightly fluorescing granules in the cytoplasm of infected cells beginning at the second day.

2. Different Rickettsiae

A suitable medium for the multiplication of rickettsiae under *in vitro* conditions is still a problem. Although various vertebrate cells *in vitro* were successfully used for their propagation, a general use of these cells such as for virus multiplication where they often substitute laboratory animals, has not been established here.

It is well known that the ticks are the biological vectors of various rickettsial diseases. We were interested in the question as to whether tick cells also *in vitro* would be able to support multiplication of different rickettsiae.

The tick *Hyalomma dromedarii* was employed for the preparation of cell cultures. The method of their establishment was the same as already described (Řeháček, 1965a). The culture was performed mostly on small cover glasses placed in test tubes which were kept at 30°. The following strains of rickettsiae were used: *Rickettsia prowazeki*, strains Breinl and W2; *R. mooseri*, a strain originated from Mexico; *R. conori*, strain M1; *R. akari*, strain Toger (the latter two isolated in USSR); and *Coxiella burneti*, strain Nine Mile, phase II. The cells were set up in the growth medium with 200 units of penicillin and 200 μg/ml of streptomycin. Between the third and seventh days after the seeding of cells, the grown cultures were twice washed with medium without antibiotics. Thereafter, 1.5 ml of the same medium and 0.2 ml of serial tenfold dilutions in saline of suspensions from infectious yolk sacs of embryonated hen eggs were added to each tube. The multiplication of rickettsiae in cells was studied at two-day intervals from the second to the tenth day following the infection. The cover glasses were taken from the tubes, quickly rinsed in saline, airdried, and stained by Giemsa. In the first group of experiments the cultures were inoculated with the *C. burneti* in phase II, using a 10^{-3} dilution of yolk sac suspension. Massive multiplication of *C. burneti* in cultures with Hanks solution without antibiotics appeared as early as on the third day (Fig. 4). The titer reached six days after inoculation of the cultures, as revealed by the yolk sac titration, was 10^6 EID_{50}/ml. When comparing these results with those obtained in our laboratory on the propagation of *C. burneti* in other substrates, the tick tissue cultures proved to be most susceptible. Their use in laboratory work on *C. burneti* appears, therefore, to have good prospects (Řeháček and Brezina, 1964).

Both Breinl and W2 strains of *Rickettsia prowazeki* were detectable from the fifth to the eighth day after infection in cells inoculated with suspensions diluted up to 10^{-5}. The propagation was manifested by an increase in the amount of single rods, chains, and threads of rickettsiae in the cytoplasm of cells. Starting with the eighth day, the chains and threads, as well as single rods, disappeared from the cells. *R. prowazeki* followed the architecture of the cytoplasm going around the cell vacuoles. The addition of 50 units of penicillin and 50 μg of streptomycin per

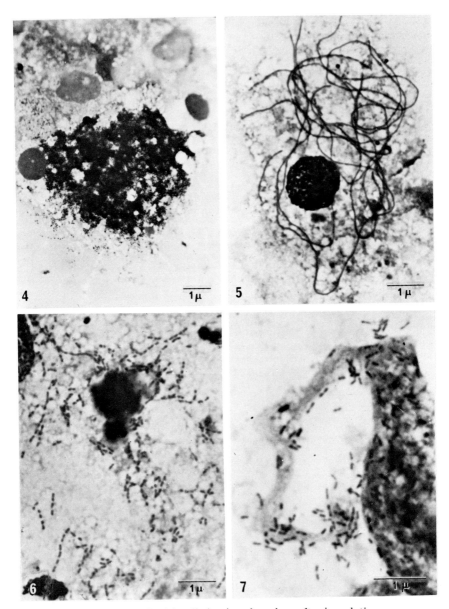

FIG. 4. *Coxiella burneti* in tick cells *in vitro*, four days after inoculation.
FIG. 5. *Rickettsia prowazeki* in tick cell *in vitro*, eight days after inoculation.
FIG. 6. *Rickettsia mooseri* in tick cells *in vitro*, five days after inoculation.
FIG. 7. *Rickettsia conori* in tick cells *in vitro*, eight days after inoculation.

milliliter of medium entirely blocked the propagation of rickettsiae. In a few cases, especially with the W2 strain, extremely long threads of rickettsiae were found in the cytoplasm of infected cells (Fig. 5). When comparing the susceptibility to *R. prowazeki* of tick cells *in vitro* and of yolk sacs, a higher dilution of rickettsiae (10^{-6}) was detected in yolk sacs than in tick cells (10^{-5}).

Rickettsia mooseri multiplied in the cells very quickly from the second day after infection. At the third day large foci of single short rods resembling dumbbells and the formation of chains showing marked segmentation were observed in the cells (Fig. 6). Starting with the eighth day, the rickettsiae began again to disappear from the cultures. The susceptibility of tick cells and yolk sacs to *R. mooseri* was of the same degree, the detection of rickettsiae having been positive in both of them in a 10^{-6} dilution of yolk sac suspension.

Also *R. conori* multiplied in tick tissue cultures very well. Short dumbbell-like rods and sometimes diplobacilli of this agent were seen in the cultures from the second to eighth days after infection. The propagation was manifested by the formation of large groups of rods without any inclination to develop chains and threads (Fig. 7). Single rods were seen also in the nuclei of infected cells. When tick cells and yolk sac were infected in parallel with serial tenfold dilutions of *R. conori,* then the agent was detected in tick cells in a 10^{-6} dilution as compared with a 10^{-4} dilution in yolk sacs.

The Toger strain of *R. akari* multiplied in tick cells from the second day and was detected in them when these cells were inoculated with dilutions of up to 10^{-5}. The highest multiplication of agent was noticed at the eighth day after infection. The propagation of this rickettsia in tick cells resembled that of *R. conori*. Single rods were also found in nuclei of the infected cells. The susceptibility of tick cells to *R. conori* was higher than that of yolk sacs in which the agent was detected only after inoculations with dilutions of up to 10^{-4} (Řeháček *et al.*, 1968).

The positive results on the multiplication of rickettsiae in tick tissue cultures open the way for further observations. When introducing *R. conori* and *R. akari* in tick cells *in vitro* and in yolk sacs of embryonated hen eggs, a higher susceptibility of tick cells to these rickettsiae was quite obvious. This finding could be promising for a possible use of tick cells for the isolation of rickettsiae from nature. We also assume that the differences in the multiplication of various rickettsiae in tick cells could serve as a preliminary criterion for their diagnosis. The multiplication of rickettsiae in tick cells *in vitro* showed many common features with that in snake tissue cultures as described by Kazár *et al.* (1966). We mention only the finding of *R. conori* and *R. akari* in cell nuclei and the higher susceptibility to these two species of tick and snake cell cultures as compared with yolk sacs. Also, the tendency of *R. prowazeki* to follow the architecture of cell cytoplasm is remarkable.

IV. USE OF STABLE CELL LINES FOR THE MULTIPLICATION
OF VIRUSES

A. Multiplication of Arboviruses in Mosquito Cell Line

1. Different Arboviruses

Mosquito stable cell line derived from *Aedes aegypti* by Grace (1966) revealed to us a quite new substrate for the proliferation of different microorganisms. Because mosquitoes are vectors of various arboviruses, some of them dangerous to man, our first effort was to use the mosquito cells for arbovirus multiplication.

First series of experiments concerned the determination of the sensitivity of these cells to different arboviruses (Řeháček, 1968a). The cells were cultured at 28° in 2-oz flat medicine bottles, and when a monolayer had formed, the cells were infected, usually, with one of three doses of viruses and without adsorbing or washing. Because of very high reproductibility of cells, at two- or four-day intervals, either ⅓ or ⅔ of the medium in the control cultures and ⅔ or all the medium with infected cells was replaced with fresh medium. Virus assay was done, from the total amount of virus in the culture up to 30th day following the infection, by means of the plaque method on chick embryonal cells or, in the case of Kokoberra virus, by intracerebral inoculation into suckling mice.

The results (Table VII) have shown that Sindbis, Semliki, Bebaru, Edge Hill, and Kokoberra viruses did not grow in the mosquito cells *in vitro* up to the 24th day after infection. The survival of virus in the presence of the mosquito cells was almost identical with virus in medium containing no cells. The medium alone showed a benign effect on virus survival; however, the low temperature of incubation also supported the survival of the virus.

On the other hand, the second group of viruses studied, i.e., Murray Valley Encephalitis (MVE), Japanese encephalitis (JE), West Nile, and Kunjin, all multiplied in the cells (Fig. 8). None of them with the exception of JE virus, caused marked specific changes in the appearance of the cells when viewed under light microscope or in their behavior. The growth of the viruses was not greater than 10^1/day. The viruses which multiplied did not show any significant differences in their propa-

TABLE VII

SURVEY ON THE CULTIVATION OF ARBOVIRUSES IN MOSQUITO CELL CULTURES

Virus	Strain	Characteristics[a]	Titrated in[a]	No. of experiments over No. of flasks	No. of controls	Growth of virus after 24 days in No. of flasks	
						Positive	Negative
Sindbis	MRM 29	Ms8	P-CEC	3/26	7	0[b]	19
Semliki	25639 Semliki	Ms20	P-CEC	2/20	5	0[b]	15
Bebaru	AMM 2354	Ms9	P-CEC	2/17	4	0[b]	13
MVE	78499-15 Plymouth brown	Un	P-CEC	5/51	5	46	0
JE	Nakayama	Ms45	P-CEC	4/38	5	33	0
West Nile	Sarafend strain	Un, HTC6, Ms2	P-CEC	4/32	7	15	10
Kunjin	MRM 16	Ms4	P-CEC	4/47	4	40	3
Kokoberra	MRM 32	Ms5	Ms	1/12	3	0[b]	9
Edge Hill	C 281	Ms4	P-CEC	2/17	4	0[b]	13

[a] The viruses were passaged and titrated in the following systems: Ms, baby white mice, by intracerebral route; CEC, chick embryo cell cultures; HTC, human tissue culture; Un, unknown number of mouse brain passages; P-CEC, plaque method on chick embryo cell cultures.

[b] Negative result.

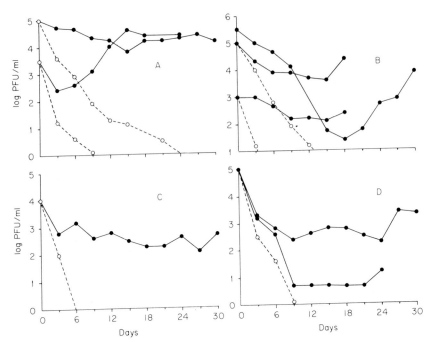

FIG. 8. Multiplication of viruses in mosquito cells line A, Murray Valley Encephalitis; B, Japanese Encephalitis; C, Kunjin; D, West Nile. —, mosquito cells; - - -, cell free medium.

gation in the cells; nevertheless, the MVE and JE viruses multiplied regularly and a little better than West Nile and Kunjin viruses, showing in 10 days of multiplication an increase in virus titer of about 10^4 PFU/ml. When West Nile virus was multiplied in most instances, especially when more than 10^4 PFU/ml of virus was used as an inoculum, a marked multiplication of virus was noticed. In other instances, the virus did not multiply (Řeháček, 1968a,b).

Converse and Nagle (1967) described the multiplication of Asibi strain of yellow fever virus in the same type of cultures. The virus in these cells exhibited a 10^2 increase in titer within 24 hours, followed by lowered titers on subsequent days. Because cell lysis in these cultures reduced cell population to approximately 50% of the original count at 24 hours, 25% at 48 hours, and 10% after four to five days of incubation, virus determinations were terminated in these cultures on the fifth day.

Singh's *Aedes albopictus* and *A. aegypti* cell lines were used extensively for the multiplication of various viruses. It was found that several viruses from different arbovirus groups multiplied in both lines (Singh and Paul, 1968a,b; Buckley, 1969; Paul et al., 1969; Yunker and Cory, 1969) and

that some of those belonging to group B produced a cytopathic effect in
A. albopictus cells. The main features of cytopathic effect were cytolysis
of the cells, development of large syncytial masses, establishment of
multinucleated giant cells, phagocytosis of dead cells, and the recovery
of the cell population in the infected cultures. In addition, these experiments demonstrated the markedly higher sensitivity of the *A. albopictus*
cell line to viruses than cells of *A. aegypti*. Suitor and Paul (1969a) described the formation of syncytia in *A. albopictus* cells infected with
dengue virus type 2 and cultured in plastic vessels. The formation of
virally induced syncytia in cultures was more sensitive than either plaque
formation in PS(Y-15) cells or inoculation in suckling mice. Suitor (1969)
first succeeded in demonstrating plaques in *A. albopictus* cells infected
with JE virus. Comparing the sensitivity of both tested cell lines to infection with various viruses with that of infant mice and the Vero cell
line, Paul and Singh (1969) found that *A. albopictus* cells were equally
or slightly more sensitive to infection with Chikungunya, West Nile, and
JE viruses than were infant mice or Vero cells, but they were slightly
less sensitive than mice and Vero cells to infection with Batai and
Chandipura viruses. These cells were 100 times more sensitive than Vero
cells and 40 times more sensitive than infant mice to infection with dengue
virus type 2.

The cell line derived from *Aedes aegypti* embryos by Peleg (1968b)
was used in virological studies by the same author (Peleg, 1968b) for the
multiplication of EEE, Semliki Forest, and West Nile viruses. All these
viruses multiplied without causing cytopathic effect and produced 100–
10 000 fold titer increases. On the other hand, the picornaviruses, encephalomyocarditis, and polio did not multiply (Peleg, 1969b).

The finding on the high sensitivity of dengue virus to mosquito cells
was used for the isolation of dengue viruses in mosquito cells (Singh
and Paul, 1969). The isolation was made from sera collected in areas in
India where dengue was endemic. The dengue viruses were isolated in
mosquito cells from 22 out of 25 samples tested. The isolation test carried
out from this material in mice and Vero cells showed the same sensitivity
of all these media to isolated dengue viruses. The mosquito cells used
for the isolation experiments served also for the identification test of
isolated dengue viruses in a complement-fixation test (Pavri and Ghosh,
1969).

2. Studies of the Morphology and Development of Murray Valley Enceph-alitis and Japanese Encephalitis Viruses in Electron Microscope

Because MVE and JE viruses multiplied very well in mosquito cells
in vitro, we chose them for the investigations on·their morphology and

replication in cultures by means of the electron microscope (Filshie and Řeháček, 1968). In these experiments the acutely, as well as persistently, infected cells (with both viruses, as mentioned above) were employed in the experiments. Every three days from 3–27 days after inoculation, cells were taken for electron microscope examination. Cells were removed from the tubes or bottles by gently pipetting the media. Following centrifuging at 600 rpm for three minutes, the cells were resuspended in a small volume of the supernatant, transferred to Spinco polyethylene microcentrifuge tubes, and again sedimented at 600 rpm for three minutes. The resulting pellet of cells was fixed with 2.5% glutaraldehyde in 0.1 M phosphate buffer at pH 7.2 for two to four hours at room temperature, washed in several changes of buffer for a total of two to four hours, post-fixed with 1% osmium tetroxide in 0.1 M phosphate buffer at pH 7.2 for two hours at room temperature, dehydrated through an acetone series, and embedded in Araldite. Sections approximately 500-Å thick were cut with a Porter-Blum MT 2 ultramicrotome, mounted on carbon films, and stained with 5% aqueous uranyl acetate for one hour at room temperature and then by lead citrate for 10–20 minutes at room temperature. The sections were examined in a Siemens Elmiskop I electron microscope operated at 80 kV with electron-optical magnifications up to 40 000 times.

The experiments to detect both the viruses in mosquito cells *in vitro* by means of this method were successful. It was found that external profiles of both MVE and JE particles in thin sections were to be either circular or elliptical, but the latter appearance could have resulted from compression introduced during sectioning. The fine structure and dimensions of MVE and JE virus particles appear identical in these sections. We have not detected any morphological differences between extracellular particles and particles found within the endoplasmic reticulum. Particles are approximately 400 Å in diameter. The outer layer of each particle consists of a single electron-dense membrane about 25 Å in thickness separated by a less dense layer about 25 Å in thickness from a central, dense core approximately 280 Å in diameter. In a few cases the outer covering of the core appears to form a distinct membrane 10–20 Å in thickness. We infer from these observations that MVE and JE virus particles consist of a nucleoid approximately 250 Å in diameter enveloped by a triple-layered unit membrane approximately 75 Å in thickness.

In the initial experiments cells were grown in medicine bottles. However, neither extra- nor intracellular particles were observed by electron microscopy during the first 24 days after inoculation. We suspected that the multiplicity and, hence, the probability of detecting particles in thin sections was low. In all further experiments involving acute infection, cells were cultured in test tubes, where higher multiplicities were achieved. MVE virus particles were first observed in acutely infected cells six days after inoculation and in persistently infected cultures in the fifth passage on the sixth day after seeding. In infected cells, mature particles are always contained within vacuoles or elements of the endoplasmic reticulum—never free in the cytoplasm. From the sixth day onwards, large vacuoles frequently contained single particles and crystalline groups of particles in addition to other debris apparently ingested from the culture medium by phagocytosis. On the ninth day, about 5% of the cells contained virus particles (Fig. 9), which were now confined mainly to the endoplasmatic reticulum. In some of these cells the reticulum was arranged in characteristic patterns, the presence of which seemed to be correlated with the presence of the relatively large numbers of particles within the reticulum. Such pattern consists essentially of areas of interconnecting smoothmembraned vesicles and cisternae, from which radiate cisternae studded with ribosomes. Numerous virus particles were found within the ribosome-covered reticulum and, to a lesser extent within the

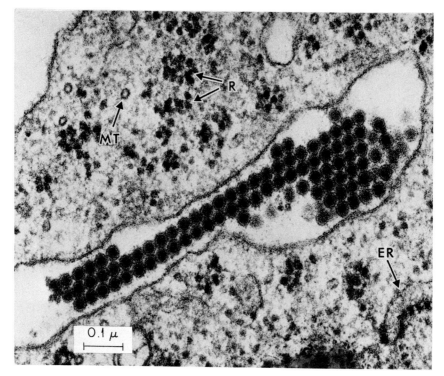

FIG. 9. Murray Valley Encephalitis Virus in mosquito cells culture nine days after infection. ER, endoplasmic reticulum; MT, microtubule; R, ribosomes.

centrally placed regions of smooth membrane. Toward the plasma membranes of infected cells, vesicles containing virus particles are devoid of a covering of ribosomes. Occasionally, the limiting membrane of these vesicles is seen to be joined to the plasma membrane. Infected cells usually carry clusters of virus particles adhering to their surfaces.

Particles of JE virus were first seen in the endoplasmic reticulum of the mosquito cells three days after inoculation. Within the cells the viruses are always enclosed either in large vacuoles or in the reticulum, but the patterns of smooth and particle-covered membranes seen in MVE infected cells were not detected in cells containing JE virus particles. In the JE virus infected cultures greater numbers of cells undergoing autolysis were observed than in either the MVE infected cultures or in the controls. Many of the moribund cells contained particles within elements of the endoplasmic reticulum.

With this additional information, it is now possible to envisage certain stages of the process of replication of MVE virus and, probably, JE virus within cultured mosquito cells. Infection appears to result from the ingestion of particles, generally in large clusters, by phagocytosis. The appearance of single particles within the cisternae of the endoplasmic reticulum suggests that this is their site of assembly; in individual cells containing large numbers of particles, the reticulum is apparently organized into a characteristic pattern for this purpose. The virus particles then appear to accumulate either singly or in small numbers within vesicles and migrate to the plasma membrane, where they are discharged from the cell by a process which is the reverse of phagocytosis. Agglomeration of particles into larger clusters probably occurs on the surface of infected cells before they are finally released into the culture medium.

It should be pointed out that, throughout our investigations, repeated difficulty was encountered in obtaining a high multiplicity of virus. Even in cultures where the heaviest infections were achieved, the ratio of PFU*:cell was never much greater than 2:1. Consequently, very few cells in individual cultures contained large numbers of virus particles, and our proposals are based on the examinations of a relatively small sample of infected cells in a very large number of thin sections.

3. Persistent Infection with Viruses

One of the interesting aspects of the relationship between viruses and cells *in vitro* is persistent infection. In our experiments we succeeded very easily in developing this with MVE and JE viruses (Řeháček, 1968c).

The MVE virus does not cause any cytopathological changes in

* Plaque forming units.

TABLE VIII

ESTIMATED MEANS OF VIRUS TITER IN PASSAGES OF PERSISTENTLY
INFECTED MOSQUITO CELLS WITH MVE AND JE VIRUSES

| Passage | Days | Virus titer in lines in log PFU/ml | |
		MVE	JE
1–4	90	4.4	4.2
5–11	48	4.2	3.0
12–21	71	3.0	1.5

mosquito cells *in vitro;* on the other hand, in the cultures infected with JE virus, degenerated cells can be seen containing viral particles (Filshie and Řeháček, 1968). For the establishment of persistent infection this appearance does not play any role because of the extremely high reproductive ability of the mosquito cells.

To enable a persistent infection to become established two different doses of both the viruses were used as inocula $10^{5.7}$ and $10^{3.7}$ PFU/ml of MVE virus and 10^6 and 10^4 PFU/ml of JE virus. Specific antisera were not added to the medium. Each of the first four passages of cells was made at 24-day intervals, and the complete volume of the medium was changed every sixth day. At this time the virus titers in the medium and cells were ascertained. After this period the cultures were passaged once weekly with a complete change of medium every third or fourth day. Experiments started on 26 May 1966 were finished, after 210 days of cultivation, on 23 December 1966, when 21 passages of cells was attained.

When comparing the virus amount in the course of 210 days of passaging (Table VIII), a slow and permanent loss of virus content was detected; however, MVE virus occurred regularly in cultures during this time. JE virus starting from the 12th passage and on the third day following the passage failed several times to be isolated. The highest titers of MVE virus observed were on the sixth day of the fifth passage ($10^{5.6}$ PFU/ml); those for JE virus were observed on the sixth day of the fourth passage ($10^{6.0}$ PFU/ml) of virus.

The growth curve of both the viruses in the mosquito cells in the 10th passage were studied by the following method. The cells were carefully washed in the medium and set up in about 2×10^5 cells/ml in 0.5-liter flat bottles. Every day 0.5 ml of the supernatant, consisting of medium and liberated cells, was withdrawn, and after sonication the suspension was titrated for the presence of virus by plaques on chick embryo cells. During this time there was no change of medium in the cultures. The MVE virus content increased about 10^5 PFU/ml from the first to the 10th day, and the JE virus increased about 10^3 PFU/ml (Fig. 10).

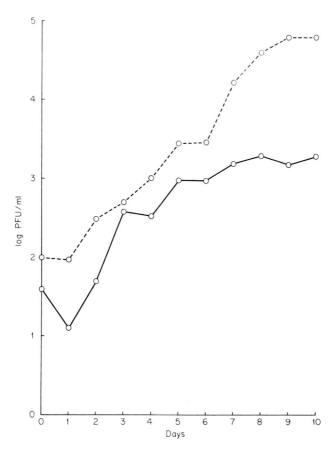

Fig. 10. Multiplication of MVE and JE viruses in mosquito cells at 10th passage. - - -, MVE virus; —, JE virus.

In the course of the experiments microscopic preparations stained by Giemsa from infected and uninfected cells were prepared. No obvious specific changes caused by MVE virus propagation were noticed. On the other hand, in cultures infected with JE virus, rounded and, on several occasions, disrupted cells were observed. When these cells were investigated under the electron microscope, viral particles were found in them (Filshie and Řeháček, 1968).

During the passages, the presence of virus in the cells and in the medium was tested. Extracellular virus was tested after carefully removing cells and debris. The intracellular virus was obtained by disrupting the cells by sonication for 40 seconds on a Mullard drill. When the samples for virus isolation were taken in several passages before the cells were washed, values about 10^1–10^2 lower (of extracellular virus) were found.

The distribution of viruses in the cells and the percentage of cells infected was studied by means of electron microscopy (Filshie and Řeháček, 1968). These results have shown that the viruses are present only in the cytoplasm of the cells and that the percentage of cells infected is very low, not exceeding 1–5% of cells.

In two experiments the growth rate of infected and uninfected cells was compared. In the course of 10 days, each day one test tube was used for virus testing and comparing the number of cells. There was no change of medium. In the first experiment about 70,000 cells/ml were set up; in the second, 520,000/ml. In the first experiment with MVE and JE viruses the infected cells multiplied at the same rate as the uninfected ones, the MVE and JE virus titers increased, up to the ninth day, about 100 times. In the second experiment, up to the sixth day, both MVE and JE virus infected cells and control cells multiplied at the same rate, but this time to the 10th day, the MVE virus infected cells increased at the same rate as the control cells. On the other hand, the number of cells in the culture infected with JE virus decreased rapidly. The MVE virus titer was $10^{4.3}$ PFU/ml and the JE virus titer was $10^{2.1}$ PFU/ml on the 10th day of introduction. Summarizing these results, it seems that when using as inoculum a small number of cells with a low dose of virus (70 000 cells/ml with 10^2 and 10^1 PFU/ml of MVE and JE viruses respectively) more rapid increase in the number of cells as well as in virus amount occurred. On the other hand, when higher amounts of cells and viruses were used in the experiment (520,000 cells/ml with 10^4 and 10^3 PFU/ml of MVE and JE viruses respectively), there was a small increase in the number of infected cells. In the case of the JE virus, a marked decrease of infected cells following the sixth day of introduction occurred, with practically no change of MVE titers and about a 10^1 decrease of JE virus on the 10th day.

In the experiments, to "cure" the persistently infected cells with both viruses, the rabbit antisera prepared and tested in 1960 by Dr. Doherty from the Queensland Institute of Medical Research in Brisbane were used. These antisera, tested immediately after their preparation, showed neutralization indexes of 3.1 (MVE) and 3.4 (JE) when the heated mixture with viruses was applied intracerebrally to infant mice.

For the experiments the cells in their 15th passage were used. To the medium at each passage a volume of 1.5% of diluted antisera was added. The viruses were tested every fourth and seventh day from the media and from twice-washed cells in medium after their disruption by sonication. In the first passage at the fourth day, the MVE virus was found at a titer $10^{4.0}$ PFU/ml in the cells and 10^3–10^4 in the medium. JE virus was found only intracellularly at $10^{3.7}$ PFU/ml. From this time the amount of virus in the cultures gradually decreased. No virus was detected in the cells infected with MVE after the third passage. In the JE infected cells, no virus was observed after the fourth passage (Table IX).

In experiments designed to determine whether the persistently infected cells could be superinfected, the heterologous Kunjin virus was chosen because it propagates readily in mosquito cells *in vitro* and develops characteristic plaques on chick embryo cells, plaques that differ in size from those established with JE and MVE viruses. This property was used also for the titration on chick fibroblasts by plaques without the use of specific antisera. This suggests that no reduction of plaques occurs when the viruses are inoculated on the monolayers simultaneously in a mixture.

The arrangement of experiments was as follows: Twenty-five bottles of monolayers of infected cells with both the viruses in the 13th passage (first experiment) and the 18th passage (second experiment) and 10 bottles of uninfected mosquito cells, after being washed five times, were challenged with 10^4 PFU/ml of Kunjin virus. After 24-hours of incubation the cells were again washed five times in fresh medium to remove the inadhered virus. On the sixth day after introduction, the virus titers in the cells and in the medium were estimated following results. There was no extracellular virus found in either of the experiments. On the other hand, the Kunjin virus was detected in the persistently infected cells with both the viruses and in the controls at an average amount of 10^2 (first experiment) and $10^{1.7}$ PFU/ml (second experiment). The per-

TABLE IX

VIRUS TITER OF MVE AND JE VIRUSES IN CULTURES GROWN IN THE SPECIFIC ANTIBODY PRESENCE IN THE MEDIUM

Number of passages	Days	MVE virus		JE virus	
		1[a]	2[b]	1[a]	2[b]
1	4	4.0	3.4	3.7	—
	7	1.0	2.0	1.5	—
2	4	1.5	—	1.2	—
	7	2.2	—	<1.0	—
3	4	1.7	—	—	—
	7	<1.0	—	—	—
4, 5	4	—	—	—	—
	7	—	—	—	—

[a] Intracellular.
[b] Extracellular.

sistently infected cells had the same titer as in normal passage (10^2–10^3 PFU/ml). Thus, the Kunjin virus very likely was adsorbed and multiplied in the cells.

During the course of the experiments, the size and appearance of plaques was studied. With neither MVE nor JE virus was a significant change observed. Sometimes the size of plaques differed a little from the original ones, but in further passages they reverted to the original form. These changes could probably be caused by the inconstant conditions occurring during the preparation of the chick embryo monolayers.

In the 4th, 10th, and 20th passages of the persistently infected cells, the virus titers by inoculation of the virus suspension from cells prepared by sonication were administered subcutaneously and intracerebrally on baby white mice. The values obtained were then compared with those obtained by the plaque method. These results showed a practically stable pathogenicity for mice of both the viruses examined after the fourth and tenth passages and a lowering of pathogenicity at the 20th passage of cells for mice inoculated subcutaneously (Table X).

These experiments were finished by identification of both viruses in the neutralization test on chick fibroblasts where the marked reduction of plaque numbers of about 10^2 occurred. Antibodies used in these tests were from the same rabbit antisera as was used in the experiments of the culture of cells in their presence to cure the infected cells.

Some of the problems studied in this work were not investigated sufficiently. Research was restricted by the limited stay of the author at the Australian institutes. Interesting and important problems such as the

TABLE X

COMPARISON OF THE TITER OF MVE AND JE VIRUSES IN 4TH, 10TH, AND
20TH PASSAGES OF PERSISTENTLY INFECTED CELLS BY PLAQUE METHOD
AND BY INTRACEREBRAL AND SUBCUTANEOUS INOCULATION OF
AGENTS INTO BABY MICE

Virus	Passage	Day	PFU/ml	Mice (intracerebral) 0.03 ml	Mice (subcutaneous) 0.1 ml
JE	4	21	2.8	2.0	2.0
	10	7	3.8	4.2	3.5
	20	8	1.6	2.5	—[a]
MVE	4	21	4.0	3.0	3.0
	10	7	4.6	4.2	4.0
	20	8	3.7	2.5	1.0

[a] Negative.

possibility of an establishment of interferonlike substances and the percentage of cells infected need more time and attention.

Carrier culture of *Aedes albopictus* cells infected with several viruses were studied by Banerjee and Singh (1968). The carrier culture infected with JE virus was, at time of publication, in its 13th passage, in the eighth passage with West Nile virus, and in the seventh passage with Chikungunya virus. When Chikungunya virus was investigated in further passages, a loss of mouse virulence was noted (Banerjee and Singh, 1969). Virus harvested at the seventh passage level and at further passages of the carrier culture failed to kill mice when inoculated intracerebrally although the titers in Vero cells were about 10^4 $TCID_{50}$/ml. These results indicate that it may be possible to prepare an attenuated tissue culture strain of virus for use as a vaccine.

A carrier state of *Aedes aegypti* embryo cells infected with Semliki Forest disease virus was developed by Peleg (1969a). He found this state was due to a small proportion of virus-producing cells in the culture and a low yield of virus per infected cell.

4. Attempts to Study Behavior of Infectious RNA from Various Viruses in Cell Line

Peleg (1969b) found that embryonal *Aedes aegypti* mosquito tissue culture inoculated with isolated RNA of Semliki Forest and West Nile arboviruses yielded virus progeny which were identical with the infectious RNA. The same type of cells failed to produce viruses after inoculation of infectious RNA derived from EMC and polio picornaviruses. Peleg suggested that this was probably due to intracellular factors rather than factors involved in virus adsorption and penetration of these cells.

B. Multiplication of Viruses in Moth Cell Line

Suitor (1966a,b) first reported the utilization of first cell line of *Antheraea eucalypti* ovarian cells, which was derived by Grace (1962) to culture Japanese encephalitis (JE) virus. These cells were adapted, by Suitor (1966a,b), to a medium consisting of Grace's original insect cell culture medium plus 10 mg/ml bovine plasma albumin (Armour, fraction V) and 2% pupal hemolymph from the moth *Philosamia cynthia pryeri* and treated according to Grace (1962). The cells were cultured at 28° in glass bottles or Falcon plastic tissue culture flasks. As the virus inoculum, the JE virus, a strain from Taiwan (T-143) originated from mosquitoes, was used in the experiments. Samples of suspended cells plus medium were taken, usually, at seven-day intervals for virus assay, which was done either in intracerebrally inoculated mice or by plaque technique in porcine kidney stable or monkey kidney stable cell monolayers.

In a series of five experiments, development of JE virus in Grace's moth ovary cells was obtained, but only at relatively low levels and only after rather lengthy incubation, as compared to growth of this virus in mammalian cell cultures; this may be due to the low temperature of incubation. In one experiment where temperatures of 22 and 28° were used, virus growth occurred only at the higher temperature. Figure 11 shows representative results of one test; the titer are shown as PFU per milliliter in porcine kidney stable cell monolayers. No apparent cytopathic effect has been seen based on comparison of infected and uninfected cells. Over the course of three to four weeks at 28°, JE virus

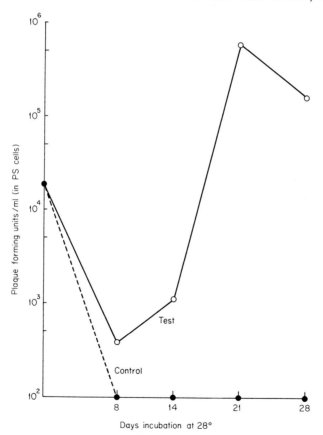

FIG. 11. Growth of Japanese encephalitis virus in Grace's moth cells line. (From Suitor, 1966b.)

has reached titers approximately 10–100 times greater than the inoculum, rising about 100- to 1000-fold after initial loss of infectivity in the first week as demonstrated by Suitor (1966a,b).

Also, the Asibi strain of yellow fever virus multiplied in these cells as shown by Converse and Nagle (1967).

In these experiments the *Antheraea eucalypti* tissue cell cultures were grown in a hemolymph-free medium. Suspension cultures of approximately 10^6 cells/ml in 250-ml Falcon plastic flasks were inoculated with approximately 10^7 mouse intracerebral LD_{50} ($MICLD_{50}$)/ml of monkey serum seed virus, incubated for one hour at 26° and at pH 6.5, and washed. Then the medium was replaced.

The authors observed 10^2–10^3 increases in virus titer in cell cultures between 24 and 72 hours after inoculation. Lower titers were obtained subsequently. Virus in medium without cells exhibited an exponential decrease in titer. The infected moth cells maintained a healthy appearance and increased in numbers throughout the incubation period. Continued incubation of the *A. eucalypti* cell cultures resulted in a gradual increase in virus titer from $10^{3.3}$ $MICLD_{50}$/ml at eight days to $10^{4.3}$ at 14 days, $10^{4.5}$ at 21 days, and an additional peak of $10^{5.5}$ at five weeks. Throughout this period, the medium was replaced and cell populations were reduced approximately 50% at 14-day intervals.

Yunker and Cory (1968) tested 34 different viruses on the *Antheraea* cell line. From this number only 10 virus strains belonging to arboviruses (two strains of JE and one strain each of yellow fever, St. Louis, Cache Valley, Bunyamwera, Tahyna, snowshoe hare, California, and Indiana vesicular stomatitis) multiplied. The viruses multiplied without developing a cytopathic effect. Viruses failing to multiply were all of the members of groups A, C, and Tacaribe, six of the 10 members of the B group, eight of the ungrouped arboviruses, and three nonarboviruses.

Suitor and Paul, (1969b), introducing JE virus in *Antheraea* cells, found two peaks of infectivity in one test at two and five weeks, respectively. Virus production plaque-assayed in porcine kidney cells showed the plaque type of the second week harvest consisted of the usual 4- to 6-mm plaque, but the fifth week's contained a small 1- to 2-mm plaque. On subsequent passages in moth cells, the two peaks persisted and the small plaque type predominated in both. No difference was seen in mouse lethality between the original inoculum and passaged virus. It is believed that a substrain of virus has been selected by passage in moth cells.

Yang *et al.* (1969), after inoculation of vesicular stomatitis virus (Indian serotype) in *Antheraea* cells, found that the virus content in cells increased for two to four days and then remained relatively constant. There was no cytopathic effect seen in infected cells. Immunofluorescence showed that only a fraction of the cell population was producing a significant amount of virus antigen. At multiplicity of 50 PFU/cell, electron microscopy of cell sections revealed the presence of large quantities of virus associated with some of the cells.

From the results on the multiplication of viruses and rickettsiae in their arthropod cells *in vitro*, which often are originated from their natural vectors or reservoirs, it is naturally impossible to draw any conclusions on the biology of these pathogens in vectors: However, we are of the opinion that the arthropod cells *in vitro* could serve as a suitable and interesting medium for the multiplication of these pathogens when investigating different problems concerning their biology (as mentioned above).

REFERENCES

Banerjee, K., and Singh, K. R. P. (1968). *Indian J. Med. Res.* 56, 812.
Banerjee, K., and Singh, K. R. P. (1969). *Indian J. Med. Res.* 57, 1003.
Benda, R. (1958). *J. Hyg., Epidemiol., Microbiol., Immunol.* 2, 314.
Blaškovič, D., and Řeháček, J. (1962). In "Biological Transmission of Disease Agents" (K. Maramorosch, ed.), pp. 135–157. Academic Press, New York.
Buckley, S. M. (1969). *Proc. Soc. Exp. Biol. Med.* 131, 625.
Converse, J. L., and Nagle, S. C., Jr. (1967). *J. Virol.* 1, 1096.
Filshie, B. K., and Řeháček, J. (1968). *Virology* 34, 435.
Filshie, B. K., Grace, T. D. C., Poulson, D. F., and Řeháček, J. (1967). *J. Invertebr. Pathol.* 9, 271.
Fujita, N., Yasui, Y., Kitamura, S., and Hotta, S. (1968). *Kobe J. Med. Sci.* 14, 241.
Grace, T. D. C. (1962). *Nature (London)* 195, 788.
Grace, T. D. C. (1966). *Nature (London)* 211, 366.
Haines, T. W. (1958). Ph.D. Thesis. Univ. Maryland (*Diss. Abstr.* 20, 1–31).
Johnson, J. W. (1967). George Washington Univ. *Diss. Abstr.* 1–60.
Johnson, J. W. (1969). *Amer. J. Trop. Med.* 18, 103.
Jones, B. N. (1962). *Biol. Rev.* 37, 512.
Kazár, J., Řeháček, J., and Brezina, R. (1966). *J. Hyg. Epidemiol. Microbiol. Immunol.* 10, 240.
Kordová, N., and Řeháček, J. (1959). *Acta Virol. (Prague), Engl. Ed.* 3, 201.
Kordová, N., and Řeháček, J. (1965). *Proc. Int. Congr. Entomol., 12th, 1964* p. 777.
Libíková, H. (1961). *Acta Virol. (Prague), Engl. Ed.* 5, 387.
Libíková, H., Řeháček, J., and Mayer, V. (1962). In "Biology of Viruses of the Tick-Borne Encephalitis Complex" (H. Libíková, ed.), pp. 201–204. Publ. House Czech. Acad. Sci., Prague.
Martin, H. M., and Vidler, B. O. (1962). *Exp. Parasitol.* 12, 192.
Nauck, E. G., and Weyer, F. (1941). *Zentralbl. Bakteriol., Parasitenk., Infektionskr. Hyg., Abt. 1: Orig.* 147, 365.
Paul, S. D., and Singh, K. R. P. (1969). *Curr. Sci.* 38, 241.
Paul, S. D., Singh, K. R. P., and Bhat, U. K. M. (1969). *Indian J. Med. Res.* 57, 339.

Pavri, K. M., and Ghosh, S. N. (1969). *Bull. W. H. O.* **40**, 984.
Peleg, J. (1968a). *Amer. J. Trop. Med. Hyg.* **17**, 219.
Peleg, J. (1968b). *Virology* **35**, 617.
Peleg, J. (1969a). *J. Gen. Virol.* **5**, 463.
Peleg, J. (1969b). *Nature (London)* **221**, 193.
Peleg-Fendrich, J., and Trager, W. (1963a). *Amer. J. Trop. Med. Hyg.* **12**, 820.
Peleg-Fendrich, J., and Trager, W. (1963b). *Ann. Epiphyt.* **14**, 211.
Pospíšil, L., Jandásek, L., and Pešek, J. (1954). *Lek. Listy* **9**, 3.
Price, W. H. (1956). *Pub. Health Rep.* **71**, 125.
Řeháček, J. (1958a). Thesis.
Řeháček, J. (1958b). *Acta Virol. (Prague), Engl. Ed.* **2**, 253.
Řeháček, J. (1962). *Acta Virol. (Prague), Engl. Ed.* **6**, 188.
Řeháček, J. (1963). *Ann. Epiphyt.* **14**, 199.
Řeháček, J. (1965a). *J. Med. Entomol.* **2**, 161.
Řeháček, J. (1965b). *Acta Virol. (Prague), Engl. Ed.* **9**, 332.
Řeháček, J. (1965c). *Proc. Int. Congr. Entomol., 12th, 1964* p. 774.
Řeháček, J. (1968a). *Proc. Int. Colloq. Invertebr. Tissue Cult., 2nd, 1967* p. 241.
Řeháček, J. (1968b). *Acta Virol. (Prague), Engl. Ed.* **12**, 241.
Řeháček, J. (1968c). *Acta Virol. (Prague), Engl. Ed.* **12**, 340.
Řeháček, J., and Brezina, R. (1964). *Acta Virol. (Prague), Engl. Ed.* **8**, 380.
Řeháček, J., and Hána, L. (1961a). *Acta Virol. (Prague), Engl. Ed.* **5**, 57.
Řeháček, J., and Hána, L. (1961b). *In* "Biology of Viruses of the Tick-borne Encephalitis Complex" (H. Líbiková, ed.), pp. 185–187. Publ. House Czech. Acad. Sci., Prague.
Řeháček, J., and Kožuch, O. (1964). *Acta Virol (Prague), Engl. Ed.* **8**, 470.
Řeháček, J., and Kožuch, O. (1969). *Acta Virol. (Prague), Engl. Ed.* **13**, 253.
Řeháček, J., and Pešek, J. (1960). *Acta Virol. (Prague), Engl. Ed.* **4**, 241.
Řeháček, J., Brezina, R., and Majerská, M. (1968). *Acta Virol. (Prague), Engl. Ed.* **12**, 41.
Řeháček, J., Rajčani, J., and Grešíková, M. (1969). *Acta Virol. (Prague), Engl. Ed.* **13**, 439.
Singh, K. R. P., and Paul, S. D. (1968a). *Curr. Sci.* **37**, 65.
Singh, K. R. P., and Paul, S. D. (1968b). *Indian J. Med. Res.* **56**, 815.
Singh, K. R. P., and Paul, S. D. (1969). *Bull. W. H. O.* **40**, 982.
Soret, M. G., and Sanders, M. (1954). *Proc. Soc. Exp. Biol. Med.* **87**, 526.
Suitor, E. C., Jr. (1966a). *Virology* **30**, 143.
Suitor, E. C., Jr. (1966b). Lect. Rev. Ser., NAMRU-2-LR-023.
Suitor, E. C., Jr. (1969). *J. Gen. Virol.* **5**, 545.
Suitor, E. C., Jr., and Paul, F. J. (1969a). *Virology* **38**, 482.
Suitor, E. C., Jr., and Paul, F. J. (1969b). *Bacteriol. Proc.* p. 230.
Trager, W. (1938). *Amer. J. Trop. Med. Hyg.* **18**, 387. (Copyright, The Williams and Wilkins Company, Baltimore.)
Weyer, F. (1952). *Zentralbl. Bakteriol., Parasitenk., Infektionskr. Hyg., Abt. 1: Orig.* **159**, 13.
Yang, Y. J., Stoltz, D. B., and Prevec, L. (1969). *J. Gen. Virol.* **5**, 473.
Yunker, C. E., and Cory, J. (1967). *Exp. Parasitol.* **20**, 267.
Yunker, C. E., and Cory, J. (1968). *Amer. J. Trop. Med. Hyg.* **17**, 889.
Yunker, C. E., and Cory, J. (1969). *J. Virol.* **3**, 631.

9

USE OF INVERTEBRATE TISSUE CULTURE
FOR THE STUDY OF PLASMODIA

Gordon H. Ball

I. THE RATIONALE OF CULTIVATING THE INSECT PHASE OF
MALARIA PARASITES

In that broad spectrum of associations between different species of organisms designated as symbiosis, parasitism is an important segment.

In obligatory parasitism, the relation between host and parasite may be so intimate that only one, or at most very few, species of host are suitable for the development of the parasite. Although in some instances this suitability is due in large part to the provision of a favorable physical environment, much more frequently it has a physiological or a biochemical basis.

Consequently, if we are to determine the nature of this dependence of parasite upon host, we must be able to supply the parasite with the essential substances provided by the host organism and be in a position to define these chemically. The obvious, but not necessarily the easiest, method would be to substitute a chemically defined medium for the host. Then, if normal development of the parasite results, one alters the medium, mainly by deletion of various constituents. If the removal of a particular compound is associated with decreased growth or development of the parasite or with complete failure of previously successful cultures, then the omitted substance may be considered essential for the parasite, and there is a high degree of probability that either this substance or its biochemical equivalent is supplied to the parasite by the host.

In nature, of course, the malaria plasmodium is found in two hosts, i.e., a vertebrate and an arthropod, the only proved arthropod host being a mosquito. In the vertebrate, the plasmodium is almost exclusively intracellular, the extracellular periods being limited to the brief interval when the parasite is in the process of transferring from one cell to another. Although *in vitro* cultivation of the vertebrate phase of malaria has had a high degree of success (Anfinsen *et al.*, 1946; Huff *et al.*, 1960; Nydegger and Manwell, 1962; Siddiqui *et al.*, 1970; Trager, 1964; Trager and Krassner, 1967; Trigg, 1969), the dependence of the parasite on host cells for completion of the life cycle makes it very difficult to determine the biochemical nature of the essential substances supplied by the host. Trager (1964, 1967) has been able to show by experiments on plasmodia removed from erythrocytes that development will not occur in the absence of various chemically defined substances. However, these experiments demonstrate that it is survival of the extracellular parasite which is interfered with if these substances are not available, and not necessarily that they are essential in the intracellular phase.

On the other hand, the parasite in the mosquito is probably extracellular throughout its development in this host except for a possible brief period when it may exist in an epithelial cell in the stomach mucosa (Garnham *et al.*, 1962). Consequently, the problem of determining the nature of the dependence of the plasmodium on the host is, at least in theory, more easily solved in the mosquito phase since one is dealing with substances present in the hemolymph rather than with the much more complex intracellular environment.

If one is able to determine the biochemical dependence of a particular species of plasmodium on a species of mosquito in which successful completion of the life cycle is the usual occurrence, then it may be possible to compare the availability of these essential substances in those mosquitoes which are not suitable for development of the same plasmodial species.

It is, of course, possible that failure of plasmodia to develop in certain kinds of mosquitoes may result not from the absence of essential materials but from an immune reaction (Weathersby, 1965, 1967), perhaps due to the presence of certain toxic factors (Weathersby and McCall, 1968). Culture experiments accompanied by accurate analyses of media may be valuable in arriving at the proper explanation.

Finally, the results of successful culture of malaria parasites, with a determination of their chemical requirements may suggest a method for the control of the plasmodia by altering the metabolism of the mosquito. Just as chemosterilants interfere with the reproductive capacity of some insects, other chemicals may prevent development of plasmodia in mosquitoes by making certain metabolites unavailable to the parasite.

II. PROBLEMS ENCOUNTERED IN INSECT TISSUE CULTURE OF PLASMODIA

In experiments involving insect tissue culture and accompanying malaria plasmodia, one important problem that must be solved is the maintenance of sterility. The digestive tract of the mosquito is not sterile, and if one uses the mosquito stomach with its attached developing oocysts, the maintenance of the preparation free from contaminating bacteria or fungi becomes of fundamental importance.

In our own investigations (Chao and Ball, 1964), we have found that the number of contaminating organisms in the stomach will be markedly reduced if the mosquito is starved for as long as possible before the culture is initiated. Mosquitoes will survive without food at a temperature of 18° for 48 hours, provided they are given access to water. At the end of this time, the mosquito's stomach is largely free of contaminants.

Additional measures for the control of contaminants are the incorporation of antibiotics, such as penicillin or streptomycin, into the medium, sterilization of the outside of the insect by immersion into a germicide of low surface tension, and thorough washing of the mosquito stomach with its attached parasites in a series of chambers containing sterile physiological salt solutions.

Axenic mosquito culture is, of course, possible (Akov, 1962; Boorman,

1967; Johnson, 1969; Lea *et al.*, 1956; Trager, 1935), but it suffers from the disadvantages of requiring considerably more equipment and care, and of providing frequently only a low yield of healthy adults. However, Boorman obtained a large number of pupae, and we are now using his method for obtaining axenic adults.

We have also employed another method of obtaining oocysts free of possible contaminants from the mosquito gut lumen, namely, dissection of oocysts free from the stomach wall (Ball and Chao, 1957). This has resulted in development of the isolated oocysts in suitable media to about the same stage as is reached with oocysts remaining attached to the intact stomach. Such a result is one that might have been expected, in view of Weathersby's findings (1954, 1960) that oocysts develop in the bodies of live mosquitoes independently of the stomach. The preparation of separate oocysts, however, is laborious and unsuitable for cultivation of very early stages.

If malaria parasites are grown in association with organs of the mosquito, an important problem, independent of the maintenance of sterility, which must be solved is the provision of an environment suitable for prolonged cultivation of the mosquito stomach or other organ as well as for the cultivation of the plasmodium. Although oocysts will continue to develop, for a period, apart from the stomach wall, one can hardly expect normal growth if they remain attached to an organ which is abnormal, dying, or dead. Despite the fact that we were able to maintain mosquito gut actively contracting in culture for as long as five weeks (Ball, 1954), many of the tissues of the gut, e.g., glandular epithelium, did not survive as long as did the muscles.

It is evident that this type of tissue culture of mosquito stomach and plasmodium is one involving the maintenance of a whole organ. Just as in the establishment of cell lines, not all of the mixed types of cells originally isolated will persist; in organ cultures, also, certain types of cells of the organ survive better than others. The establishment of cell lines of mosquito tissue has been accomplished in the last few years, and attempts to grow malaria plasmodia in cultures of this type are discussed in Section IV of this chapter.

Since malaria plasmodia develop in culture at a slower rate than they do in the live mosquito, it is essential that the constituents of the medium be available over a period of two to three weeks. This is a condition that cannot always be fulfilled, especially with organic solutions; so transfer to fresh medium at frequent intervals is usually necessary. Insect hemolymph, a constituent of various successful media for insect tissue culture, is highly unstable if untreated when it is removed from the host.

An additional complication is the probable requirement of trace elements for the complete development of plasmodia in culture. Since the nature of these substances is virtually unknown, there is no way of finding out their concentrations or stability. Nevertheless, the depletion of such trace elements in culture probably explains why the cultivation of malaria plasmodia, both in the insect phase and in the blood phase of vertebrates, has so far proved to be possible only over a relatively short period of time.

III. SPECIAL TECHNIQUES

One method of investigation which has been pursued in order to produce satisfactory culture media for maintaining organs of insects over a two- to three-week period is to analyze the hemolymph or the whole bodies of the particular species which serves as a successful host for the plasmodium being studied. Although this type of research is capable of producing useful information, it must be pointed out that, even if one knew the exact proportions of every substance available to the parasite in the living mosquito, it is very doubtful that an identical mixture would prove successful *in vitro* where reaction rates are different and such phenomena as active transport, mitochondrial activity, etc., may be greatly altered.

Nevertheless, it is useful to analyze the chemical make-up of a suitable host to determine the relative concentrations of different constituents as well as to ascertain if some previously unsuspected compounds are present in a particular species of host.

The major inorganic constituents of the bodies of a very favorable host, i.e., *Culex tarsalis* for *Plasmodium relictum,* as well as of those of a less suitable one, i.e., *Culex stigmatosoma,* were determined by microanalytical techniques (Clark and Ball, 1954). The amino acid composition of the whole bodies of *C. tarsalis* and of other mosquitoes were investigated by means of paper chromatography (Table I) (Clark and Ball, 1952), and a partial analysis of *C. tarsalis* proteins was made by paper electrophoresis (Clark and Ball, 1956). The results of these determinations were used to modify and to improve various types of media which had been employed previously by our laboratory for the attempted cultivation of insect tissues (Ball, 1948, 1954) or by others for the vertebrate phase of malaria plasmodia (Anfinsen *et al.,* 1946). The net result was a considerable improvement in the development of the parasite and of the persistence of the mosquito tissue in culture. Table II gives the composition of the medium which up to the present has proved very successful

TABLE I
CHROMATOGRAPHIC ANALYSES OF FREE AMINO ACIDS IN THREE
SPECIES OF *Culex*—SOUTHERN CALIFORNIA

Amino acid	*Culex tarsalis*	*Culex stigmatosoma*	*Culex quinquefasciatus*
Alanine	+	+	+
Arginine	+	+	+
Aspartic acid	+	−	+
Cysteic acid	+	+	−
Glutamic acid	+	+	+
Glycine	+	+	+
Histidine	+	+	+
Hydroxyproline	−	−	−
Isoleucine and/or leucine	+	+	+
Lysine	+	+	+
Methionine	+	+	+
Phenylalanine	−	−	−
Proline	+	+	+
Serine	+	+	+
Threonine	+	+	+
Tryptophan	+	+	+
Tyrosine	+	+	+
Valine	+	+	+
β-Alanine	+	+	+
α-Amino-n-butyric acid	+	+	+
Asparagine	−	−	−
Glutamine	+	+	+
Taurine	+	+	+
Glutathione	+	+	+

in the cultivation of the mosquito phase of *Plasmodium relictum*. Its composition and the concentration of the various components are the result of studies, such as those which have been described above, of analyses of the body of the insect host. More recent work on the amino acid composition of mosquito hemolymph has been carried by Chen and his co-workers (Chen, 1960, 1963; Chen and Briegel, 1965).

Since this is a highly artificial medium as well as being a very complex one, it is very probable that it contains certain chemical antagonists which may prevent utilization of essential materials by the parasite in culture. With this in mind, we modified the medium by the omission of various constituents. This type of experiment is very tedious since the results may not be obvious except after careful study of the relative degree of development over a considerable period of time. Sometimes the examination of fixed tissue is required. The experiments showed that the medium which was being used was more suitable than a less complex one

TABLE II

Composition of Medium for *in vitro* Cultivation of the
Mosquito Phase of *Plasmodium relictum*[a]

Milligrams/1000 ml			
NaCl	6800	DL-Threonine	240
KCl	400	DL-Tryptophan	160
$Ca(NO_3)_2 \cdot 4H_2O$	200	L-Tyrosine	80
$MgCl_2 \cdot 6H_2O$	100	DL-Valine	100
Na_2HPO_4	60		
KH_2PO_4	60	Thiamine–HCl	1.30
$NaHCO_3$	1000	Calcium pantothenate	0.25
		Riboflavin	0.13
Glucose	1000	Pyridoxine–HCl	0.25
		Niacin	0.20
L-Arginine–HCl	140	Biotin	0.05
DL-Aspartic acid	240	Folic acid	0.05
L-Cystine	40	*p*-Aminobenzoic acid	0.01
L-Glutamic acid	200	Choline	0.25
L-Glutamine	100	Cholesterol	0.10
Glycine	200		
L-Histidine–HCl	140	Adenine	1.40
L-Isoleucine	100	Guanine	1.50
L-Leucine	100	Xanthine	1.50
L-Lysine–HCl	200	Hypoxanthine	1.40
DL-Methionine	200	Cytosine	1.10
L-Phenylalanine	100	Uracil	1.10
DL-Proline	200	Thymine	1.30

[a] The final medium consisted of 77.5% of the above nutrient solution, 20% chicken serum, and 2.5% chick embryo extract. pH was adjusted to 6.8.

and that no antagonistic action, at least of blocks of components, was demonstrable (Ball, 1964).

More recently, we have cultured plasmodial oocysts in the presence of insect cell lines, using the media which have proved successful for mosquito cells. The results are discussed later in this section.

Still another type of study aiming to improve the composition of culture media has been employed in insect culture but not, so far as I am aware, in the cultivation of the insect phase of the malaria parasite. This is the use of depletion studies of a culture medium, i.e., analyzing it before and after tissues have been grown in it. Experiments of this nature with insect cells in culture have been those of Clements and Grace (1967) and of Grace and Brzostowski (1966) with the cells of *Antheraea eucalypti*. Their results have shown what sugars and amino acids are utilized by tissue culture cells of this insect. Such studies for determining the constituents used by the malaria oocyst *in vitro* would be difficult because of its relatively small size. On the other hand, it should be possible to extend the type of investigation described above, by comparative analysis of media which have contained uninfected stomachs and of media which previously held stomachs bearing numerous oocysts.

The use of isotope tracers is discussed in Section IV below.

If the malaria parasite, either attached to the stomach wall or dissected away from it, is transferred to an artificial culture medium, it is subjected to a considerable amount of stress. This is true also for the mosquito tissue. Since the stress is usually accompanied by a decreased rate of metabolism in artificial media, both of the parasite and of the tissues, neither the host nor the parasite may be able to respond adequately to the new environment. This is reflected by the loss of an exceptionally large number of early culture preparations, while those that survive the first day or so are much more apt to last for some time. This difficulty may be overcome in part by decreasing the metabolic stress immediately after a culture is started and keeping it at a low level until the preparation has become acclimated to the new conditions.

One of the most obvious methods of mitigating the shock of surgery and of transfer to the artificial conditions of culture is to maintain the preparation for a period of time at low temperature. This may be below that which is required for the development of the parasite but may still not be lethal or even harmful to it. In the case of *P. relictum* cultures, transfer of freshly prepared cultures to a temperature of 18° for the first 24 hours produces a marked increase in the number of successful cultures since the decreased rate of metabolism enables the preparations to get past the period of initial stress (Ball, 1964).

IV. RESULTS

Not only has it been possible to obtain the successful cultivation of a malaria plasmodium in the insect phase, as will be discussed in more detail below, but certain other relationships between the parasite and its insect host have been elucidated as a result of studies of this type. In some instances, these relationships had been suspected previously but not proved; in others, there had been considerable dispute about the relative roles of the mosquito and of the plasmodium.

One of the findings resulting from the *in vitro* studies is that of the low oxygen requirement by the plasmodium in the mosquito. In a series of experiments using different mixtures and concentrations of ambient gases, it was discovered that high concentrations of oxygen were harmful to both the parasite and the mosquito tissue. Since the material being cultured must be immersed in fluid, it might be hypothesized that gassing with a mixture high in oxygen would be required to prevent asphyxiation. However, as shown in Table III, the most favorable gas mixture proved to be 95% air and 5% CO_2 for the greater part of the life history of the plasmodium (Ball and Chao, 1963; Chao and Ball, 1964). The older oocysts and the sporozoites did best in culture in an atmosphere of air alone. According to Martignoni (1960), the concentration of oxygen is low in insect blood; consequently, the developing parasite must be capable of living in this type of environment. Tissue culture studies indicate that this is an inherent property of the insect phase of the parasite and probably is important in the utilization of the insect hemocoele as a region of parasite development.

Cultivation experiments also afforded proof that the developing parasite is relatively independent of specific organs of the mosquito for completion of various stages in its life history. Weathersby (1954, 1960) had shown by injection of different stages of plasmodia into hemocoeles of mosquito hosts that oocysts would develop to maturity if attached to other parts of the body than the stomach or even if they were floating freely in the hemocoele fluid. Culture techniques used in some of our experiments also proved that the mosquito stomach was not essential for the continued development of the malaria plasmodium. If oocysts of various sizes, even as small as 25 μ in diameter, were dissected away from the mosquito stomach wall and transferred to suitable culture media, they continued to grow and develop for as long as 14 days, very much as if they were still attached to the stomach (Ball and Chao, 1957). It is evident that in life the stomach provides a substrate to hold the oocyst in place and is the organ most immediately available for invasion and attachment, but culture and injection experiments show that it is not essential for oocyst development.

The literature of malariology includes a series of investigations purporting to show the necessity of a sojourn in the salivary glands by sporozoites before they become infective for a suitable vertebrate host. It is not necessary to review these experiments, the most recent of which, that of Trembley, Greenberg, and Coatney (1951), indicated that sporozoites are infective before they enter the mosquito's salivary glands. But the use of mature oocysts from live mosquitoes has resulted in contradictory interpretations, mainly because of the impossibility of demonstrating that a few sporozoites had not been released prematurely from the ripe oocysts and had already invaded the salivary glands.

The employment of the tissue culture technique has probably answered this question by permitting the experimenter to obtain sporozoites from

TABLE III
RESULTS WITH VARIOUS AERATION MIXTURES

Gas mixtures	Lens paper support		Agar sublayer		After 96 hours (one subculture)	
	Initial pH of medium	pH after 48 hours	Initial pH of medium	pH after 48 hours	Contraction of hindgut	Increase of oocyst diam.
5% CO_2 and 95% air	6.9	6.9	—	—	++	50%+
5% CO_2 and 95% N_2	6.8	6.6	—	—	0	0
5% CO_2 and 95% O_2	7.0	7.4	7.0	7.3	0	0
50% O_2 and 50% N_2	6.7	8.3	6.7	7.2	0	0
50% O_2 and 50% CO_2	7.0	7.2	7.0	7.3	+	29%[a]
Air (20% O_2, 78% N_2, 0.03% CO_2)	6.9	8.3	6.9	8.1	++	40%−

[a] Vacuolated.

oocysts developing in culture. In these cases, the sporozoites will never be in contact with salivary glands. If *P. relictum* sporozoites, which have developed from cultured oocysts, are injected into susceptible canaries (Table IV), infection will result (Ball and Chao, 1961), and with about the same prepatent period that is found in natural infections. Further-

TABLE IV

RESULTS OF INJECTIONS OF CANARIES WITH SPOROZOITES AND
MATURE OOCYSTS OF *P. relictum* DEVELOPED *in Vitro*

Age and size of oocysts at beginning of culture	Days oocysts cultured	Conditions of culture	Infection in canary
13 days; 65 μ	1	22°	Negative 5–34 days; positive later from infected blood
13 days; 65 μ	2	Room temp.	Negative 10–25 days
10 days; 55 μ	2	22°; aeration with 5% CO_2–95% air	Negative 5–27 days
7 days; 50 μ	3	3 stomachs at 22°; 1 stomach at room temp.	Negative 6–24 days; positive later from infected blood
7 days	2	22°; aeration with 5% CO_2–95% air	Negative 9–23 days
10 days; 61 μ	1	Room temp. (26.5°)	Negative 5–19 days; positive later from infected blood
10 days; 61 μ	2	Room temp. (26°)	Positive 10th day; mosquito biting this bird had sporozoites in salivary glands at 10 days
5 days; 44 μ	4	Room temp. aerated as above (1 culture) Rocked but not aerated (4 cultures)	Negative 5–25 days
9 days; 50 μ	3	22°	Negative 6–25 days
9 days; 50 μ	3	22°	Negative 6–25 days
10 days; 50 μ	4	22°	Positive 9th day; mosquito biting this bird had sporozoites in salivary glands at 11 days
10 days	—	—	Positive 12th day
10 days	—	—	Positive 8th day
10 days	—	—	Positive 8th day
13 days room temp.; 16 days at 4°	—	—	Positive 8th day

more, the intensity of infection does not seem to be diminished, and the resulting gametocytes are capable of infecting *C. tarsalis* and, through this vector, other canaries. Although in the body of the mosquito the salivary glands probably provide the sporozoites with nutrition, prevent excessive motility, and offer access to the salivary ducts for passage into the proboscis at biting, these cultivation experiments show that the glands contribute nothing essential to the initial infectivity of the sporozoites.

In recent years, there has been some discussion of the normal shape and motility of malarial ookinetes. According to Howard (1962), the zygote of *Plasmodium gallinaceum* and, presumably, of other plasmodia is spherical and nonmotile, while the elongated ookinete is really a degenerate gametocyte. His observations were made through the wall of the mosquito stomach while the latter was flattened and subjected to considerable pressure. Contrary observations of ookinete motility were explained by Howard as the result of passive rolling over of this stage. This may occur along with a flowing of formed particles in the stomach fluid past the ookinete. The resultant picture, according to Howard, would be an optical illusion of active movement and of change of shape by the ookinete.

By cultivating zygotes removed from the stomachs of mosquitoes which had bitten infected hosts, Freyvogel (1966), using media developed by Chao and Ball, was able to observe marked activity for as long as 43 hours in ookinetes belonging to four species of plasmodia. *In vitro* culture was the means of establishing the fact that the ookinetes are motile, and it permitted a detailed study of their method of locomotion. It also showed that the ookinete is a normal elongated stage of the malaria plasmodium and is derived from the spherical zygote.

Plasmodia of the mosquito phase either obtained directly from culture or transferred from the mosquito to suitable culture media have been used to study both locomotor patterns and the mechanics of locomotion. The stage most thoroughly studied has been the sporozoite, and high-speed cinematography has been used to investigate the locomotor behavior. The success in cultivating the plasmodia in media in which they remain viable and normal over considerable periods of time has made possible this type of investigation (Jahn and Bovee, 1968).

In our own experiments (Ball, 1964, 1968, 1969; Ball and Chao, 1963; Chao and Ball, 1964), we have been able to grow *Plasmodium relictum* in association with the stomach of its vector, *Culex tarsalis*, in suitable media and under the favorable physical conditions described in the earlier portions of this paper. All stages of the life cycle were grown *in vitro*, starting with mature gametocytes from the blood of a canary to the formation of sporozoites which were capable of producing the infection in clean canaries. All cultures were kept at 18° for the first 24 hours; subsequent development took place at an optimum temperature of 22–25°. Since 27° is the optimum for development of the plasmodium in live *C. tarsalis,* the development in culture proceeded more slowly. Up to the present, it has not been possible to maintain a single preparation which will develop continuously from zygote to sporozoites. Although growth and development go on for several days, these processes gradually slow down and stop, yet the parasites may remain in a viable condition for several weeks (Ball, 1948). It is very probable that development ceases because of the depletion of some essential trace substance or substances, the nature of which has not yet been discovered. Consequently, it has been necessary to obtain the complete cycle by overlapping successive developmental stages. For example, a two-day oocyst may develop *in vitro* to a stage corresponding to that of a five-day oocyst in the living mosquito although it may take five days in culture to reach a point of development comparable to what occurs in the host in three days. Since oocysts in this type of culture usually cease development after about five days, larger and more mature oocysts are obtained by taking a four-day oocyst and culturing it up to a stage corresponding to a seven-day oocyst. Similar overlapping is required at about five-day intervals throughout the cycle.

In initiating cultures with gametocytes from the blood of infected canaries, it was found that fertilization took place *in vitro* only in blood recovered from a mosquito's stomach. Exflagellation occurred if mature microgametocytes were taken directly from the bird, but no zygotes or ookinetes were ever seen. Our culture experiments indicate that the mosquito stomach probably possesses some substance essential either for maintaining the viability of the gametes or for insuring fertilization. The favorable effect of the stomach is exhibited very quickly since zygotes may be seen in blood removed from the mosquito stomach just after the mosquito has completed its blood meal. Yoeli and Upmanis (1968), by examining blood discharged from the anus of mosquito biters, found that this activation could be demonstrated in 50 seconds after the mosquito's proboscis was inserted.

If the entire preparation of a mosquito stomach filled with parasitized blood was maintained in culture for about five days, the zygotes developed into ookinetes (Fig. 1), penetrated the stomach wall, and gave rise to small oocysts, which might reach an eventual size of 31 μ in diameter (Fig. 2). As noted above, a second culture was then initiated with oocysts smaller than 31 μ, and these were grown until they ceased development at a stage larger and more mature than that reached by the first

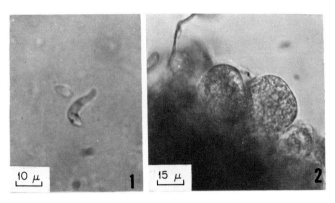

FIG. 1. Ookinete of *Plasmodium relictum*. Blood taken from mosquito stomach and cultured 24 hours.

FIG. 2. Young oocysts of *P. relictum* on stomach of *Culex tarsalis*. Culture started day of biting. Fourth day of culture.

culture (Figs. 3 and 4). This procedure was repeated until mature oocysts were obtained with sporozoites visible within. Oocysts reached about the same stage of development regardless of whether they were attached to the stomach wall or had been dissected off under aseptic conditions. The separated oocysts grew more slowly but did not differ essentially otherwise from those attached.

If nearly mature oocysts, but with no visible sporozoites, were used to initiate a culture, they would give rise to sporozoites after three to

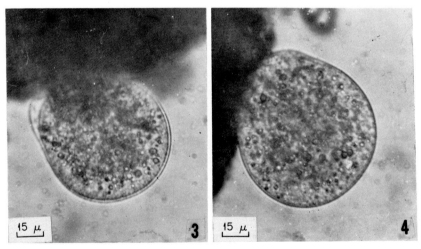

FIG. 3. Oocyst of *P. relictum* on stomach of *C. tarsalis*. Nine days after biting. First day of culture.

FIG. 4. Same oocyst as in Fig. 3. Second day of culture.

four days *in vitro* (Figs. 5 and 6). It was these sporozoites which proved infective by intravenous injection into susceptible birds.

A study of cultured plasmodia growing at lower temperatures than do comparable stages in the insect indicates that no essential biochemical mechanisms are seriously interfered with by growth at 18° (Ball and Chao, 1964). The optimum for *P. relictum* in *C. tarsalis* is 27°.

In the normal life cycle, the plasmodia of warm-blooded vertebrates are physiologically and biochemically triggered to develop at radically different temperatures when ingested by a mosquito or when injected back from the mosquito into the homoiothermous vertebrate. We have demonstrated in our laboratory (Ball and Chao, 1964; Chao and Ball, 1962) that the plasmodia are already partly prepared for these temperature changes some time before the transfer actually occurs and that these adaptations are gradually acquired as the parasite develops.

As indicated in Table I, amino acid analyses of whole bodies of different species of *Culex* show different distribution patterns for individual amino acids (Ball and Clark, 1953). This study was initiated with the aim of preparing a culture medium with an amino acid picture resembling that of the mosquito vector. The discovery that differences in amino acid distribution are associated with differences in suitability of the species of *Culex* to transmit *P. relictum* suggests a causal relationship between the two findings. More extensive investigation will be necessary to test the validity of this hypothesis.

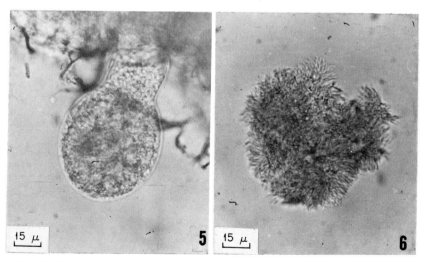

Fig. 5. Oocyst of *P. relictum* on stomach of *C. tarsalis*. Thirteen days after biting. First day of culture.

Fig. 6. Sporozoites from oocyst shown in Fig. 5. Fourth day of culture.

In view of the extensive use of insect cell lines for the growth of viruses *in vitro* (see Chapters 8 and 10), it is surprising that relatively few culture studies have been carried out with mosquito cell lines and malaria parasites. Mosquito cell lines have been established by Grace (1966), Kitamura (1964, 1965, 1966), Peleg (1965), Singh (1967), Singh and Bhat (1969), Schneider (1969), and others. These lines have been derived from the genera *Aedes, Anopheles,* and *Culex* and comprise cell lines of vectors of both mammalian and avian plasmodia. Originally, the maintenance of mosquito cell lines required the presence of insect hemolymph, but this can now be replaced by various substances, such as fetal bovine serum (Gubler, 1968; Hsu *et al.*, 1969; Mitsuhashi and Grace, 1969; Sohi, 1969; Varma and Pudney, 1969), greatly simplifying the maintenance of cultures. Although some mosquito cell line cultures show a high degree of polyploidy (Hink and Ellis, 1970; McHale *et al.*, 1970; Suitor *et al.*, 1966), diploid cultures have also been obtained (Nichols *et al.*, 1970; Schneider, 1969; Stevens, 1970; Varma and Pudney, 1969).

There is no certainty as to the type or types of mosquito cells which survive in culture; it is improbable that intestinal epithelial cells have been maintained. However, this is not believed to be crucial for malaria development in view of Weathersby's experiments (1954, 1960) referred to in an earlier portion of this chapter.

Schneider (1968a), starting with eight- to nine-day-old oocysts of *Plasmodium gallinaceum* obtained sporozoites in a modified Grace's mosquito medium, without mosquito cells. Extracts of whole mosquitoes or organs were added; the sporozoites survived seven days but were infective for only 24 hours. Later (1968b), she cultured oocysts of the

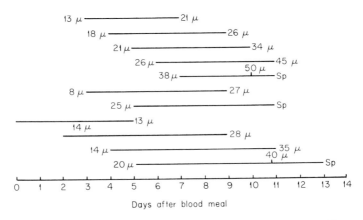

Fig. 7. Increase in size of oocysts of *P. relictum* on stomach of *C. tarsalis* in culture with Grace's cell line of *Aedes aegypti.* Sp: sporozoites.

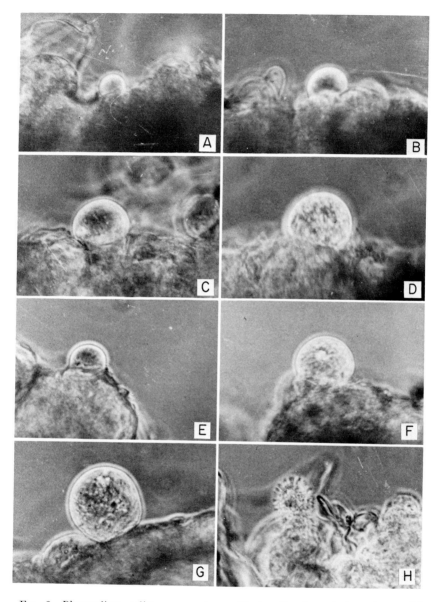

Fig. 8. *Plasmodium relictum* oocysts on *C. tarsalis* stomach in culture with Grace's cell line of *A. aegypti*. (A)–(D) Culture started four days after blood meal. (A) Day 1, 0.14 μ. (B) Day 2, 22 μ. (C) Day 4, 31 μ. (D) Day 5, 35 μ. (E)–(H) Culture started five days after blood meal. (E) Day 0, 20 μ. (F) Day 3, 33 μ. (G) Day 6, 40 μ. (H) Day 8, sporozoites attached to sporoblast core.

same species with Grace's cell line of *Aedes aegypti* and found that eight-day oocysts grew more rapidly than did comparable stages grown in the absence of mosquito cells. In those mature oocysts which did not rupture spontaneously, sporozoites remained active within the oocysts in the cultures for 15 days. In some instances, it was necessary to restrict the rate of growth of the mosquito cells lest they overgrow and eventually kill the oocysts.

Fink and Schicha (1969) used Grace's medium to maintain sporozoites of *Plasmodium berghei yoelii* and *P. cathemerium* in an infective state, and Walliker and Robertson (1970) were able to obtain development of *P. berghei* six- to nine-day oocysts from *Anopheles stephensi* to infective sporozoites in Grace's medium supplemented with various mosquito organs. The most satisfactory were salivary glands, toward which the sporozoites moved.

In our laboratory, we have attempted to cultivate the oocysts of *P. relictum* in *Aedes aegypti* cell lines, both those developed by Grace and those by Singh (Ball and Chao, 1971). This mosquito is a vector for *P. relictum* but a less efficient one than *Culex tarsalis*. The combination of Grace's cell line and the parasite has proved more successful for the development of *P. relictum* than the media used previously and discussed above. Young oocysts only 6 to 9 μ in diameter reached a size of 24–31 μ before ceasing to grow. A second culture, started with a five-day oocyst measuring 20–28 μ, developed to a stage of motile sporozoites after six days of cultivation, approximately the time required by oocysts of this size to develop to the sporozoite stage *in vivo*. With this type of culture, we have approximately doubled the extent of *in vitro* development of a single preparation of the mosquito phase of *P. relictum* (Figs. 7 and 8).

We are now carrying on investigations of the nature of the stimulus or stimuli supplied by the mosquito cell line to the developing parasite. This involves the use of radioactive isotopes which are supplied to the plasmodium from the mosquito cells or the medum in which they are growing. Hayashi and Sohi (1970) have studied the incorporation of ^{3}H-uridine and of ^{14}C-leucine into *A. aegypti* cells grown *in vitro*.

V. FUTURE STUDIES ON PLASMODIA USING INVERTEBRATE TISSUE CULTURE

One of the lines of research to be pursued in the future is the perfection of culture conditions so that the products arising from a single zygote can be grown continuously *in vitro* until infective sporozoites have developed from it. This will eliminate the present necessity of overlapping successive stages. Continuous culture of a plasmodium depends upon the perfection of the medium, probably by supplying it with presently unknown essential trace elements.

When a completely successful medium for continuous culture has been obtained, this medium should be studied analytically with the aim of producing one that is entirely chemically defined. We already have a partially defined medium; it will be necessary to substitute known substances for chicken serum [cf. Siddiqui *et al.* (1967), using the erythrocyte phase of *P. knowlesi*] and for chick embryo extract. The next step will be to carry out systematic changes in the composition of this medium to determine which constituents are essential. It is very possible that the deletion of one compound may result in an altered effect on the culture by another or that alteration in the concentration of a substance may reveal some inhibitory influence on the tissue being cultured.

The use of defined media can very well be combined with a series of depletion studies using the culture preparations. This type of work will certainly be facilitated if the present rather complex medium can be simplified. Otherwise, an investigation of this sort is certain to be tedious and to stretch over several years.

Another type of investigation which should be pursued with plasmodial cultures is the determination of the biochemical and physiological changes which occur in the parasite during its developmental cycle in the mosquito. The changes in the temperature parameters at different stages in the life history have been referred to earlier (Ball and Chao, 1964; Chao and Ball, 1962). These temperature limits have been observed in the living mosquito. The biochemical mechanisms which are responsible can be examined by cultivating the plasmodia in insect tissue culture and obtaining a large number of organisms which are in identical stages of development. Investigation of incorporation of isotopically labeled compounds into the plasmodia during culture should be continued, using mosquito cell lines as the donors.

It is possible that the nutritive requirements of the parasites as supplied by the mosquito change during the plasmodial life history. Comparison of the demands of different stages on media of known constitution should provide evidence for or against this assumption.

Possibly, primary cultures of mosquito cell lines might be more suitable for plasmodial development than longer established and presumably more highly selected and specialized cell lines (Schneider, 1968b).

Parasites might be cultured in sufficient numbers in chemically defined media to serve as test antigens for possible immune reactions in resistant mosquitoes. Plasmodia from culture could be used at various stages of development to determine if there are changes in the antigenic expression by the parasite during its development. Immunological studies on mosquito cell lines have already begun (Ibrahim *et al.*, 1970).

It is known that failure of malaria parasites to develop in resistant mosquitoes may be expressed at different stages in the life history. This is also the case with drugs acting on the developing plasmodia. For example, sometimes zygotes fail to penetrate the stomach wall; in other instances, the oocysts may fail to grow or to form sporozoites; or the sporozoites may be produced but may not be infective (Gooding, 1966; Huff, 1934; Hunninen, 1953; Terzian *et al.*, 1949; Ward, 1963; Whitman, 1948). *In vitro* experiments using different stages of the plasmodia would offer an opportunity of explaining such differences in susceptibility or in its expression.

Finally it may be possible at some future date to dispense with all of the hosts of the malaria parasite and to obtain complete development in culture media, for both the vertebrate and the insect phases. If this occurs, then the net result, at least as far as the cycle in the mosquito is concerned, will be that we have employed invertebrate tissue culture to eliminate the need for invertebrate tissue in the cultivation of plasmodia.

ACKNOWLEDGMENTS

Most of the investigations carried out in my laboratory were in collaboration with Dr. E. W. Clark and Dr. Jowett Chao. The research was supported by Grant 00087 National Institutes of Health and Zoology Grant 254, University of California. Figures 1–6 and Tables I–III are reprinted by permission of the Oregon State University Press. Figures 7 and 8 and Table IV are published by permission of the Editor of the Journal of Parasitology.

REFERENCES

Akov, S. (1962). *J. Insect Physiol.* **8**, 319.
Anfinsen, C. B., Geiman, Q. M., McKee, R. W., Ormsbee, R. A., and Ball, E. G. (1946). *J. Exp. Med.* **84**, 607.
Ball, G. H. (1948). *Amer. J. Trop. Med.* **28**, 533.
Ball, G. H. (1954). *Exp. Parasitol.* **3**, 358.
Ball, G. H. (1964). *J. Parasitol.* **50**, 3.
Ball, G. H. (1968). *Abstr. 8th Int. Congr. Trop. Med. Malar., 1968* p. 1283.
Ball, G. H. (1969). *In* "Progress in Protozoology," p. 325. Nauka, Leningrad.
Ball, G. H., and Chao, J. (1957). *J. Parasitol.* **43**, 409.
Ball, G. H., and Chao, J. (1961). *J. Parasitol.* **47**, 787.
Ball, G. H., and Chao, J. (1963). *Ann. N. Y. Acad. Sci.* **113**, 322.

Ball, G. H., and Chao, J. (1964). *J. Parasitol.* **50**, 748.
Ball, G. H., and Chao, J. (1971). *J. Parasitol.* **57**, 391.
Ball, G. H., and Clark, E. W. (1953). *Syst. Zool.* **2**, 138.
Boorman, J. (1967). *Nature (London)* **213**, 197.
Chao, J., and Ball, G. H. (1962). *J. Parasitol.* **48**, 252.
Chao, J., and Ball, G. H. (1964). *Amer. J. Trop. Med. Hyg.* **13**, 181.
Chen, P. S. (1960). *Proc. Int. Congr. Entomol., 11th, 1960* Symposium 3, p. 201.
Chen, P. S. (1963). *J. Insect Physiol.* **9**, 453.
Chen, P. S., and Briegel, H. (1965). *Comp. Biochem. Physiol.* **14**, 463.
Clark, E. W., and Ball, G. H. (1952). *Exp. Parasitol.* **1**, 339.
Clark, E. W., and Ball, G. H. (1954). *Physiol. Zool.* **27**, 334.
Clark, E. W., and Ball, G. H. (1956). *Physiol. Zool.* **29**, 206.
Clements, A. N., and Grace, T. D. C. (1967). *J. Insect Physiol.* **13**, 1327.
Fink, E., and Schicha, E. (1969). *Z. Parasitenk.* **32**, 93.
Freyvogel, T. A. (1966). *Acta Trop.* **23**, 201.
Garnham, P. C. C., Bird, R. G., and Baker, J. R. (1962). *Trans. Roy. Soc. Trop. Med. Hyg.* **56**, 116.
Gooding, R. H. (1966). *Comp. Biochem. Physiol.* **17**, 115.
Grace, T. D. C. (1966). *Nature (London)* **211**, 366.
Grace, T. D. C., and Brzostowski, H. W. (1966). *J. Insect Physiol.* **12**, 625.
Greene, A. E., and Charney, J. (1971). *Curr. Topics Microbiol. Immun.* **55**, 51.
Gubler, D. J. (1968). *Amer. J. Epidemiol.* **87**, 502.
Hayashi, Y., and Sohi, S. S. (1970). *In Vitro* **6**, 148.
Hink, W. F., and Ellis, B. J. (1970). *In Vitro* **6**, 230.
Howard, L. M. (1962). *Amer. J. Hyg.* **75**, 287.
Hsu, S. H., Liu, H. H., and Suitor, E. C., Jr. (1969). *Mosquito News* **29**, 439.
Huff, C. G. (1934). *Amer. J. Hyg.* **19**, 123.
Huff, C. G., Pipkin, A. C., Weathersby, A. B., and Jensen, D. V. (1960). *J. Biophys. Biochem. Cytol.* **7**, 93.
Hunninen, A. V. (1953). *J. Parasitol.* **39**, 28.
Ibrahim, A. N., Cupp, E. W., and Sweet, B. H. (1970). *In Vitro* **6**, 231.
Jahn, T. L., and Bovee, E. C. (1968). *In* "Infectious Blood Diseases of Man and Animals" (D. Weinman and M. Ristic, eds.), Vol. 1, pp. 393–436. Academic Press, New York.
Johnson, J. W. (1969). *Amer. J. Trop. Med. Hyg.* **18**, 103.
Kitamura, S. (1964). *Kobe J. Med. Sci.* **10**, 85.
Kitamura, S. (1965). *Kobe J. Med. Sci.* **11**, 23.
Kitamura, S. (1966). *Kobe J. Med. Sci.* **12**, 63.
Lea, A. O., Dimond, J. B., and DeLong, D. M. (1956). *J. Econ. Entomol.* **49**, 313.
McHale, J. S., Cupp, E., Unthank, H. D., and Sweet, B. H. (1970). *In Vitro* **6**, 230.
Martignoni, M. E. (1960). *Experientia* **16**, 125.
Mitsuhashi, J., and Grace, T. D. C. (1969). *Appl. Entomol. Zool.* **4**, 121.
Nichols, W. W., Bradt, C., and Bowne, W. (1970). *In Vitro* **6**, 238.
Nydegger, L., and Manwell, R. D. (1962). *J. Parasitol.* **48**, 142.
Peleg, J. (1965). *Nature (London)* **206**, 427.
Schneider, I. (1968a). *Exp. Parasitol.* **22**, 178.
Schneider, I. (1968b). *Proc. Int. Colloq. Invertebr. Tissue Cult., 2nd. 1967* p. 247.
Schneider, I. (1969). *J. Cell Biol.* **42**, 603.
Siddiqui, W. A., Schnell, J. V., and Geiman, Q. M. (1967). *Science* **156**, 1623.
Siddiqui, W. A., Schnell, J. V., and Geiman, Q. M. (1970). *Amer. J. Trop. Med. Hyg.* **19**, 586.

Singh, K. R. P. (1967). *Curr. Sci.* **36,** 506.

Singh, K. R. P., and Bhat, U. K. M. (1969). *Indian J. Med. Res.* **57,** 52.

Sohi, S. S. (1969). *Can. J. Microbiol.* **15,** 1197.

Stevens, T. M. (1970). *Proc. Soc. Exp. Biol. Med.* **134,** 356.

Suitor, E. C., Jr., Chang, L. L., and Liu, H. H. (1966). *Exp. Cell Res.* **44,** 572.

Terzian, L. A., Stahler, N., and Weathersby, A. B. (1949). *J. Infec. Dis.* **84,** 47.

Trager, W. (1935). *Amer. J. Hyg.* **22,** 18.

Trager, W. (1964). *Amer. J. Trop. Med. Hyg.* **13,** 162.

Trager, W. (1967). *J. Protozool.* **14,** 110.

Trager, W., and Krassner, S. M. (1967). *In* "Research in Protozoology" (T. T. Chen, ed.), Vol. 2, pp. 357–382. Pergamon, Oxford.

Trembley, H. L., Greenberg, J., and Coatney, G. R. (1951). *J. Nat. Malar. Soc.* **10,** 68.

Trigg, P. I. (1969). *Parasitology* **59,** 925.

Varma, M. G. R., and Pudney, M. (1969). *J. Med. Entomol.* **6,** 432.

Wallicker, D., and Robertson, E. (1970). *Trans. Roy. Soc. Trop. Med. Hyg.* **64,** 5.

Ward, R. A. (1963). *Exp. Parasitol.* **13,** 328.

Weathersby, A. B. (1954). *Exp. Parasitol.* **3,** 538.

Weathersby, A. B. (1960). *Exp. Parasitol.* **10,** 211.

Weathersby, A. B. (1965). *Mosquito News* **25,** 44.

Weathersby, A. B. (1967). *J. Georgia Entomol. Soc.* **2,** 31.

Weathersby, A. B., and McCall, J. W. (1968). *J. Parasitol.* **54,** 1017.

Whitman, L. (1948). *J. Infec. Dis.* **82,** 251.

Yoeli, M., and Upmanis, R. (1968). *Exp. Parasitol.* **22,** 122.

Note added in proof: According to Greene and Charney (1971), Grace's cell line of *Aedes aegypti,* distributed in this country originally from Dr. Suitor's laboratory, is really a cell line of the moth *Antheraea eucalypti.* This is the cell line used by us. On the contrary, Singh's cell line consists of mosquito cells.

10

USE OF INVERTEBRATE CELL CULTURE FOR THE STUDY OF PLANT VIRUSES

Jun Mitsuhashi

I. INTRODUCTION

Recently, cell cultures of some plant virus transmitting insects became possible. Attempts to use insect cell cultures for the study of viruses have been made in the fields of insect viruses and insect-borne animal viruses. Such a type of study on plant viruses had also been desired by plant pathologists as well as by insect virologists, but the difficulty of culturing the cells of insect vectors hampered the progress of studies in this particular field until recently. In 1963 the first growing cells were obtained from leafhoppers which had been known as vectors of plant viruses (Vago and Flandre, 1963), and since then, cells of other leaf-hopper vectors have been successfully cultivated (Mitsuhashi and Mara-morosch, 1964b; Mitsuhashi, 1965a; Chiu *et al.*, 1966). The obstacle to study plant viruses in vector cell cultures was overcome, and some experiments have already been carried out, by using the primary cultures as well as the established cell lines of vector insects. This chapter sum-marizes the results hitherto obtained from the study of plant virus multiplication in the cultured vector cells.

II. METHODS FOR CULTURING LEAFHOPPER TISSUES

Growing cells can be obtained from any stage of leafhoppers (Mitsu-hashi and Maramorosch, 1964b), but when the tissues of nymphs or adults are cultivated, the leafhoppers which provide explants have to be reared aseptically. For this purpose, methods for rearing leafhoppers aseptically have been developed (Mitsuhashi and Maramorosch, 1963; Mitsuhashi, 1965c). The use of embryonic tissues as explants is prefer-able because the eggs of leafhoppers can be easily surface-sterilized, and therefore, aseptic embryos can be obtained from septically reared stock colonies. Another advantage of using embryos is the high reproducibility of the growth of cells from their tissues. This is important for the use of the primary cultures in any type of experiment. Consequently, all the experiments on plant virus multiplication in vector cell cultures have been carried out using cultivated cells from leafhopper embryos.

The details of leafhopper tissue culture are given by Maramorosch and Hirumi in Volume I, Chapter 10. However, it might be worthwhile here to describe some points important for infections and to take as example the preparation of materials for culturing leafhopper embryonic cells.

For setting up the primary cultures from leafhopper embryos, the eggs excised from the oviposition site are surface-sterilized by submersion in a sterilizing agent, followed by two rinses with sterilized distilled water. The sterilizing agent should be carefully chosen because embryos of many species of leafhoppers are extremely sensitive to some sterilizing agents. The embryos of *Nephotettix cincticeps*, for instance, are easily killed by treatment with 0.1% mercuric chloride for one minute, while embryos of *Macrosteles fascifrons* can tolerate this treatment. After surface-sterilization, the eggs are placed in sterilized Rinaldini's salt solution (NaCl, 0.8 gm; KCl, 0.02 gm; $NaH_2PO_4 \cdot H_2O$, 0.005 gm; glucose, 0.1 gm; $NaHCO_3$, 0.1 gm; sodium citrate, 0.0676 gm; all made up to 100 ml with distilled water), and the embryos are squeezed out through cuts made on tips of the eggs. The embryos are freed from the yolk, then cut into small pieces to provide adequate size of explants, and finally subjected to trypsinization. The trypsinization is carried out with 0.1% trypsin (Trypsin 1:250, Difco, Detroit, Mich.) at 25° for several minutes This trypsinization aimed to loosen the connection of cells and enhance the adhesiveness of the tissues to the glass surface but not to obtain separate cells. Prolonged trypsinization resulted in the formation of very viscous material and damaged the cells. The trypsinized tissues should be transferred into sterilized Ringer–Tyrode's salt solution (Carlson, 1946) placed on the bottom of the culture vessels when the surface of the tissues began to become sticky. The tissues adhere to the glass surface instantaneously when they are released from a pipette into Ringer–Tyrode's solution. Finally, the Ringer–Tyrode's solution is replaced with culture medium.

For the observation of infected cells, the glass which constitutes the vessels should be very thin and plain, and the height of vessels be within a working distance of the condenser lens. To satisfy these conditions, the present author made a culture vessel with two pieces of cover glass and a glass ring (Fig. 1). The components of the culture vessel were sterilized before assembly, which was made under sterile conditions. One piece of cover glass was fastened to one side of the ring with melted paraffin to

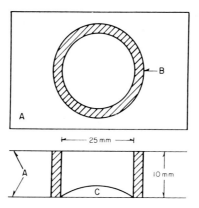

25 mm

10 mm

Fig. 1. Schematic representation of culture vessel. A, cover glass; B, glass ring; C, culture medium.

make the bottom of the vessel, and the opposite side of the ring was closed with another cover glass by means of grease after the culture was set up. This culture vessel is suitable for observation of cells with inverted phase contrast microscope. No optical distortion was produced with this vessel even by using 100 times magnification, oil immersion objective lens. The vessel is most efficient for photomicrograph as well as cinematography. This vessel also has advantages in changing the medium and in handling the cultured materials because direct manipulation is possible by simply removing the top cover glass.

The culture medium used is free of insect hemolymph. The composition of NCM-2B medium, one of the simple and efficient culture media for leafhopper cells, is shown in Table I. In this culture medium the cells of the following leafhopper species have been successfully cultivated; *Macrosteles fascifrons, Agallia constricta, Dalbulus maidis, Nephotettix cincticeps, N. apicalis,* and *Inazuma dorsalis.*

In the primary cultures of leafhopper embryos, several types of cells

TABLE I
Composition of NCM-2B Medium

Substances	Concentrations	
	(gm)	(ml)
NCM-2 mixture (see below)		40
TC-199 medium		20
Fetal bovine serum		20
Bidistilled water		19
Dihydrostreptomycin sulfate (10 mg/ml)		1
NCM-2 mixture		
Ringer–Tyrode's stock solution A (see below)		10
Ringer–Tyrode's stock solution B (see below)		10
Glucose	0.4	
Lactalbumin hydrolysate	1.3	
Bacto-peptone	1.3	
TC-Yeastolate	0.5	
Bidistilled water		80
Ringer–Tyrode's solution A		
NaCl	7	
$CaCl_2 \cdot 2H_2O$	0.2	
NaH_2PO_4	0.2	
KCl	0.2	
$MgCl_2 \cdot 6H_2O$	0.1	
Bidistilled water		100
Ringer–Tyrode's solution B		
$NaHCO_3$	0.12	
Bidistilled water		100

migrate from the explants, and some of them multiply by mitoses. The detailed morphology and behavior of cells have been reported elsewhere (Mitsuhashi and Maramorosch, 1964b; Mitsuhashi, 1965a). By changing the culture medium once a week, the cells usually continued to multiply for three or four months and then gradually degenerated. This survival period is long enough to carry out some experiments with plant viruses.

Chiu and Black (1967) succeeded to establish lines of leafhopper cells. The primary cultures of embryonic tissues of *Agallia constricta* were set up in small petri dishes in the same manner as described above. The first subculture was made when extensive cell growth was attained about three to four weeks after the culture had been set up. The culture medium developed by Mitsuhashi and Maramorosch (1964b) was used successfully for starting primary cultures, but this medium was found unsatisfactory for subcultures. Successful subculturing was achieved with a medium based on Schneider's salt solution, (D-glucose, 400 mg; NaCl, 105 mg; KCl, 80 mg; MgSO$_4$·7H$_2$O, 185 mg; CaCl$_2$, 30 mg; KH$_2$PO$_4$, 30 mg; NaHCO$_3$, 35 mg; lactalbumin hydrolysate, 650 mg; yeastolate, 500 mg; fetal bovine serum, 17.5–20.0 ml; penicillin, 10 000 IU, streptomycin, 10 000 μg; neomycin, 5000 μg; fungizone, 250 μg; and water added to make 100 ml). Substitution of 2.5 ml of insect or lobster hemolymph for a like quantity of fetal bovine serum supported vigorous cell growth after subculturing. But once adapted, the cultures grew well without insect or lobster hemolymph. Subculturing was carried out by a procedure which involved trypsinization for six to eight minutes in 0.05% trypsin to dissociate the cells, centrifugation for two minutes at 200 g, and dispersion of the sedimented cells in medium for distribution to new containers.

III. CULTURING CELLS CARRYING VIRUSES

Virus carrying cells can be obtained only from viruliferous insects. If a plant virus is not transmissible transovarially, aseptic viruliferous insects may be needed to obtain virus carrying tissues from larvae or adults. For this purpose a method has been devised to allow an aseptically reared leafhopper to acquire a plant virus without contamination (Mitsuhashi and Maramorosch, 1964a). If embryos are already virus infected when the eggs are laid, such embryos are suitable mateiral for the culture of virus carrying cells. The egg laid *by Nephotettix cincticeps* carrying rice dwarf virus is such an instance. Rice dwarf virus can be transmitted from the mother leafhopper to her eggs through the ovaries.

Therefore, the cells from viruliferous embryos of *N. cincticeps* were cultured in the same manner as the virus-free embryonic cells (Mitsuhashi, 1965b).

Cell migration and multiplication from the explanted viruliferous tissues occurred almost normally. All the cell types observed in the culture of virus-free embryonic cells were also obtained. Epithelial-like cells were predominant and multiplied by forming monolayer cell sheets in a large area around the explants. These cells sometimes contained many granules as compared with the identical type of cells obtained from virus-free embryonic tissues. But the appearance of granules is not a characteristic feature of the virus carrying cells because similar granules also appeared in virus-free cells when cell growth became unfavorable. During the culture considerable histolysis of the explants occurred simultaneously with the development of cell sheets. But this phenomenon is also not usual in the culture of viruliferous tissues. No appreciable difference in growth of fibroblast-like cells and wandering cells was observed. Cell degeneration seemed to begin somewhat earlier in the viruliferous cell cultures than in the virus-free cell cultures although there was no clear difference. In the earliest case, viruliferous cells began

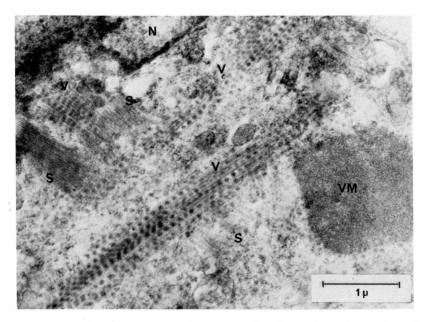

FIG. 2. Rice dwarf virus in the viruliferous cell culture. The cell grew from the viruliferous embryonic tissues of *Nephotettix cincticeps*. N, nucleus; S, spindle-like fibrous structure; V, rice dwarf virus particles; VM, viral matrix.

to degenerate 30 days after the culture had been set up, while cells were still multiplying after three months in some cases.

Electron microscopic observation was made on the cultured viruliferous cells. The cells were fixed with osmium tetroxide solution buffered at pH 7.2 on the 27th day of culture and embedded in Epon 812. The ultra-thin sections revealed the presence of rice dwarf virus particles as well as viral matrix in the cytoplasm. The virus particles were found in abundance in almost all the cells examined. Most of the particles were arranged in lines and were accompanied by a sheath-like structure (Fig. 2). Some virus particles were seen at the periphery of the viral matrix. In intact eggs, viruliferous embryos contained only small numbers of rice dwarf virus particles, and the particles were seen only in mycetomes (Nasu, 1965). It is therefore evident that rice dwarf virus multiplied in the cells grown *in vitro*.

IV. INOCULATION OF CULTURED VECTOR CELLS WITH PLANT VIRUSES

Cultures of vector cells of plant viruses have been successful in leaf-hoppers and aphids. By using the primary cultures of embryonic cells or the established cell lines, some experiments on intracellular multi-plication of plant virus have been carried out.

Inoculation of cells from *Nephotettix cincticeps* with rice dwarf virus has been carried out (Mitsuhashi, 1965b; Mitsuhashi and Nasu, 1967). Inoculum for *N. cincticeps* cells were prepared in two different ways. In the first experiment the inoculum containing rice dwarf virus was prepared from viruliferous female adults of *N. cincticeps* (Mitsuhashi, 1965b). The viruliferous leafhoppers were first surface-sterilized by sub-mersion in 70% ethyl alcohol for one minute and then in 0.1% mercuric chloride for five minutes. After rinsing with sterilized distilled water, the leafhoppers were cut open and the fat bodies were removed aseptically. The fat bodies from five viruliferous female adults were homogenized in 0.5 ml of the culture medium and the homogenate was centrifuged at 3000 rpm for five minutes. The clear part between the lipid layer and the sediment was used as inoculum. As recipient of the virus, 14-day-old virus-free embryonic cell cultures were used. At this age of culture, the epithelial-like cells already formed monolayer cell sheets around the explants, and also, all the other types of cells were seen. Inoculation of virus was carried out simply by replacing the culture medium with the inoculum. The inoculated cultures were kept at 25° for 24 hours with the inoculum, and then the inoculum was thoroughly washed away by repeating the changes of the culture medium. The inoculated cultures were thereafter maintained by changing the medium once a week.

Some morphological changes of the inoculated cells were recognized within 24 hours. Granules appeared first around the nuclei, and later they increased in number and filled the whole area of the cytoplasm (Fig. 3). This granulation finally resulted in destruction of the cells. From the time lapse cinematographic analysis the granules seemed to be substances which were brought into the cells from the environment by active phagocytosis of the cells. The granules disappeared when the cells were treated with xylene after fixation with alcohol. The granules were stained with Sudan Black B after fixation with formal-calcium. Therefore, the granules seemed to be of lipidic nature. The granulation was quite marked in the epithelial-like cells. It proceeded from the periphery of the cell sheets to the inner parts. During the first three days after the inoculation, mitoses could be seen even in the granulated cells at the periphery of the cell sheets. But with the progress of the granulation, the cells shrank and the cell sheets were destroyed. The explants themselves were also subjected to some changes. The contracting movement of the explants lasted for a week after the inoculation but ceased later. The cells constituting the explants also became granulated.

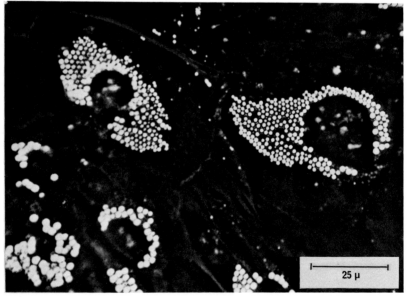

Fig. 3. Granules appeared in the *Nephotettix cincticeps* cells inoculated with rice dwarf virus.

In the control experiments, the inoculum was prepared from the fat bodies of virus-free adult leafhoppers in the same manner as the viral inoculum. When this virus-free inoculum was brought into the virus-free embryonic cell cultures, the cells and the explants did not show marked changes as observed in the cultures inoculated with the virus. Sometimes, granulation of cells occurred also in the control cultures, but not severely. In this case, the granulation usually did not cover a wide area and resulted in local destruction of the cell sheets. Similar granules were also seen in noninoculated virus-free cell cultures when the culture conditions became unfavorable for the growth of cells. It is therefore reasonable to consider that the granulation of cells after the inoculation of the virus was a secondary effect of virus multiplication or merely effects of substances in the inoculum other than the virus. This point will be discussed further in the second experiment.

At the ninth day post-inoculation, the inoculated cells were fixed for electron microscopic examination. The ultrathin sections revealed the presence of many small particles which looked like rice dwarf virus particles but far smaller (30 mμ in diameter) than the rice dwarf virus particles (70 mμ in diameter) in the cytoplasm. The small particles appeared in crystalline aggregates or in linear alignment (Fig. 4). Some-

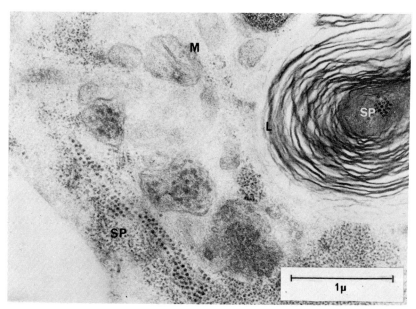

FIG. 4. Small particles appeared in the *Nephotettix cincticeps* cells inoculated with rice dwarf virus. L, lamellar structure; M, mitochondria; SP, small particles.

times the crystalline aggregates of the small particles were found in lamellar structures (Fig. 4, upper right). Besides the small particles, rice dwarf virus particles of normal size were also seen in some sections, but not many. The small particles seemed to be virus particles from their geometrical alignment, but it is not certain whether the small particles are polymorphism or developmental stages of rice dwarf virus particles. The possibility that the small particles are unknown virus particles which have contaminated the cultures cannot be neglected. Nasu (1965) has also reported the presence of similar small particles in the cells of intact viruliferous leafhoppers (Fig. 5). But the infectivity of these small particles has not been determined and further investigations will be required to clarify their nature.

In the first experiment, enough evidence for the multiplication of rice dwarf virus in the inoculated cells could not be obtained although some ultrathin sections revealed the presence of rice dwarf virus particles. The second experiment was undertaken to get more conclusive evidence of the intracellular multiplication of the virus. In order to exclude the deleterious effects of substances other than the virus in the inoculum, the inoculum was prepared from the viruliferous cell cultures (Mitsuhashi

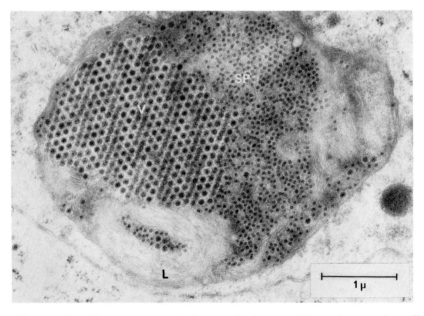

FIG. 5. Crystalline aggregate of rice dwarf virus particles and mass of small particles in the mycetome of the viruliferous *Nephotettix cincticeps*. L, lamellar structure; SP, small particles; V, rice dwarf virus particles. (From Nasu, 1965.)

and Nasu, 1967). For the preparation of the inoculum the cultured cells from viruliferous embryos were harvested on 57th day of culture. At that time in the viruliferous cultures, epithelial-like cells developed mono-layer cell sheets around the explants which were still pulsating; fibroblast-like cells and wandering cells were seen in abundance. These cells were still multiplying. From the results of viruliferous cell cultures, it was assumed that rice dwarf virus multiplied well enough in these cells. The harvested cells were homogenized with a small amount of culture medium and centrifuged for 30 minutes at 4000 rpm. No lipid layer was obtained, and the supernatant was used as inoculum. This inoculum was supposed to have less deleterious effects to the cells to be inoculated, as compared with the inoculum prepared from viruliferous fat bodies. The 37-day-old virus-free embryonic cell cultures were inoculated with this inoculum in the same manner as in the first experiment. The status of cell growth in the recipient cultures before the inoculation was about the same as in the first experiment.

Contrary to the results obtained in the first experiment, granulation of the cells did not occur after the inoculation. It may therefore be

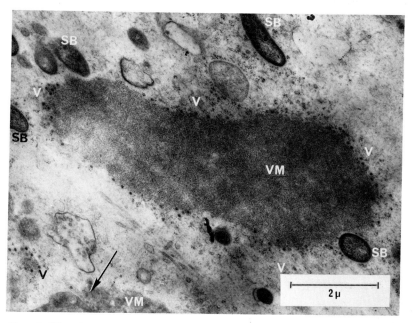

FIG. 6. Rice dwarf virus particles and viral matrix appeared in the *Nephotettix cincticeps* cells inoculated with rice dwarf virus. SB, symbiotic bacteria; V, rice dwarf virus particles; VM, viral matrix; arrow, rice dwarf virus particles accompanied by sheath-like structure. (From Mitsuhashi, 1969.)

concluded that the granulation of the cells observed in the first experiment was not cytopathic effects of the rice dwarf virus but the effect of some other substances in the supernatant of fat body homogenate from viruliferous insects.

No appreciable changes occurred in the inoculated cells. The cells continued to multiply, and the explants kept contracting for long periods after the inoculation. The growth of the inoculated cells began to stop after the 25th day, and later the cells gradually degenerated.

Electron microscopic examinations of the inoculated cells was made on the samples obtained at the 28th day post-inoculation. Abundant viral matrix and rice dwarf virus particles were observed in the cytoplasm of the inoculated cells. In the inoculated cells, rice dwarf virus particles appeared as individual separated particles in the viral matrix (Fig. 6, lower left), at the periphery of the matrix (Fig. 6, center), in the linear alignment accompanied by sheath-like structure (Fig. 6, arrow), and in crystalline array (Fig. 7). Sometimes, the virus particles were found in lamellar structures. The small particles which were observed in the first experiment were also seen in the viral matrix. The electron dense area, which is designated as viral matrix here, was ascertained to be a mass of

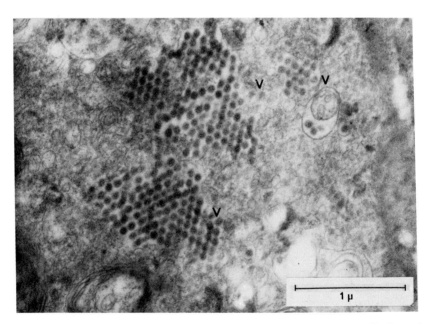

Fig. 7. Crystalline array of rice dwarf virus particles appeared in the *Nephotettix cincticeps* cells inoculated with rice dwarf virus. V, rice dwarf virus particles. (From Mitsuhashi, 1969.)

viral material by means of the ferritin-conjugated antibody technique (Nasu and Mitsuhashi, 1968) (Fig. 8).

It is evident that rice dwarf virus multiplied in the cultured vector cells when artificially inoculated.

Another example of study on intracellular growth of plant virus was given by Chiu *et al.* (1966). Working with wound-tumor virus they succeeded in infecting primary cultures of the vector leafhoppers *Agallia constricta*. Cells obtained from explanted embryonic tissues dissected from virus-free eggs were cultivated in the primary cultures. For setting up the cultures, essentially the same techniques as Mitsuhashi's were employed. The basic culture medium of Mitsuhashi and Maramorosch (1964b) gave satisfactory growth of cells.

The inoculum was prepared from adult leafhoppers which had acquired the virus by feeding for two or three days as young nymphs on crimson clover infected with wound-tumor virus. Thirty to sixty viruliferous insects were homogenized in 2 ml of the culture medium, and the homogenate was centrifuged for five minutes at 3100 g. The supernatant was then sterilized by passage through Millipore filters of 0.65 and 0.30 μ pore diameters.

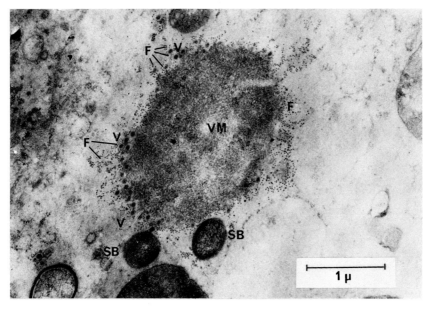

Fig. 8. Demonstration of rice dwarf virus particles and viral matrix by ferritin-conjugated antibody technique. The cell grew from the virus-free embryonic tissues of *Nephotettix cincticeps* and was inoculated with rice dwarf virus. F, ferritin particles; SB, symbiotic bacteria; V, rice dwarf virus particles; VM, viral matrix.

Ten-day-old cultures were used for inoculation. In these cultures the explants were surrounded by well-developed cell sheets of epithelial-like cells. Other types of cells as described by Mitsuhashi and Maramorosch (1964b) were also seen in the mixed cell population.

Inoculation was carried out by replacing culture medium with inoculum and allowing an adsorption period of two or three hours at room temperature. The inoculated cultures were subsequently maintained at 24° by replacing the medium every two or three days.

The inoculated cultures were examined for evidence of wound-tumor virus infection by the fluorescein-conjugated antibody technique. On the third day after the inoculation, the viral antigens were detected as discrete brilliant yellow greenish spots in the cytoplasm of a few cells in the inoculated cultures. The intensity of staining and the number of stained cells increased on later days. The viral antigens were not observed in nuclei. In the control cultures, which had been exposed to the homogenate prepared from virus-free leafhoppers, the specific staining was not observed.

Evidence of wound-tumor virus infection was also given by infectivity tests made on the inoculated cells or the medium in which they had been cultivated. The relative concentration of infective wound-tumor virus was calculated from the results of precipitin ring-time tests on extracts from injected leafhoppers. An increase in infectivity was obtained from the inoculated cells although the relative concentration of infective virus did not exceed the level present in the original inoculum. The relative concentration of infective virus in the medium reached a plateau level in about two days, after which no significant increases in infectivity was obtained. This was explained by the equality of the rate of virus release from the infected cells and the rate of virus inactivation at 24°.

A serial passage experiment was carried out to obtain further evidence of wound-tumor virus multiplication in tissue cultures. In this experiment, the inoculum was prepared from root tumors on sweet clover infected by wound-tumor virus. Serial passages were carried out at 8- to 11-day intervals. In each passage of the inoculated cultures, infection of the cells was obtained as revealed by the development of specific antigens. The passages were repeated seven times. At the end of the seventh passage, the virus concentration should have been reduced to $10^{-2.5}$ from the original inoculum, and this concentration was well below that detectable by the infectivity test. However, the experimental results did not show such a gradual virus reduction, and the virus was recovered at about the same concentration from the cells at each passage.

Transfer of the inoculated explants was also attempted. The explants were transplanted from cell sheets at 4- and 8-day post-inoculations to establish new cultures. The transplanted explants developed new cell sheets until 45 days after the inoculation. The newly developed cell sheets took the specific staining with fluorescein-conjugated antiserum, suggesting the presence of infective wound-tumor virus in the cells.

After establishment of cell lines of *Agallia constricta* Chiu and Black (1967) examined the susceptibility of the cells to wound tumor virus. All the established cell lines were susceptible to wound tumor virus. Virus antigen was detected as early as 12 hours at 30° after inoculation by the fluorescent antibody staining. Infected lines, with 75% or more of the cells positively stained, showed a growth rate comparable with the healthy lines.

The particles of wound tumor virus which are required to cause infection in the monolayer cells of *A. constricta* were studied quantitatively (Gamez and Chiu, 1968). The infection occurred by inoculation of $10^{6.1}/$ml of the virus particles but not by $10^{5.4}/$ml of the particles.

The assay method for wound tumor virus by using *A. constricta* cell line was devised (Chiu and Black, 1969). Monolayers of the cell were prepared in circles of about 30 mm² on the coverslips of 15 mm in diameter, to which wound tumor virus was inoculated. The adsorption of the virus was maximum for three hours at 24° or two hours at 30°. The optimal time for counting the cells was about 27 hours after inoculation at 30° although the specific fluorescence was detected after 12 hours. The distribution of infected cells in the monolayer was found to follow Poisson distribution. The number of cells which showed specific fluorescence increased with time, and the linear relationship was obtained between the relative concentration of the virus in the inoculum and the number of cells showing the specific fluorescence. This seems to indicate that a single virus particle produced each infection. The cell-infecting unit was estimated as 405 virions.

Chiu *et al.* (1970) established a cell line from *Aceratagallia sanguinolenta* which is a vector of New York potato yellow dwarf virus (NY-PYDV). This cell line was found to be susceptible to NY-PYDV. At the early stages of the infection, only nuclei reacted with fluorescent antibody, and later cytoplasm became to react with the antibody to some extent. The percentage of infection has been reported to be enhanced by the incorporation of 25 μg/ml of DEAE-dextran to the inoculum. A linear relationship was established between the virus concentration in the inoculum and the incidence of infected cells in the inoculated monolayers. The regression lines for NY-PYDV, which is transmitted by *A. sanguinolenta* but not by *A. constricta,* and New Jersey potato yellow dwarf virus (NJ-PYDV), which is transmitted by *A. constricta* but not by *A. sanguinolenta,* were compared on the monolayers of the respective vector cells, and the slopes of these lines were virtually identical. On the other hand, when wound tumor virus and NJ-PYDV were both determined on *A. constricta* cells, the slopes were definitely different from each other.

Attempts to inoculate nonvector cell culture with viruses have been made. Chiu and Black (1969) succeeded in infecting the cultured cells of *A. sanguinolenta* with wound tumor virus. The number of infected cells was 1 or 2 log units lower than with the vector cells, *A. constricta* cell line. When NJ-PYDV and NY-PYDV were inoculated to their non-vector cell cultures, *A. sanguinolenta* cells and *A. constricta* cells respectively, the infectivity was about 10% of that on the vector cells (Liu and Black, after Black, 1969). Infection of the primary culture of *Macrosteles fascifrons*, which is nonvector leafhopper of wound tumor virus, by wound tumor virus has been reported also (Hirumi and Maramorosch, 1968).

Recently, Peters and Black (1970) succeeded in culturing the cells from ovarian and embryonic tissues of the aphid, *Hyperomyzus lactucae*, which is a vector of sowthistle yellow vein virus. When primary cultures of the cells were inoculated with purified samples of the virus, infection of the cells was demonstrated by fluorescent antibody staining. The first infection was noticed 37 hours after the inoculation, and the number of infected cells reached maximum after 48 hours. As is the case with potato yellow dwarf virus infections, staining was prominent in the nuclei.

V. CONCLUSION

Up to the present, methods for cultivation of leafhopper cells have been established, and the studies on plant virus in the cultivated vector cells have started. Multiplication of rice dwarf virus, wound tumor virus, potato yellow dwarf virus, and sowthistle yellow vein virus in respective vector cells *in vitro* has already been proved. But cell cultures of many other vector insects should be made. Although some viruses have been proved to multiply in the cultured nonvector cells, specificity of virus and vector relationship may require cell cultures of specific vector insects when one attempts to study a plant virus in cell cultures. Some plant viruses may multiply in the cell cultures of nonvector insects or even in the established vertebrate cell lines, but the use of cultures of its own vector cells is much more preferable for the study of the virus because the virus in its own vector cells *in vitro* will be much closer to the natural status. Very few attempts have been made on the cell cultures of vector insects other than leafhoppers. Methods for culturing the cells of vector insects other than homopterous insects should also be studied in future because the same culture method cannot necessarily be applied to all the insect species. Besides insect tissue culture, culture methods for the tissues of mites and nematodes should be established in order to study mite- and nematode-borne viruses in their vector cell cultures.

Establishment of a vector cell line which forms monolayer cell sheets will be preferable for the titration of the virus by means of plaque formation or by focus formation. Much improvement of the culture media and culture techniques will be required for their establishment of a cell line from the primary cultures of leafhopper cells although the Chiu and Black (1967) method for establishing cell lines is applicable to other species of leafhoppers.

In the inoculation experiments of rice dwarf virus and wound tumor virus, the growth of the inoculated cells did not seem to be affected by the virus multiplication. This may be considered as a kind of carrier culture. If a cell line can be established from such infected cultures, the resulting cell line may maintain the carrier state. Establishment of plant virus carrier cultures may also facilitate the preservation of the virus.

Although establishment of a cell line from vector animals is most desirable, there are many experiments which can be done at the pirmary culture levels. Among them, it is interesting to study the relationship between plant pathogenic virus and symbiotes of vector insects. Most of homopterous insects are known to have intracellular symbiotes. And some intracellular symbiotic bacteria can multiply in the cultivated host cells *in vitro* (Mitsuhashi and Maramorosch, 1964b). In addition, Nasu (1965) has reported that some intracellular symbiotes of *Nephotettix cincticeps* play some role in the transovarial passage of the rice dwarf virus, and that bacterial symbiotes of *N. cincticeps* seem to have some relation to the virus multiplication in the vector cells. Such relationship between virus and symbiotes can be studied by a use of the primary cultures of vector leafhoppers.

Interaction of viruses can also be studied by the use of the primary cultures of vector insects. *Nephotettix cincticeps* is a vector of rice dwarf virus as well as a vector or rice yellow dwarf mycoplasma. Interaction of both these plant pathogenic agents, if any, will be studied in the primary cultures of *N. cincticeps* by means of electron microscopy, immunological techniques, infectivity tests, etc. This type of study may throw more light on the problem of interference in case of mixed infections.

Apart from viruses, some plant pathogenic microorganisms which can be transmitted by insect vectors could be studied in the same manner as the study of viruses in the vector cell cultures. Recently, considerable numbers of plant pathogenic mycoplasmas have been reported. It will be desirable that plant pathogenic mycoplasma are studied in the vector cell cultures.

In conclusion, the present author would like to emphasize the necessity of establishing more cell lines from vector animals. This will promise further development of the studies in this particular field.

REFERENCES

Black, L. M. (1969). *Annu. Rev. Phytopathol.* **7,** 73.

Carlson, J. G. (1946). *Biol. Bull.* **90,** 109.

Chiu, R.-J., and Black, L. M. (1967). *Nature (London)* **215,** 1076.

Chiu, R.-J., and Black, L. M. (1969). *Virology* **37,** 667.

Chiu, R.-J., Reddy, D. V. R., and Black, L. M. (1966). *Virology* **30,** 562.

Chiu, R.-J., Liu, H.-Y., McLeod, R., and Black, L. M. (1970). *Virology* **40,** 387.

Gamez, R., and Chiu, R.-J. (1968). *Virology* **34,** 356.

Hirumi, H., and Maramorosch, K. (1968). *Proc. Colloq. Invertebr. Tissue Cult., 2nd, 1967* p. 203.

Mitsuhashi, J. (1965a). *Jap. J. Appl. Entomol. Zool.* **9,** 107.

Mitsuhashi, J. (1965b). *Jap. J. Appl. Entomol. Zool.* **9,** 137.

Mitsuhashi, J. (1965c). *Kontyu* **33,** 271.

Mitsuhashi, J. (1969). "Viruses, Vectors and Vegetation" (K. Maramorosch, ed.), p. 475. Wiley (Interscience), New York.

Mitsuhashi, J., and Maramorosch, K. (1963). *Contrib. Boyce Thompson Inst.* **22,** 165.

Mitsuhashi, J., and Maramorosch, K. (1964a). *Virology* **23,** 277.

Mitsuhashi, J., and Maramorosch, K. (1964b). *Contrib. Boyce Thompson Inst.* **22,** 435.

Mitsuhashi, J., and Nasu, S. (1967). *Appl. Entomol. Zool.* **2,** 113.

Nasu, S. (1965). *Jap. J. Appl. Entomol. Zool.* **9,** 225.

Nasu, S., and Mitsuhashi, J. (1968). *Ann. Meet. Jap. Soc. Appl. Entomol. Zool.*

Peters, D., and Black, L. M. (1970). *Virology* **40,** 847.

Vago, C., and Flandre, O. (1963). *Ann. Epiphyt.* **14,** 127.

III

Cell Lines

11

A CATALOG OF INVERTEBRATE CELL LINES

W. Fred Hink

I. INTRODUCTION

This chapter provides information on cell lines from 35 species of insects from four different orders and from one species of mollusk. Diptera are represented by 14 species, Lepidoptera by 10 species, Homoptera by seven species, and Dictyoptera by four species. The species represented by the largest number of lines is *Drosophila melanogaster*, from which 23 separate lines have been established. There are seven lines from *Aedes aegypti* and five from *Anopheles stephensi*.

The total number of cell lines from all species is 85. The recent successes in the field of long-term insect cell culture are evident by the fact that of these 85 lines, 78 have been described during the years 1967–1971.

II. CHRONOLOGICAL LIST OF CELL LINES

Invertebrate species	Order	Primary explant	Year of publication	References
Bombyx mori	Lepidoptera	ovary	1958	Vago and Chastang
Helix aspersa	Mollusca	foot	1958	Vago and Chastang
Bombyx mori	Lepidoptera	gonad	1959	Gaw, Liu, and Zia
Antheraea eucalypti	Lepidoptera	ovary	1962	Grace
Aedes aegypti	Diptera	larva	1966	Grace
Blattella germanica	Dictyoptera	embryo	1966	Landureau
Drosophila melanogaster	Diptera	embryo	1966	Horikawa, Ling, and Fox
Aedes aegypti	Diptera	larva	1967	Singh
Aedes albopictus	Diptera	larva	1967	Singh
Agallia constricta	Homoptera	embryo	1967	Chiu and Black
Agallia quadripunctata	Homoptera	embryo	1967	Chiu and Black
Bombyx mori	Lepidoptera	ovary	1967	Grace
Chilo suppressalis	Lepidoptera	hemocytes	1967	Mitsuhashi
Blabera fusca	Dictyoptera	embryo (2 lines)	1968	Landureau
Periplaneta americana	Dictyoptera	embryo (2 lines)	1968	Landureau
Aceratagallia sanguinolenta	Homoptera	embryo	1969	Chiu and Black
Aedes aegypti	Diptera	embryo (2 lines)	1969	Peleg
Aedes aegypti	Diptera	larva (3 lines)	1969	Varma and Pudney
Anopheles stephensi	Diptera	larva	1969	Schneider
Drosophila melanogaster	Diptera	embryo	1969	Kakpakov, Goosdev, Platova, and Polukarova
Leucophaea maderae	Dictyoptera	heart	1969	Vago and Quiot
Aedes vexans	Diptera	pupa	1970	Sweet and McHale

Invertebrate species	Order	Primary explant	Year of publication	References
Aedes vittatus	Diptera	larva	1970	Bhat and Singh
Culiseta inornata	Diptera	adult	1970	Sweet and McHale
Culex molestus	Diptera	ovary	1970	Kitamura
Culex quinquefasciatus	Diptera	ovary (2 lines)	1970	Hsu, Mao, and Cross
Drosophila melanogaster	Diptera	embryo (8 lines)	1970	Echalier and Ohanessian
Heliothis zea	Lepidoptera	ovary	1970	Hink and Ignoffo
Samia cynthia	Lepidoptera	hemocytes	1970	Chao and Ball
Trichoplusia ni	Lepidoptera	ovary	1970	Hink
Aedes novo-albopictus	Diptera	larva (2 lines)	1971	Bhat (unpublished)
Aedes w-albus	Diptera	—	1971	Singh (unpublished)
Anopheles gambiae	Diptera	larva	1971	Varma and Pudney (unpublished)
Anopheles stephensi	Diptera	larva	1971	Pudney and Varma
Anopheles stephensi	Diptera	larva (3 lines)	1971	Varma (unpublished)
Bombyx mori	Lepidoptera	ovary	1971	Grace (unpublished)
Carpocapsa pomonella	Lepidoptera	embryo (2 lines)	1971	Hink and Ellis
Choristoneura fumiferana	Lepidoptera	larva	1971	Sohi
Colladonus montanus	Homoptera	embryo (3 lines)	1971	Richardson and Jensen
Culex salinarius	Diptera	larva	1971	Schneider
Culex tritaeniorhynchus	Diptera	ovary	1971	Hsu (unpublished)
Culex tritaeniorhynchus	Diptera	larva	1971	Schneider
Drosophila melanogaster	Diptera	embryo (3 lines)	1971	Dolfini (unpublished)
Drosophila melanogaster	Diptera	embryo (7 lines)	1971	Richard-Molard (unpublished)
Drosophila melanogaster	Diptera	embryo (2 lines)	1971	Schneider (unpublished)
Drosophila melanogaster	Diptera	imaginal discs	1971	Schneider (unpublished)
Leucophaea maderae	Dictyoptera	heart	1971	Quiot and Vago (unpublished)
Leucophaea maderae	Dictyoptera	ovary	1971	Quiot and Vago (unpublished)
Macrosteles sexnotatus	Homoptera	embryo	1971	Peters and Spaansen
Malacosoma disstria	Lepidoptera	hemocytes (5 lines)	1971	Sohi (unpublished)
Nephotettix apicalis	Homoptera	embryo	1971	Mitsuhashi (unpublished)
Nephotettix cincticeps	Homoptera	embryo	1971	Mitsuhashi (unpublished)
Spodoptera frugiperda	Lepidoptera	ovary	1971	Vaughn (unpublished)

III. CHARACTERISTICS OF CELL LINES

A. Lepidoptera

Invertebrate species: *Antheraea eucalypti*
Common name: Australian emperor gum moth
Primary explant from which line was developed: Trypsinized pupal
    ovaries
Date primary culture was initiated: August 1960
Population doubling time: 36–72 hours
Maximum population per milliliter of medium: 2–3 $\times$ 10⁶ cells
Subculture interval: 7–10 days
Growth medium: G.M.A. plus 5% *A. eucalypti* hemolymph
Characteristics of line: Three distinct cell types; polygonal with finely
    granulated cytoplasm and 20–40 μ diameter, round to fibroblast-
    like in outline and 15–20 μ diameter, small and round with a clear
    cytoplasm and 10–15 μ diameter. The chromosome numbers range
    from 2n to 128n
Reference: Grace (1962)

Invertebrate species: *Bombyx mori*
Common name: Silkworm
Primary explant from which line was developed: Larval ovaries
Date primary culture was initiated: 1956
Number of subcultures to date: 12 as of original 1958 publication +57
Growth medium: Modified Vago and Chastang, 1958, medium
Characteristics of line: Fibroblast-like and epithelial-like cells
Reference: Vago and Chastang (1958)

Invertebrate species: *Bombyx mori*
Primary explant from which line was developed: Minced larval gonads
Number of subcultures to date: 22 as of original 1959 publication
Growth medium: Trager's (1935) solution A plus 10% *B. mori*
    hemolymph
Reference: Gaw, Liu, and Zia (1959)

Invertebrate species: *Bombyx mori*
Primary explant from which line was developed: Trypsinized larval
    ovaries
Date primary culture was initiated: November 1964
Population doubling time: 48 hours

Maximum population per milliliter of medium: 1.6×10^6 cells
Subculture interval: 6 days
Growth medium: G.M.A. plus 5% *Antheraea eucalypti* hemolymph
Characteristics of line: Most common cell is spindle-shaped and 12–25 μ wide and 50–70 μ long; another cell type is slightly spindle-shaped and 18–30 μ in diameter. Cells contain many more than 100 chromosomes
Reference: Grace (1967)

Invertebrate species: *Bombyx mori*
Primary explant from which line was developed: Ovary
Date primary culture was initiated: March 1968
Population doubling time: 60 hours
Growth medium: G.M.A. plus 1% *A. pernyi* hemolymph
Reference: Grace (unpublished)

Invertebrate species: *Carpocapsa pomonella*
Common name: Codling moth
Designation of cell line: CP-1268
Primary explant from which line was developed: Minced embryos
Date primary culture was initiated: December 1968
Number of subcultures to date: 119 as of May 1971
Population doubling time: 24 hours
Maximum population per milliliter of medium: 1×10^7 cells
Subculture interval: 7 days
Growth medium: TNM-FH
Characteristics of line: Cell population consists of spindle-shaped, round, and cells with single extensions. Chromosome distribution is bimodal with 51% of cells having 51–57 chromosomes and 35% having 102–110 chromosomes
Reference: Hink and Ellis (1971)

Invertebrate species: *Carpocapsa pomonella*
Designation of cell line: CP-169
Primary explant from which line was developed: Minced embryos
Date primary culture was initiated: December 1968
Number of subcultures to date: 116 as of May 1971
Population doubling time: 24 hours
Maximum population per milliliter of medium: 5×10^6 cells
Subculture interval: 7 days
Growth medium: TNM-FH

Invertebrate species: *Carpocapsa pomonella, continued*
Characteristics of line: Most cells (70%) are spherical with diameters
    of 11.0–20.2 μ, 22% have single protoplasmic extensions, and 7%
    are spindle-shaped. Line is highly heteroploid with about 9% of
    population diploid and 72% contain more than 100 chromosomes
Reference: Hink and Ellis (1971)

Invertebrate species: *Chilo suppressalis*
Common name: Rice stem borer
Primary explant from which line was developed: Larval hemocytes
Date primary culture was initiated: February 1965
Number of subcultures to date: 42 as of June 1967
Population doubling time: 108 hours
Maximum population per milliliter of medium: 4×10^5 cells
Subculture interval: 1–2 weeks
Growth medium: CSM-2F
Characteristics of line: Prohemocytes were predominate cell type. Ac-
    cidentally infected with CIV and carrier cultures were established
Reference: Mitsuhashi (1967)

Invertebrate species: *Choristoneura fumiferana*
Common name: Spruce budworm
Primary explant from which line was developed: Minced larvae
Date primary culture was initiated: May 1970
Number of subcultures to date: 15 as of March 1971
Subculture interval: 7 days
Growth medium: 85% G.M.A., 10% FBS, and 5% *B. mori* hemolymph
Reference: Sohi (1971)

Invertebrate species: *Heliothis zea*
Common name: Corn earworm or cotton bollworm
Designation of cell line: IMC-HZ-1
Primary explant from which line was developed: Minced adult ovaries
Date primary culture was initiated: March 1967
Number of subcultures to date: 175 as of May 1971
Population doubling time: 33 hours
Maximum population per milliliter of medium: $1–2 \times 10^6$ cells
Subculture interval: 7 days
Growth medium: Modified Yunker, Vaughn, and Cory medium, desig-
    nated IMC-1
Characteristics of line: Most cells (90%) are spherical or ellipsoidal and

9.9–26.2 μ diameter, 6% of cells have a single protoplasmic extension, and 4% are oblong binucleate cells. Cells are heteroploid.
Reference: Hink and Ignoffo (1970)

Invertebrate species: *Malacosoma disstria*
Common name: Forest tent caterpillar
Designation of cell lines: 5 separate cell lines
Primary explant from which line was developed: Larval hemocytes
Date primary culture was initiated: September 1969
Number of subcultures to date: 40 as of March 1971
Maximum population per milliliter of medium: 4–8 $\times$ 10⁵ cells
Subculture interval: 7 days
Growth medium: 92% G.M.A., 5% FBS, and 3% *B. mori* hemolymph
Reference: Sohi (unpublished)

Invertebrate species: *Samia cynthia*
Common name: Cynthia moth
Primary explant from which line was developed: Pupal hemocytes
Date primary culture was initiated: September 1968
Number of subcultures to date: 70 as of 5 March 1971
Subculture interval: 14 days
Growth medium: G.M.A., 10% FBS, and 0.5% *S. cynthia* hemolymph
Reference: Chao and Ball (1970)

Invertebrate species: *Spodoptera frugiperda*
Common name: Fall armyworm
Designation of cell line: IPRL-21
Primary explant from which line was developed: Pupal ovaries treated
    with trypsin and hyaluronidase
Date primary culture was initiated: October 1968
Number of subcultures to date: 80 as of March 1971
Population doubling time: 24 hours
Growth medium: 100 parts G.M.A., 5 parts FBS, 5 parts EU, and 3 parts
    B. mori hemolymph
Characteristics of line: The cells are polyploid and grow attached to surface of culture vessel
Reference: Vaughn (unpublished)

Invertebrate species: *Trichoplusia ni*
Common name: Cabbage looper
Designation of cell line: TN-368
Primary explant from which line was developed: Minced adult ovaries
Date primary culture was initiated: March 1968

Invertebrate species: *Trichoplusia ni, continued*

Number of subcultures to date: 357 as of May 1971

Population doubling time: 16 hours

Maximum population per milliliter of medium: 2–3 $\times$ 10^6 cells

Subculture interval: 2–3 days

Growth medium: TNM-FH

Characteristics of line: Cells are round, oval, or possess one, two, or three protoplasmic extensions. Protoplasmic extensions are up to 105 μ long. A majority of the cells (90%) have 82–95 chromosomes but some have 160–180 chromosomes

Reference: Hink (1970)

B. Diptera

Invertebrate species: *Aedes aegypti*

Common name: Yellow-fever mosquito

Primary explant from which line was developed: Minced last instar larvae

Date primary culture was initiated: June 1963

Population doubling time: 51 hours

Subculture interval: 7–10 days

Growth medium: G.M.A. plus 5% *A. eucalypti* hemolymph

Characteristics of line: Most common cell type is spindle-shaped and 40–50 μ long by 8–10 μ wide and less common are round cells with 20-μ diameters. Small numbers of large round cells, 55–60 μ in diameter and cells with long extensions are also present. Most cells are 32n while next most frequent chromosome number is 16n

Reference: Grace (1966)

Invertebrate species: *Aedes aegypti*

Primary explant from which line was developed: Minced trypsinized larvae

Number of subcultures to date: 115 as of July 1969

Subculture interval: 7–14 days

Growth medium: Mitsuhashi and Maramorosch (1964) culture medium

Characteristics of line: Mainly epithelial-like cells

Reference: Singh (1967)

Invertebrate species: *Aedes aegypti*

Designation of cell line: Two separate lines, 59 and 364

Primary explant from which line was developed: Homogenized embryos

Number of subcultures to date: 80 as of 1969 publication (see below)
Subculture interval: 4–6 days
Growth medium: Kitamura's (1965) medium plus 10% FBS, 10% conditioned medium, and 1% chick embryo extract
Characteristics of line: Grow in more than one layer
Reference: Peleg (1969)

Invertebrate species: *Aedes aegypti*
Designation of cell line: Mos. 20
Primary explant from which line was developed: Minced trypsinized larvae
Date primary culture was initiated: March 1968
Number of subcultures to date: 129 as of March 1971
Subculture interval: 7 days
Growth medium: Mitsuhashi and Maramorosch (1964) medium
Reference: Varma and Pudney (1969)

Invertebrate species: *Aedes aegypti*
Designation of cell line: Mos. 20A
Primary explant from which line was developed: Minced trypsinized larvae
Date primary culture was initiated: March 1968
Number of subcultures to date: 124 as of March 1971
Subculture interval: 7 days
Growth medium: MM/VP_{12}
Reference: Varma and Pudney (1969)

Invertebrate species: *Aedes aegypti*
Designation of cell line: Mos. 29
Primary explant from which line was developed: Minced trypsinized larvae
Date primary culture was initiated: May 1968
Number of subcultures to date: 135 as of March 1971
Population doubling time: 29 hours
Maximum population per ml medium: 5×10^6 cells
Subculture interval: 7 days
Growth medium: Modified Kitamura's (1965) medium
Characteristics of line: The majority of cells are diploid
Reference: Varma and Pudney (1969)

Invertebrate species: *Aedes albopictus*
Common name: Mosquito

Invertebrate species: *Aedes albopictus, continued*
Primary explant from which line was developed: Minced trypsinized
    larvae
Number of subcultures to date: 112 as of July 1969
Subculture interval: 7 days
Growth medium: Mitsuhashi and Maramorosch (1964) culture medium
Characteristics of line: Predominant cell type is round and 6–20 μ in
    diameter, next common is spindle-shaped and 7–10 μ wide by 15–
    90 μ long, and third type is binucleated
Reference: Singh (1967)

Invertebrate species: *Aedes novo-albopictus*
Common name: Mosquito
Designation of cell lines: Two separate lines, ATC-170 and ATC-173
Primary explant from which the line was developed: Newly hatched
    larval tissues
Date primary cultures were initiated: March 1971
Number of subcultures to date: 12 and 5 as of May 1971
Growth medium: Mitsuhashi and Maramorosch (1964) culture medium
Reference: Bhat (unpublished)

Invertebrate species: *Aedes vexans*
Common name: Mosquito
Primary explant from which line was developed: Minced pupae
Number of subcultures to date: 18 as of August 1969
Maximum population per milliliter of medium: 5×10^5 cells
Subculture interval: 7 days
Growth medium: G.M.A. plus 10% FBS
Characteristics of line: Chromosome numbers range from 156 to 216 with
    a mean of about 190
Reference: Sweet and McHale (1970)

Invertebrate species: *Aedes vittatus*
Common name: Mosquito
Primary explant from which line was developed: Minced newly hatched
    larvae
Date primary culture was initiated: April 1969
Number of subcultures to date: 21 as of May 1970
Subculture interval: 12–14 days
Growth medium: Schneider (1969) medium with 10% FBS
Reference: Bhat and Singh (1970)

Invertebrate species: *Aedes w-albus*
Common name: Mosquito
Designation of cell line: ATC. No. 136
Number of subcultures to date: 26 as of March 1970
Growth medium: Mitsuhashi and Maramorosch (1964) culture medium
Reference: Singh (unpublished)

Invertebrate species: *Anopheles gambiae*
Common name: Mosquito
Designation of cell line: Mos. 55
Primary explant from which line was developed: First stage larvae
Date primary culture was initiated: August 1970
Number of subcultures to date: 19 as of March 1971
Subculture interval: 7 days
Reference: Varma and Pudney (unpublished)

Invertebrate species: *Anopheles stephensi*
Common name: Mosquito
Primary explant from which line was developed: Minced larvae
Date primary culture was initiated: March 1968
Number of subcultures to date: 120 as of April 1971
Population doubling time: 65 hours
Growth medium: Modified Grace (1962) culture medium
Characteristics of line: Most cells are epithelial in appearance and range
    from 4–9 μ in diameter and 12–20 μ in length. Most cells are diploid
Reference: Schneider (1969)

Invertebrate species: *Anopheles stephensi*
Designation of cell line: Mos. 43
Primary explant from which line was developed: Minced trypsinized
    first instar larvae
Date primary culture was initiated: April 1969
Number of subcultures to date: 90 as of March 1971
Population doubling time: 16 hours
Maximum population per milliliter of medium: 8×10^6 cells
Subculture interval: 7 days
Growth medium: Mod Kit/VP$_{12}$
Characteristics of line: The majority of cells are diploid and fibroblast-
    like
Reference: Pudney and Varma (1971)

Invertebrate species: *Anopheles stephensi*
Designation of cell line: Mos. 44, Mos. 45, and Mos. 46; three separate
    cell lines
Primary explant from which line was developed: First stage larvae
Date primary cultures were initiated: April 1970
Number of subcultures to date: 23–28 as of March 1971
Subculture interval: 7 days
Reference: Varma (unpublished)

Invertebrate species: *Culex molestus*
Common name: Mosquito
Primary explant from which line was developed: Adult ovaries
Date primary culture was initiated: October 1967
Number of subcultures to date: 87 as of February 1970
Population doubling time: 30 hours
Maximum population per milliliter of medium: 1.1×10^6 cells
Subculture interval: Twice a week
Growth medium: Kitamura (1970) medium
Characteristics of line: Most cells are diploid with an indication of
    heteroploidy in a few cells
Reference: Kitamura (1970)

Invertebrate species: *Culex quinquefasciatus*
Common name: Mosquito
Designation of cell line: Two separate cell lines
Primary explant from which line was developed: Adult ovaries
Number of subcultures to date: 92 as of December 1969
Growth medium: 721 medium
Characteristics of line: Most cells are diploid. The predominant type of
    cell is spindle-shaped measuring about $21.5 \times 12.9 \, \mu$ and the other
    major cell type is spherical and about $12.1 \, \mu$ in diameter
Reference: Hsu, Mao, and Cross (1970)

Invertebrate species: *Culex salinarius*
Common name: Mosquito
Primary explant from which line was developed: Neonate larvae
Date primary culture was initiated: March 1970
Number of subcultures to date: 46 as of April 1971
Population doubling time: 30 hours
Growth medium: Hsu's medium
Reference: Schneider (1971)

Invertebrate species: *Culex tritaeniorhynchus*
Common name: Mosquito
Primary explant from which line was developed: Ovarian tissue
Date primary culture was initiated: December 1970
Number of subcultures to date: 17 as of March 1971
Population doubling time: 44 hours
Growth medium: Modified 721 medium
Reference: Hsu (unpublished)

Invertebrate species: *Culex tritaeniorhynchus*
Primary explant from which line was developed: Neonate larvae
Date primary culture was initiated: March 1970
Number of subcultures to date: 53 as of April 1971
Population doubling time: 24 hours
Growth medium: Hsu's medium
Reference: Schneider (1971)

Invertebrate species: *Culiseta inornata*
Common name: Mosquito
Primary explant from which line was developed: Minced adult
Number of subcultures to date: 27 as of August 1969
Maximum population per milliliter of medium: 1.5×10^6 cells
Subculture interval: 7 days
Growth medium: G.M.A. plus 10% FBS
Characteristics of line: Extremely polyploid, at least 50–70n
Reference: Sweet and McHale (1970)

Invertebrate species: *Drosophila melanogaster*
Common name: Fruit fly
Primary explant from which line was developed: Homogenized embryos
Number of subcultures to date: 43 as of 1966 publication
Maximum population per milliliter of medium: 1×10^6 cells
Subculture interval: 10 days
Growth medium: H-6
Characteristics of line: Modal chromosome number is diploid with
    evidence of heteroploidy
Reference: Horikawa, Ling, and Fox (1966)

Invertebrate species: *Drosophila melanogaster*
Designation of cell line: Eight independent lines of diploid cells
Primary explant from which line was developed: Homogenized 6–12 hours
    embryos

Invertebrate species: *Drosophila melanogaster, continued*
Date primary cultures were initiated: February 1968
Number of subcultures to date: 125 line K, 110 line C, as of 17 March
    1971
Population doubling time: 20 hours
Subculture interval: 7 days
Growth medium: D-20
Characteristics of line: The karyotype of most lines is fundamentally
    diploid. Cells are small, about 10 μ in diameter
Reference: Echalier and Ohanessian (1970)

Invertebrate species: *Drosophila melanogaster*
Primary explant from which line was developed: Embryo
Reference: Kakpakov, Goosdev, Platova, and Polukarova (1969)

Invertebrate species: *Drosophila melanogaster*
Designation of cell line: Seven distinct lines
Date primary cultures were initiated: 1969
Reference: Richard-Molard (unpublished)

Invertebrate species: *Drosophila melanogaster*
Primary explant from which line was developed: Embryo
Date primary culture was initiated: August 1969
Number of subcultures to date: 50 as of April 1971
Population doubling time: 28 hours
Growth medium: Modified Schneider's medium
Reference: Schneider (unpublished)

Invertebrate species: *Drosophila melanogaster*
Primary explant from which line was developed: Embryo
Date primary culture was initiated: December 1969
Number of subcultures to date: 44 as of April 1971
Population doubling time: 24 hours
Growth medium: Modified Schneider's medium
Reference: Schneider (unpublished)

Invertebrate species: *Drosophila melanogaster*
Primary explant from which line was developed: Imaginal discs
Date primary culture was initiated: February 1970
Number of subcultures to date: 31 as of April 1971
Population doubling time: 35 hours
Growth medium: Modified Schneider's medium
Reference: Schneider (unpublished)

Invertebrate species: *Drosophila melanogaster*
Designation of cell line: GM_1, GM_2, GM_3
Date primary culture was initiated: February and September 1970
Number of subcultures to date: 11–22 as of July 1971
Subculture interval: 7 days
Growth medium: D-225: The components of D-20 are dissolved in 1150 ml
Characteristics of line: GM_1 is composed of 60% X + F cells that have one X and a centric heterochromatic fragment (a portion of Y), 10% X + F cells having two IV chromosomes, 8% XO cells, and 22% tetraploid cells. Most cells of GM_2 are XO with two IV chromosomes. GM_3 karyological analysis has not been done
Reference: Dolfini (unpublished)

C. Dictyoptera

Invertebrate species: *Blabera fusca*
Common name: Cockroach
Designation of cell line: Two separate cell lines
Primary explant from which line was developed: Minced trypsinized embryos
Date primary cultures were initiated: March 1965
Reference: Landureau (1968)

Invertebrate species: *Blattella germanica*
Common name: Cockroach
Primary explant from which line was developed: Minced trypsinized embryos
Date primary culture was initiated: March 1965
Characteristics of line: Most cells are diploid
Reference: Landureau (1966)

Invertebrate species: *Leucophaea maderae*
Common name: Cockroach
Designation of cell line: LM 42
Primary explant from which line was developed: Larval heart
Date primary culture was initiated: 1968
Number of subcultures to date: 31 as of September 1970
Subculture interval: 8–10 days
Growth medium: SFM 72 of Vago and Quiot (1969)
Characteristics of line: Very elongated cells attaches to surfaces
Reference: Vago and Quiot (1969)

Invertebrate species: *Leucophaea maderae*
Designation of cell line: LM 75
Primary explant from which line was developed: Larval heart
Date primary culture was initiated: October 1969
Number of subcultures to date: 67 as of May 1971
Subculture interval: 8–10 days
Growth medium: SFM 72 of Vago and Quiot (1969)
Characteristics of line: Spherical and fibroblast type cells 30–50 μ
    diameter, in suspension and attached to surfaces
Reference: Quiot and Vago (1971)

Invertebrate species: *Leucophaea maderae*
Designation of cell line: LM 112
Primary explant from which line was developed: Larval ovaries
Date primary culture was initiated: July 1970
Number of subcultures to date: 32 as of June 1971
Subculture interval: 8–10 days
Growth medium: SFM 72 of Vago and Quiot (1969)
Characteristics of line: Spherical and spindle-shaped cells 20–40 μ diam-
    eter, in suspension, rarely attached to surfaces
Reference: Quiot and Vago (1971)

Invertebrate species: *Periplaneta americana*
Common name: Cockroach
Designation of cell line: EPa
Primary explant from which line was developed: Minced trypsinized
    embryos
Date primary culture was initiated: September 1965
Number of subcultures to date: More than 200 as of March 1971
Subculture interval: Twice a week
Growth medium: Landureau, J. C. (1966)
Characteristics of line: Chitinase is produced by the cultured cells and
    cells are fundamentally euploid
Reference: Landureau (1968)

Invertebrate species: *Periplaneta americana*
Primary explant from which line was developed: Minced trypsinized
    embryos
Date primary culture was initiated: September 1965
Number of subcultures to date: 32 as of March 1971
Subculture interval: Once a fortnight

Characteristics of line: Karyotype is polyploid with many cells approximately 16n
Reference: Landureau (1968)

D. Homoptera

Invertebrate species: *Aceratagallia sanguinolenta*
Common name: Clover leafhopper
Designation of cell line: AS
Primary explant from which line was developed: Embryos
Growth medium: Chiu and Black's (1967) medium
Reference: Chiu and Black (1969)

Invertebrate species: *Agallia constricta*
Common name: Leafhopper
Designation of cell line: AC20
Primary explant from which line was developed: Minced embryos
Date primary culture was initiated: December 1965
Number of subcultures to date: About 55 as of May 1967
Population doubling time: 72 hours
Growth medium: Chiu and Black's (1967) medium
Characteristics of line: Predominately epithelial-type cells
Reference: Chiu and Black (1967)

Invertebrate species: *Agallia quadripunctata*
Common name: Leafhopper
Primary explant from which line was developed: Minced embryos
Number of subcultures to date: 12 as of May 1967
Growth medium: Chiu and Black's (1967) medium
Reference: Chiu and Black (1967)

Invertebrate species: *Colladonus montanus*
Common name: Leafhopper
Designation of cell line: Three separate cell lines
Primary explant from which the line was developed: Fragmented trypsinized embryos
Subculture interval: Weekly
Growth medium: Chiu and Black (1967) medium

Invertebrate species: *Colladonus montanus, continued*
Characteristics of line: Monolayers of large epithelial-like cells, some of
  which are elongate and spindle-shaped. Also present are larger cells
  with many processes and giant cells 3–4 times the size of usual cell
  types
Reference: Richardson and Jensen (1971)

Invertebrate species: *Macrosteles sexnotatus*
Common name: Leafhopper
Primary explant from which the line was developed: Embryos
Number of subcultures to date: 31 as of June 1971
Growth medium: Chiu and Black (1967) medium
Characteristics of line: Cells of epithelial-like and fibroblast-like
  morphology
Reference: Peters and Spaansen (1971)

Invertebrate species: *Nephotettix apicalis*
Common name: Leafhopper
Primary explant from which line was developed: Embryonic tissue
Date primary culture was initiated: May 1970
Number of subcultures to date: 23 as of March 1971
Growth medium: MGM401 (unpublished)
Reference: Mitsuhashi (unpublished)

Invertebrate species: *Nephotettix cincticeps*
Common name: Leafhopper
Primary explant from which line was developed: Embryonic tissues
Date primary culture was initiated: February 1970
Number of subcultures to date: 52 as of March 1971
Population doubling time: 48 hours
Growth medium: MGM401 (unpublished)
Reference: Mitsuhashi (unpublished)

E. Mollusca

Invertebrate species: *Helix aspersa*
Designation of cell line: HA 31
Primary explant from which line was developed: Foot
Date primary culture was initiated: 1967
Number of subcultures to date: 8 as of original 1958 publication plus 48
Growth medium: HA medium of Vago and Chastang (1958)
Characteristics of line: Fibroblast-type cells
Reference: Vago and Chastang (1958)

IV. CULTURE MEDIA USED FOR CELL LINES*

G.M.A.: Grace (1962) insect tissue culture medium (mg/100 ml)

Salts			L-Threonine	17.5
$NaH_2PO_4 \cdot 2H_2O$	114		L-Valine	10
$NaHCO_3$	35		Sugars	
KCl	224		Sucrose (in grams)	2.668
$CaCl_2$	100		Fructose	40
$MgCl_2 \cdot 6H_2O$	228		Glucose	70
$MgSO_4 \cdot 7H_2O$	278		Organic acids	
Amino acids			Malic	67
L-Arginine–HCl	70		α-Ketoglutaric	37
L-Aspartic acid	35		Succinic	6
L-Asparagine	35		Fumaric	5.5
L-Alanine	22.5		Vitamins	
β-Alanine	20		Thiamine–HCl	0.002
L-Cystine–HCl	2.5		Riboflavin	0.002
L-Glutamic acid	60		Calcium pantothenate	0.002
L-Glutamine	60		Pyridoxine–HCl	0.002
L-Glycine	65		p-Aminobenzoic acid	0.002
L-Histidine	250		Folic acid	0.002
L-Isoleucine	5		Niacin	0.002
L-Leucine	7.5		Isoinositol	0.002
L-Lysine–HCl	62.5		Biotin	0.001
L-Methionine	5		Choline chloride	0.02
L-Proline	35		Antibiotics	
L-Phenylalanine	15		Penicillin G, Na salt	3
DL-Serine	110		Streptomycin sulfate	10
L-Tyrosine	5			
L-Tryptophan	10			

Vago and Chastang (modified 1958) insect tissue culture medium (gm/liter)

KCl	3.0	Lactalbumin hydrolysate	5.0
$CaCl_2$	0.5	Organic fraction of TC 199 dried	2.38
$NaH_2PO_4 \cdot H_2O$	1.0		
$MgSO_4 \cdot 7H_2O$	3.5	Penicillin	200 000 IU
$MgCl_2 \cdot 6H_2O$	3.0	Streptomycin	0.05
Glucose	1.0	B. mori hemolymph	10%
$NaHCO_3$	to pH 6.4	FBS	10%

* The composition of media effectively used to maintain cell lines may differ slightly from the original formula. For this reason the composition of some media is mentioned here although the original formulas are given in different chapters of this book.

Gaw, Liu, and Zia (1959) insect tissue culture medium

Trager (1935) Solution A	moles/liter	gm/liter
Maltose	0.06	20.538
NaCl	0.015	0.8766
$MgCl_2 \cdot 6H_2O$	0.001	0.2033
$CaCl_2$	0.001	0.111
$NaH_2PO_4 \cdot H_2O$	0.0015	0.207
K_2HPO_4	0.0015	0.2613
B. mori hemolymph	10%	
Penicillin	100 units/ml	
Streptomycin	100 μg/ml	

TNM-FH: Hink (1970) insect tissue culture medium

G.M.A.	90.0 ml
Fetal bovine serum (FBS)	8.0 ml
Chicken egg ultrafiltrate	8.0 ml
TC yeastolate	0.3 gm
Lactalbumin hydrolysate	0.3 gm
Bovine plasma albumin, crystallized	0.5 gm

CSM-2F: Mitsuhashi (1967) insect tissue culture medium (mg/100 ml)

$NaH_2PO_4 \cdot H_2O$	50	Lactalbumin hydrolysate	520
$MgCl_2 \cdot 6H_2O$	120	Bacto-peptone	520
$MgSO_4 \cdot 7H_2O$	160	TC-yeastolate	200
KCl	120	Choline chloride	40
$CaCl_2 \cdot 2H_2O$	40	TC-199	20 ml
Glucose	80	Fetal bovine serum	20 ml
Fructose	80	Dihydrostreptomycin sulfate	10

IMC-1: Hink and Ignoffo (1970), modified Yunker, Vaughn, and Cory (1967) insect tissue culture medium

G.M.A.	90.0 ml
Fetal bovine serum	10.0 ml
Egg ultrafiltrate	10.0 ml
Bovine plasma albumin, crystallized	1.0 gm

Yunker, Vaughn, and Cory (1967) insect tissue culture medium

G.M.A.	79%
Fetal bovine serum	10%
Whole chicken egg ultrafiltrate	10%
Bovine plasma albumin, fraction V	1%

Mitsuhashi and Maramorosch (1964) insect tissue culture medium (mg/100 ml)

NaH$_2$PO$_4$·H$_2$O	20	D-Glucose	400
MgCl$_2$·6H$_2$O	10	Lactalbumin hydrolysate	650
KCl	20	Yeastolate	500
CaCl$_2$·2H$_2$O	20	FBS	20%
NaCl	700	Penicillin	100 units/ml
NaHCO$_3$	12	Streptomycin	100 μg/ml

Kitamura (1965) insect tissue culture medium (gm/100 ml)

NaCl	0.65	Sucrose	1.0
KCl	0.05	Lactalbumin hydrolysate	2.0
CaCl$_2$	0.01	TC 199	various amounts
KH$_2$PO$_4$	0.01	Calf serum	various amounts
NaHCO$_3$	0.01		

Modified Kitamura (1965) medium; Varma and Pudney (1969) insect tissue culture medium (mg/100 ml)

NaCl	650	D-Glucose	400
KCl	50	Lactalbumin hydrolysate	650
CaCl$_2$·2H$_2$O	10	Yeastolate	500
KH$_2$PO$_4$	10	FBS	20%
NaHCO$_3$	10	Penicillin	1000 units/ml
		Streptomycin	1 mg/ml

VP$_{12}$: Varma and Pudney (1969) insect tissue culture medium (mg)

NaCl	390	Inositol	40
NaH$_2$PO$_4$·2H$_2$O	55	Lactalbumin hydrolysate	500
MgCl$_2$·6H$_2$O	110	Bovine plasma albumin, fraction V	100
MgSO$_4$·7H$_2$O	120	5% Glutamine	0.6 ml
KCl	55	Basal medium Eagle vitamin	
CaCl$_2$·2H$_2$O	40	mixture (100X)	2.0 ml
D-Glucose	200	H$_2$O	97.4 ml
NaHCO$_3$	50	FBS	10%
Choline chloride	25		

ModKit/VP$_{12}$: Pudney and Varma (1971) insect tissue culture medium

A 1:1 ratio of modified Kitamura (1965) medium and VP$_{12}$ medium

MM/VP$_{12}$: Varma and Pudney (1969) insect tissue culture medium

A 1:1 ratio of Mitsuhashi and Maramorosch (1964) medium and VP$_{12}$ medium

Schneider (1969) medium: modified Grace (1962) insect tissue culture medium (gm/liter)

NaCl	3.0	Sucrose	16.0
KCl	1.1	Glucose	1.0
$MgCl_2 \cdot 6H_2O$	1.14	Trehalose	0.5
$MgSO_4 \cdot 7H_2O$	0.4	Cholesterol	2.0 mg
$NaHCO_3$	0.35	Phenol red	0.01%
$CaCl_2$	0.40	FBS	15%

Amino acids as in Grace's (1962) medium
Supplemented with 1% 10X solution of abbreviated NCTC 135 that lacks sugars, amino acids, and salts

Kitamura (1970) insect tissue culture medium (mg)

NaCl	650	Glucose	200
KCl	50	Lactalbumin hydrolysate	1000
$CaCl_2 \cdot 2H_2O$	10	H_2O	100 ml
KH_2PO_4	10	TC 199	2 parts
$NaHCO_3$	10	Calf serum	1 part

721 medium: Hsu, Mao, and Cross (1970) insect tissue culture medium (mg/100 ml)

KCl	80	Lactalbumin hydrolysate	2000
NaCl	450	Bacto-peptone	500
$CaCl_2$	35	TC-yeastolate	200
$MgSO_4 \cdot 7H_2O$	40	L-Malic acid	60
KH_2PO_4	25	α-Ketoglutaric acid	40
$NaHCO_3$	100	Succinic acid	6
D-Glucose	160	Fumaric acid	6
Sucrose	600	Medium No. 199 (1X)	20%
		FBS	10%

H-6: Horikawa, Ling, and Fox (1966) insect tissue culture medium (mg/liter)

$NaH_2PO_4 \cdot 2H_2O$	200	Lactalbumin hydrolysate	7500
$NaHCO_3$	350	L-Tryptophan	80
KCl	200	L-Cysteine–HCl	20
$CaCl_2 \cdot 2H_2O$	20	Yeast extract	1200
$MgCl_2 \cdot 6H_2O$	100	Phenol red	10
NaCl	7000	Penicillin G	30
Glucose	5500	Streptomycin sulfate	100
Sucrose	5500	Calf serum	10%

D-20: Echalier and Ohanessian (1970) insect tissue culture medium

Glutamic acid	7.35 gm	Neutralize with 10 N KOH and add H_2O to 100 ml,	
Glycine	3.74 gm	use 54 ml of this solution	
Glutamic acid	7.35 gm	Neutralize with 10 N NaOH and add H_2O to 100 ml,	
Glycine	3.74 gm	use 94 ml of this solution	
$MgCl_2 \cdot 6H_2O$	1.0 gm	Sodium acetate·$3H_2O$	25 mg
$MgSO_4 \cdot 7H_2O$	3.7 gm	Glucose	2 gm
$NaH_2PO_4 \cdot 2H_2O$	0.47 gm	Lactalbumin hydrolysate	15 gm
$CaCl_2$	0.89 gm	Grace's (1962) vitamins	
Yeastolate	1.5 gm	H_2O to 1000 ml	
Malic acid	670 mg	Adjust pH to 6.7	
Succinic acid	60 mg	FBS 10–20% supplement	

Landureau (1966) insect tissue culture medium (mmoles/liter)

L-Arginine	11.5	NaCl	113
L-Aspartic acid	1.5	KCl	12
L-Glutamic acid	13.5	$CaCl_2$	4.4
α-Alanine	1.35	$MgSO_4 \cdot 7H_2O$	5.6
β-Alanine	0.5	$NaHCO_2$	4.3
L-Cysteine–HCl	0.83	PO_3H_3	10
L-Glutamine	4.1	Glucose	5.6
L-Glycine	23.5	Trehalose	25
L-Histidine	2.6	α-Ketoglutaric acid	2.5
L-Leucine	1.91	Citric acid	0.8
L-Lysine–HCl	0.88	Fumaric acid	0.5
L-Methionine	3.35	Malic acid	5.0
L-Proline	6.53	Succinic acid	0.5
L-Serine	0.76	Yeast extract	0.5 gm/liter
L-Threonine	1.68	Lactalbumin hydrolysate	3.5 gm/liter
L-Tyrosine	2.0	Fetal calf serum	10%
L-Valine	1.28	Grace's (1962) vitamins	

S.F.M. 72 of Vago and Quiot (1969) (modified) insect tissue culture medium (gm/liter)

TC 199 organic fraction (dried)	2.38	L-Tyrosine	0.10
TC 199 mineral fraction (dried)	8.89	L-Arginine	0.76
Special amino acid complement		I-Lysine	0.09
L-Proline	0.67	L-Threonine	0.08
L-Cysteine	0.26	$NaHCO_3$ to pH	6.8
L-Glycine	0.65	Antibiotics	
L-Phenylalanine	0.10	Bacitracine	0.10
L-Methionine	0.44	Penicillin	0.12
L-Tryptophan	0.16	Streptomycin	0.05
L-Histidine	0.26	Colimycine	0.025
		FBS	15%

Chiu and Black (1967) insect tissue culture medium (mg/100 ml)

D-Glucose	400	Lactalbumin hydrolysate	650
NaCl	105	Yeastolate	500
KCl	80	Penicillin	10,000 U
MgSO₄·7H₂O	185	Streptomycin	10,000 µg
CaCl₂	30	Neomycin	5,000 µg
KH₂PO₄	30	Fungizone	250 µg
NaHCO₃	35	FBS	17.5–20.0 ml

HA: Vago and Chastang (1958) (modified) mollusk tissue
culture medium (gm/liter)

NaCl	6.5	Glucose	1.0
KCl	0.14	Lactalbumin hydrolysate	5.0
CaCl₂·2H₂O	0.12	Organic fraction of TC 199 dried	2.38
NaH₂PO₄	0.01		
NaHCO₃	pH to 7.6–7.9	Penicillin	200,000 IU
		Streptomycin	0.05
		Helix hemolymph	10%
		FBS	10%

REFERENCES

Bhat, U. K. M., and Singh, K. R. P. (1970). *Curr. Sci.* **39**, 388.
Chao, J., and Ball, G. H. (1970). Symp. Arthr. Cell Cult. Bethesda.
Chiu, R., and Black, L. M. (1967). *Nature* **215**, 1076.
Chiu, R., and Black, L. M. (1969). *Virology* **37**, 667.
Echalier, G., and Ohanessian, A. (1970). *In Vitro* **6**, 162.
Gaw, Z., Liu, N. T., and Zia, T. U. (1959). *Acta Virol.* **3**, 55.
Grace, T. D. C. (1962). *Nature (London)* **195**, 788.
Grace, T. D. C. (1966). *Nature (London)* **211**, 366.
Grace, T. D. C. (1967). *Nature (London)* **216**, 613.
Hink, W. F. (1970). *Nature (London)* **226**, 466.
Hink, W. F., and Ellis, B. J. (1971). *Curr. Topics Microbiol. Immunol.* **55**, 19.
Hink, W. F., and Ignoffo, C. M. (1970). *Exp. Cell Res.* **60**, 307.
Horikawa, M., Ling, L., and Fox, A. S. (1966). *Nature (London)* **210**, 183.
Hsu, S. H., Mao, W. H., and Cross, J. H. (1970). *J. Med. Ent.* **7**, 703.
Kakpakov, V. T., Goosdev, V. A., Platova, T. P., and Polukarova, L. G. (1969). *Genetika* **5**, 67.
Kitamura, S. (1965). *Kobe J. Med. Sci.* **11**, 23.
Kitamura, S. (1970). *Kobe J. Med. Sci.* **16**, 41.
Landureau, J. C. (1966). *Exp. Cell Res.* **41**, 545.
Landureau, J. C. (1968). *Exp. Cell Res.* **50**, 323.
Mitsuhashi, J. (1967). *Nature (London)* **215**, 863.
Mitsuhashi, J., and Maramorosch, K. (1964). *Contrib. Boyce Thompson Inst.* **22**, 435.
Peleg, J. (1969). *J. Gen. Virol.* **5**, 463.

Peters, D., and Spaansen, C. H. (1971). Proc. III Internat. Coll. Invert. Tissue Cult.

Pudney, M., and Varma, M. G. R. (1971). *Exp. Parasit.* **29,** 7.

Quiot, J. M., and Vago, C. (1971). Unpublished.

Richardson, J., and Jensen, D. D. (1971). *Ann. Ent. Soc. Am.* **64,** 722.

Schneider, I. (1969). *J. Cell. Biol.* **42,** 603.

Schneider, I. (1971). Proc. III Internat. Coll. Invert. Tissue Cult.

Singh, K. R. P. (1967). *Curr. Sci.* **36,** 506.

Sohi, S. S. (1971). Proc. III Internat. Coll. Invert. Tissue Cult.

Sweet, B. H., and McHale, J. S. (1970). *Exp. Cell. Res.* **61,** 51.

Trager, W. (1935). *J. Exp. Med.* **61,** 501.

Vago, C., and Chastang, S. (1958). *Experientia* **14,** 110.

Vago, C., and Quiot, J. M. (1969). *Ann. Zool. Ecol. Anim.* **1** (3), 281.

Varma, M. G. R., and Pudney, M. (1969). *J. Med. Ent.* **6,** 432.

Yunker, C. E., Vaughn, J. L., and Cary, J. (1967). *Science* **155,** 1565.

AUTHOR INDEX

Numbers in italics refer to the pages on which the complete references are listed.

SUBJECT INDEX

A

Aceratagallia sanguinolenta
  cell line from, 364
    characteristics, 379
  tissue culture, of plant viruses, 357–358
Acetylcholine, effects on explanted dorsal
  vessel physiology, 236, 238, 239
Acheta, dorsal vessel physiology of, 228,
  231, 232, 233
Acheta domestica, dorsal vessel physiology of, 218
Adocia elegans, organ culture of, 115
Adrenaline, effects on explanted dorsal
  vessel physiology, 236–237, 238
Aedes, see also Mosquito
  cell lines of, 336
Aedes aegypti
  cell lines from, 364
    characteristics, 370–371
  tissue culture of, in genetic studies, 166
  organ culture of, 6, 7, 10, 12, 23, 31
  plasmodia culture in cells of, 336, 338
  virus culture in tissues of, 281–284, 289–
    291
    stable cell lines, 304–307
Aedes albopictus
  cell lines from, 364
    characteristics, 371–372

  tissue cultures, in genetic studies, 166
  virus culture in cells of, 306–307, 316
Aedes communis, organ culture of, 7, 12,
  31
Aedes hexodontus, organ culture of, 7,
  12, 31
Aedes iridescent virus, 262–263
Aedes novo-albopictus, cell line, 365
  characteristics, 372
Aedes triseriatus tissue cultures, VEE
  and EEE viruses propagation in, 284
Aedes vexans cell line, 364
  characteristics, 372
Aedes vittatus cell line, 365
  characteristics, 372
Aedes w-albus cell line, 365
  characteristics, 373
Aeschna, dorsal vessel physiology of,
  214–215
Aeschna cyanea, organ culture of, 9, 14,
  16
Agallia constricta
  cell lines from, 364
    characteristics, 379
  organ culture of, 6, 13, 14
  tissue cultures of, 346
    of wound tumor virus, 355–357
Agallia quadripunctata cell lines, 364
  characteristics, 379

399

of platyhelminths, 83–91
of Polychaetes, 71–77, 124
in protochordates, 44–45
of Sipunculoidea, 81
of sponges, 105–119
terminology of, 4
of trematodes, 85–86
Organophosphorous compounds, effects
  on explanted dorsal vessel physiology, 238
Orthoptera
  cell cultures from, in virology, 250
  dorsal vessel physiology of, 218
  organ culture studies on, 14
Oscarella, organ culture of, 115, 116
Osmic acid, as fixative for electron microscopy of viruses, 251
Ovarian sheath fibroblast cultures, for
  pathology studies, 247
Ovaries, of insects
  culture of, 6, 8, 31–35
  of tubules, 6
Ovogenesis, in insects, organ culture
  studies on, 8
Owenia fusiformis, organ culture of, 76
Oysters, phagocytosis in, 273

P

Pappenheim method of virus detection,
  250
Paracentrotus lividus, organ culture of,
  46
Patella vulgata
  organ culture of, 62, 64–65
  hormone and sexuality studies using,
    193, 196–197
Pathology, of invertebrates, cell and
  organ culture studies on, 245–278
Patiria miniata, spawning studies on, 50
Pediculus humanus, organ culture of, 6
Pelmatohydra, organ culture of, 95–96,
  125
Penicillin, effects on explanted dorsal
  vessel physiology, 236
Pennaria tiarella, organ culture of, 92–93
Peridroma saucia, plasmatocyte cultures
  from, 247
Perinereis organ culture, hormone and
  sexuality studies on, 192

Perinereis cultrifera, organ culture of, 72–
  77
Periplaneta, dorsal vessel physiology of,
  213, 215, 223, 227, 228, 229, 230, 232,
  236–239
Periplaneta americana
  cell lines from, 364
    characteristics, 378–379
  dorsal vessel physiology of, 217, 218,
    219, 233
  fat body, NDV studies on, 289
  organ culture of, 8, 10, 14, 29, 32
Phagocytosis, in cell culture, 272–273
Philosamia cynthia, organ culture of, 13
Phormia, dorsal vessel physiology of, 232
Phormia regina, dorsal vessel physiology
  of, 219
Phyllody, mycoplasma agents of, 270–271
Picornaviruses, mosquito cell studies on,
  307, 316
Pieris brassicae
  granulosis virus of, tissue culture
    studies, 261
  protozoan-infected, organ culture
    studies on, 249
Pinctada radiata, organ culture of, 58
Planaria, organ culture studies of
  blastemas of, 124
Planaria dorotocephala, regeneration
  studies on, 86
Planaria gonocephala, regeneration
  studies on, 86–87
Planaria polychroa, regeneration studies
  on, 87–88
Planaria torva, regeneration studies on,
  86, 89
Plasma coagulum, cell culture in, 247
Plasmatocyte cultures, for pathology
  studies, 247
Plasmodia, tissue culture studies of, 321–
  342
Plasmodium berghei yoelii, mosquito cell
  culture of, 338
Plasmodium cathemerium, mosquito cell
  culture of, 338
Plasmodium gallinaceum
  mosquito cell culture of, 336, 338
  zygote of, 332
Plasmodium knowlesi, tissue culture
  studies on, 339